T0256179

BIOLOGICAL SYSTEMATICS

BIOLOGICAL SYSTEMATICS

Principles and Applications

Third Edition

Andrew V. Z. Brower

National Identification Services, Plant Protection and Quarantine, Animal and Plant Health Inspection Service, United States Department of Agriculture, Riverdale, Maryland; Research Associate, Division of Invertebrate Zoology, American Museum of Natural History, New York, New York; and Department of Entomology, National Museum of Natural History, Washington, District of Columbia

Randall T. Schuh

George Willett Curator of Entomology Emeritus, Division of Invertebrate Zoology, and Professor Emeritus, Richard Gilder Graduate School, American Museum of Natural History, New York, New York; Adjunct Professor Emeritus, Department of Entomology, Cornell University, Ithaca, New York; and Department of Biology, City College, City University of New York, New York

COMSTOCK PUBLISHING ASSOCIATES

AN IMPRINT OF CORNELL UNIVERSITY PRESS   ITHACA AND LONDON

First edition first published 2000 by Cornell University Press. Second edition 2009. Third edition 2021.

Library of Congress Cataloging-in-Publication Data

Names: Brower, Andrew V. Z. (Andrew Van Zandt), 1962– author. | Schuh, Randall T., author.
Title: Biological systematics : principles and applications / Andrew V. Z. Brower, Randall T. Schuh.
Description: 3rd edition. | Ithaca [New York] : Comstock Publishing Associates, an imprint of Cornell University Press, 2021. | Includes bibliographical references and index.
Identifiers: LCCN 2020028724 (print) | LCCN 2020028725 (ebook) | ISBN 9781501752773 (hardcover) | ISBN 9781501752780 (epub) | ISBN 9781501752797 (pdf)
Subjects: LCSH: Biology—Classification.
Classification: LCC QH83 .B76 2021 (print) | LCC QH83 (ebook) | DDC 578.01/2—dc23
LC record available at https://lccn.loc.gov/2020028724
LC ebook record available at https://lccn.loc.gov/2020028725

Contents

Preface to the First Edition

All fields of science have undergone revolutions, and systematics is no exception. For example, the discovery of DNA structure fundamentally altered our conception of the mechanisms of inheritance. One might assume that the most recent revolution in systematic biology would have come about through the proposal of a coherent theory of organic evolution as the basis for recovering information on the hierarchic relationships observed among organisms. Such was not the case, however, no matter the frequency of such claims. Rather, it was the realization by Willi Hennig—and others—nearly one hundred years after the publication of the *Origin of Species* by Charles Darwin, that homologies are transformed *and* nested, and that phylogenetic relationships can best be discovered through the application of what have subsequently come to be called cladistic methods. The fact that the theory of evolution allowed for the explanation of a hierarchy of descent was seemingly not sufficient to arrive at a method for consistent recovery of genealogical relationships. It can further be argued that neither was it necessary.

The revolutionary changes did not stop there, however. At the same time that the methods of cladistics were changing taxonomic practice on how to recognize natural groupings, the issue of quantification was being discussed with equal fervor. Whereas systematics was long a discipline marked by its strong qualitative aspect, the analysis of phylogenetic relationships is now largely quantitative.

The introduction of quantitative methods to systematics began with the "numerical taxonomists." Their approach to grouping was based on overall similarity concepts, and the attendant assumption of equal rates of evolutionary change across phyletic lines. Establishment of systematic relationships is now dominated by cladistic methods, which form groups on the basis of special similarity and allow for unequal rates of evolutionary change. The logic and application of quantitative cladistics were in large part developed by James S. Farris.

The overall approach of this book is to present a coherent and logically consistent view of systematic theory founded on cladistic methodology and the principle of parsimony. Some of its subject matter is in a style that would commonly be found in research papers, that is, argument and critique. This approach allows material to be presented in its unadulterated form rather than in the abstract, such that sources of ideas at which criticism is being directed are not obscured and can be found readily in the primary literature. The tradition of critical texts in biological systematics was established by Blackwelder, Crowson, Hennig, Sokal and Sneath, and others. I hope that the style of this book will help students see argumentation in science for what it is, a way of developing knowledge and understanding ideas. The alternative would be to obscure historical fact by pretending that the formulation of a body of critical thought has proceeded in a linear fashion, without sometimes acrimonious debate.

Organization of the Text. This work is divided into three sections, representing more or less logical divisions of the subject matter. Section 1, Background for the Study of Systematics, comprises three chapters, which offer, respectively, an introduction to biological systematics, binominal nomenclature, and the philosophy of science as applied to systematics. Section 2, Cladistic Methods, outlines the methods of phylogenetic analysis, with chapters on homology and outgroup comparison, character analysis, computer-implemented phylogenetic analysis, and evaluation of phylogenetic results. Section 3, Application of Cladistic Results, comprises chapters on the preparation of formal classifications, historical biogeography and coevolution, testing evolutionary scenarios, and biodiversity and conservation. A terminal glossary provides definitions for the specialized terminology of systematics used in this book.

Each chapter ends with lists of Literature Cited and Suggested Readings. The references cited in the text are those actually needed to validate an argument, but do not in all cases necessarily represent the most useful available sources. The Suggested Readings are intended to augment the material presented in the text with more detailed knowledge to challenge the more sophisticated and inquiring student. The readings are chosen for their breadth and quality of coverage, with consideration also being given to their accessibility. Most should be available in major university libraries, and thus be readily available to most students and professors using this book.

R. T. Schuh, 2000

Preface to the Second Edition

Nearly a decade has passed since the publication of the first edition of *Biological Systematics*. Computers have become faster, phylogenetic data matrices have become larger, and presentation of phylogenetic trees has become commonplace, even in literature outside the traditional realm of systematics. The exponential growth of DNA sequence data production has led to the emergence of the new disciplines of genomics and bioinformatics. During this interval, however, the core principles of systematics—discovery and interpretation of characters, construction of data matrices, search for most parsimonious trees—have remained largely unaltered. Therefore, our revision incorporates philosophical and technical advances of the past ten years, but also elaborates and enhances with additional examples the ideas that have formed the basis of modern systematics since its origins nearly fifty years ago.

Although likelihood-based methods of phylogenetic inference have increased in popularity, perhaps due to their implementation in easy-to-use software packages, our book retains its cladistic emphasis. As we have each found in our respective empirical research on Hemiptera and Lepidoptera, the cladistic approach is the most transparent, flexible, and direct means to interpret patterns of character-state transformation as evidence of hierarchical relationships among taxa. The most vociferous advocates of alternative methods are not biologists, but statisticians and computer programmers. We have been accused of "bias" in our preference for cladistic methods over alternatives, but we think—and endeavor to explain in the book—that our methodological choices are based on a clear and objective understanding of the problem being addressed. Systematics is not just about tree-building algorithms; our book devotes just one of its ten chapters to that aspect of the discipline. It is, rather a world view, nothing less than a coherent approach for organizing and understanding information about the natural world. It is with that idea in mind that we have chosen our subject matter and organized our overall presentation.

Reorganization of the Text. We have revised and expanded the entire book, although its overall structure remains largely the same as the first edition. Chapter 1 reviews the history of modern systematics and philosophical differences among various schools. Chapter 2 addresses philosophical underpinnings. An extensively reorganized discussion of character coding and homology is addressed in Chapters 3 and 4. Chapter 5 covers tree-building methods and offers an expanded discussion and critique of the rationale and methods of maximum likelihood, and Chapter 6 describes methods for assessing support for resultant topologies. The discussion of biological nomenclature has been moved to Chapter 7, and merged with an expanded critique of "phylogenetic nomenclature" and the Phylocode. Chapters 8, 9, and 10 examine applications of cladistic results to biogeography, ecology, and biodiversity, respectively. All of the works cited are listed at the end of the book in a comprehensive Literature Cited section, rather than at the end of individual chapters, as in the first edition. Each chapter is accompanied by a supplementary list of Suggested Readings, which represents a cross-section of classic and recent articles and books intended to provide background and deeper understanding of relevant issues. The glossary of terms at the back of the book has been expanded and revised.

R. T. Schuh and A. V. Z. Brower

Preface to the Third Edition

Another decade has passed since the second edition of *Biological Systematics* appeared, and the changes to systematic biology we described in our 2009 preface continue to unfold. Our discipline has proceeded into an era of incomprehensibly large molecular data sets, with automated pipelines to assemble matrices for comparative genomics, and a growing skepticism in some quarters that relationships among the diversity of living things are straightforwardly represented by "a tree." Although it is hard to object to larger data sets as more comprehensive evidentiary bases for phylogenetic inference, and although we appreciate that the scope of "big data" means that some automation is inevitable and perhaps benign (just as tree-searching has been facilitated by computers), we find that data enormity comes at a price: we are alarmed by the degree to which defective systematic methods, discarded long ago, have been resurrected in the workflows of contemporary phylogenomics. It is clear from the literature that many contemporary workers embrace a bioinformatic operationalism that no longer concerns itself with the fundamental principles that have underlain systematics during preceding centuries, and particularly since the Hennigian revolution of the 1970s. Researchers may be adept at pushing buttons, but we think they ought also to understand why they push the ones they do and what assumptions underlie those choices.

The aim of this book, as it has been through the previous editions, is to offer a theoretically coherent roadmap to aid navigation of this vast data landscape—one that not only advises the reader which turns to take but also explains why some routes are better than others. Our goal remains to explicate the theoretical grounds for interpreting the form and meaning of biosystematic evidence, for understanding how that evidence is used to infer patterns of relationship among taxa, and for applying those patterns to inform other aspects of comparative biology. To this end, now more than ever, we maintain and continue to advocate the cladistic approach.

We are cladists, and we do not refrain from advocating our methodological preferences and noting contrasts with alternative viewpoints. A number of the methods described in the book have been in use for several decades, but venerability is not per se a legitimate criticism of a methodology's philosophical soundness and ongoing utility. Fundamental concepts such as homology, the irregularly bifurcating hierarchy, and the principle of parsimony have been with us for centuries or millennia, yet remain critical elements of the conceptual framework of biological systematics. For that matter, we might observe that popular alternative frameworks are hardly recent innovations: maximum likelihood was conceived by Ronald Fisher nearly 100 years ago and was applied to phylogenetic questions in the early 1960s. Bayes' Theorem was published in 1763.

You may have read—perhaps on a social media site—that cladistics is old-fashioned, Luddite, or even utterly irrelevant to modern phylogenetic studies, and that the people who still employ its methods are irrational zealots, like acolytes of a religious cult. We are prepared—indeed, enthusiastic—to defend cladistics against sober and legitimate criticisms, but naturally, we find such ad hominem stuff to be puerile and without substance. A religion is a system of metaphysical beliefs without a firm empirical foundation. This book is all about the empirical foundations of systematics and about questioning metaphysical suppositions. One of its take-home messages is that quantitative complexity does not equate to explanatory robustness. In fact, as any statistician can tell you, just the opposite is true. The approach we endorse values empirical clarity and methodological transparency between evidence and inference—in short, parsimony.

What's new in the third edition? We have updated the entire book, with major revision to Chapters 1 and 2. We have added two new chapters, one addressing species concepts and issues related to phylogenetic inference at its lower bound, and another on understanding molecular clocks. We have significantly expanded the glossary, as well. (Note that, as has been the case through all editions of the book, terms italicized in the text are defined in the glossary). The numerous systematics resources available on the web that we cite in the text are listed in the reference section with current URLs. These are indicated to be web resources in the text with the parenthetical statement "(online)."

Since the publication of the previous edition of *Biological Systematics*, new or revised editions of several other systematics-related texts have appeared. Because this book will not be to everyone's taste, and because a circumspect systematist should always strive for a clear understanding of the breadth of opinions in the field, we offer the following list of books for the reader's awareness, edification, and/or amusement:

Baum, D.A., and S.D. Smith. 2013. Tree Thinking: An Introduction to Phylogenetic Biology. Greenwood Village, CO: Roberts.
Bromham, L. 2016. An Introduction to Molecular Evolution and Phylogenetics, 2nd ed. Oxford: Oxford University Press.
Chen, M.H., L. Kuo, and P. Lewis, eds. 2014. Bayesian Phylogenetics: Methods, Algorithms, and Applications. Boca Raton, FL: Chapman and Hall/CRC Press.
Warnow, T. 2018. Computational Phylogenetics: An Introduction to Designing Methods for Phylogeny Estimation. Cambridge: Cambridge University Press.

Wheeler, W.C. 2012. Systematics: A Course of Lectures. Chichester, UK: Wiley–Blackwell.

Wiley, E.O., and B.S. Lieberman. 2011. Phylogenetics: Theory and Practice of Phylogenetic Systematics. Hoboken, NJ: John Wiley and Sons.

Yang, Z. 2014. Molecular Evolution: A Statistical Approach. Oxford: Oxford University Press.

Zander, R.H. 2013. A Framework for Post-Phylogenetic Systematics. St. Louis: Zetetic Publications.

It is the responsibility of all scholars to understand the premises and assumptions of their chosen methodologies. Even if you disagree with our approach to systematics, we hope that this book provokes you to think about the reasons why.

A. V. Z. Brower and R. T. Schuh

Acknowledgments to the First Edition

Several colleagues and friends provided discussion, assistance, advice, reviews, and encouragement during the course of writing this book. For reviews of an early version of the manuscript, or parts thereof, I thank James Ashe, Gerasimos Cassis, David Lindberg, Steven Keffer, Norman Platnick, James Slater, Christian Thompson, Quentin Wheeler, and Ward Wheeler. For reviews of the complete manuscript, I offer special thanks to Andrew Brower, James Carpenter, Eugene Gaffney, Pablo Goloboff, Dennis Stevenson, and John Wenzel. Dennis Stevenson gave me much advice on botanical examples and nomenclature, and offered some very timely encouragement as this project progressed. My conception of issues of philosophy and systematic theory, as presented in this volume, has been influenced by discussions with Andrew Brower, James Carpenter, Eugene Gaffney, Pablo Goloboff, and Norman Platnick. Pablo Goloboff was immensely helpful in clarifying my presentation of the quantification of cladistics. Gregory Edgecombe offered suggestions on relevant literature. The students and auditors in my Spring 1998 Principles of Systematics course at the City University of New York field-tested a version of the manuscript. Christine Johnson read and commented on the final manuscript and prepared the figures. Whatever the inputs from others, in the end, I am solely responsible for the final form of all arguments presented in the text.

The development of my views on the nature of systematics was shaped by two people in particular, my long-time friends and colleagues James S. Farris and Gareth Nelson. Since 1967, they, more than any other individuals, have profoundly affected our understanding of systematic theory. Thus, in an indirect way, they have greatly influenced the way I have written this book.

The encouragement of my wife, Brenda Massie, and Steven Keffer caused me to go to work on this project. Their confidence that I could produce a useful final product spurred me on. My young daughter, Ella, has been a patient helper during the preparation of the manuscript. The term 'systematics' is now indelibly imprinted in her mind.

Randall T. Schuh
New York, October 1998

Acknowledgments to the Second Edition

This edition of *Biological Systematics* is co-authored by Andrew Brower, a systematic entomologist whose research is focused on the phylogenetic relationships of nymphalid butterflies. Andy's training as a systematist began at Cornell University and continued at the American Museum of Natural History and the Smithsonian Institution. He is extremely grateful to his colleagues and mentors at these institutions for providing a collegial and scholarly environment that gave him the opportunity to develop his perspectives on the discipline. Access to a free copier in a great library is a wonderful thing. As the Rice Professor of Systematic Entomology, Andy taught a graduate course in Principles of Systematics at Oregon State University between 1998 and 2005, an experience that helped him develop an organized framework for training systematics students. He thanks Harold and Leona Rice for their generous support of his systematics research and training program. He would like to acknowledge Darlene Judd for her systematic insight and moral support. He would also like to thank Randall "Toby" Schuh for the opportunity to contribute to this revision of the first edition. Both authors are grateful to Marc Allard and two anonymous reviewers of the revised manuscript for their thoughtful comments. We thank also Gerasimos Cassis, Dimitri Forero, James S. Miller, Mark E. Siddall, F. Christian Thompson, Ward C. Wheeler, and David M. Williams for discussion of our approach, comments on portions of the manuscript, or for other assistance. Once again we acknowledge Pablo Goloboff for his contributions to issues relating to the quantification of cladistics in the first edition, because we continue to use much of that material in the revised version.

Randall Schuh, New York
Andrew Brower, Murfreesboro, Tennessee
December 2008

Dedication and Acknowledgments to the Third Edition

As we were checking the copyedited version of the manuscript for the third edition, we learned that our longtime colleague and friend, Norman I. Platnick, had suffered a mortal injury that eventually ended his remarkable life at the age of sixty-eight. Norman was a person of prodigious intellect who joined the curatorial staff of the American Museum of Natural History in New York City, Harvard Ph.D. in hand, at the age of twenty-one. Over the course of the next 40 years he became one of the most influential spider specialists of all time. As readers of this work will find, he also had a profound impact on the relationship of the philosophy of science to systematics, the theory and practice of phylogenetic systematics, and historical biogeography. It is in recognition of his seminal contributions to the field that we dedicate this third edition of *Biological Systematics* to his memory.

We thank Kitty Liu and the staff of Cornell University Press for their willingness to publish a third edition of our book and two anonymous reviewers who provided frank opinions and valuable suggestions on the manuscript, many of which we have incorporated into the revision. We thank Jennifer Savran Kelly and Eva Silverfine for meticulous copyediting, reference checking, and thoughtful suggestions to improve the flow of the text.

Andy is grateful to former colleagues at Middle Tennessee State University and new colleagues at the United States Department of Agriculture, Animal and Plant Health Inspection Service, Plant Health Programs for their friendship and support, and in particular to the National Identification Service for bringing him aboard as a supervisor of the National Taxonomists (the specialists responsible for final authoritative identification of potential pests and pathogens intercepted at US ports of entry by US Customs and Border Protection inspectors). Precise, accurate, and timely identification of potential quarantine pests is where the systematic rubber hits the road, and after an academic career in pursuit of butterfly phylogeny, Andy is excited to be a part of this practical endeavor to protect global agriculture via applied regulatory biogeography. Many of the changes to the new edition of the book were composed on the Brunswick Line of the MARC train (not on "government time"). Nevertheless, it is prudent to assert, "The opinions expressed in this book do not necessarily represent the policies or views of the US Department of Agriculture or the United States Government.

Randall Schuh, New York, and Andrew Brower, West Virginia

Section I
HISTORICAL AND PHILOSOPHICAL BACKGROUND FOR SYSTEMATICS

INTRODUCTION TO SYSTEMATICS
First Principles and Practical Tools

Historical Précis: 400 BC–1950

Systematics is the science of biological classification. It embodies the study of organic diversity and provides the comparative framework to study the historical aspects of the evolutionary process. In this chapter we will explore the nature of systematics as an independent discipline and briefly survey the literature sources most frequently used by systematists.

Beginning in about 400 BC, the ancient Greeks produced the first writings in the Western world that might be classed as scientific by modern standards. Many of the contributions of Plato, his student Aristotle, and others were translated into Latin by the Romans, and later into Arabic, whereby they received wider distribution and by which means many of them survived to modern times. Even though these important works had great influence in their day, they remained obscure to European scholars for about 10 centuries until being "rediscovered" in the Middle Ages. It was that rediscovery, of Aristotle's work in particular, that rekindled interest among Europeans in the thought processes that led to the development of modern science.

The exact nature of Aristotle's contribution to the field of systematics is subject to varied interpretation, parts of it positive, parts of it negative, as we will see later on. What is not in dispute is that Aristotle made some of the most detailed observations of the living world during his time, particularly with regard to animals. For example, he documented that whales are mammals, not fish.

Systematics—what is often called taxonomy—as currently practiced has its beginning in the work of the Swedish botanist and naturalist Carolus Linnaeus (Carl von Linné) and his contemporaries in the mid-eighteenth century. Linnaeus's work built on the earlier contributions of authors such as the sixteenth-century Italian physician Andrea Caesalpino and the mid-seventeenth-century English naturalist John Ray.

The detailed history of systematics is a fascinating subject in its own right and would shed much light on how current systematic knowledge, and the methods used to acquire that knowledge, have achieved their current form. However, much of that history is beyond the practical scope of this book. References dealing with the subject are included at the end of the chapter under Suggested Readings. For our purposes, most of the history critical to understanding the current state of affairs in systematic methods dates from about 1950.

The scope of systematic research may be divided into three basic activities that have changed surprisingly little over the last 250 years. First is the recognition of fundamental units of biological diversity in nature, which are usually called species. Our understanding of the perpetuation of species has advanced greatly since the time of Linnaeus, primarily because of improved knowledge of the mechanisms of inheritance. Yet, with more than 2 million described species of plants, animals, and other taxa, it has not been possible to study all of them in detail. Consequently, most species are still recognized on the basis of morphological and other characteristics observable in preserved, dead specimens, much as they were in the time of Linnaeus. The details of how species are actually identified and circumscribed in mammalogy, entomology, bryology, and other fields of taxonomic specialization are discussed in further detail in Chapter 7.

Second is the classification of those species in a hierarchic scheme that reflects our understanding of their phylogenetic relationships. The existence of a hierarchical pattern of groups within groups in the living natural world has been recognized at least since the time of Aristotle, and formal, hierarchic classifications of plants and animals, such as those published by Linnaeus (1753, 1758), existed well before—and provided evidence to support—the proposal of a now widely accepted theory of organic evolution by Charles Darwin and Alfred Russel Wallace in the mid-nineteenth century. Linnaeus and other authors of early classifications were content to describe organic diversity as the handiwork of a manifestly beneficent God, and although the divine purpose of these creations might be inscrutable to mortal scientists, their relationships as revealed by similarities and differences of form were not. For example, Linnaeus placed humans and orangutans together in the genus *Homo* and (following Aristotle) categorized whales and dolphins as an order of mammals, viewing their lungs and warm blood as stronger evidence of kinship than their flukes, flippers, and other adaptations to a marine lifestyle. The introduction of the Darwinian theory of organic evolution changed the explanation for the observed relationships among organisms from a revelation of the plan of divine creation to a representation of the results of evolutionary processes. This change in the explanation of the underlying causal mechanisms did not contribute, however, to the development of a well-articulated set of methods for discovering the relationships that most investigators began to assume were phylogenetic. That development had to wait nearly one hundred years and is the subject matter of much of this book.

Third is the application of information about species and their classification to broader contexts of geography, time, and evolutionary interactions, subjects to which we will return in Section 3.

The Schools of Taxonomy: The Development of Systematics in the Twentieth Century

The fundamental challenge of biological systematics is the management and interpretation of vast amounts of potentially conflicting information: which organismal features are important as evidence of grouping, which are not? While today we have computers that can perform billions of calculations per second, systematists in the nineteenth and early twentieth centuries were forced to document and organize their evidence by hand. This inevitably resulted in classifications proposed on the bases of taxonomic judgment and experience and accepted by colleagues on the basis of the classifier's reputation. By the 1930s, nearly all biologists agreed that an ideal classification ought to reflect patterns resulting from the historical process of evolutionary divergence, generally referred to as *phylogeny* (following the works of Ernst Haeckel), but there was no agreement on how those might be determined, and many systematists despaired of ever attaining that phylogenetic nirvana with their chosen group of study. At the same time, systematics, once "the Queen of Sciences" (physics was "King"), was being eclipsed by the new and exciting field of genetics, which promised to reveal actual mechanisms of heredity and evolutionary change within populations.

The Evolutionary Taxonomic Point of View

By the early 1950s, taxonomic theory, referred to as the "new systematics," had become heavily influenced by recent successes in theoretical genetics, from which perspective the study of populations and infraspecific variability represented the crucial element for understanding biological diversity. Under this new paradigm, efforts to infer phylogenetic relationships were viewed with growing skepticism as speculative flights of fancy, unless anchored by evidence of ancestor-descendant relationships inferred from the fossil record. The "new systematic" approach can be appreciated by examining the textbook, *Methods and Principles of Systematic Zoology*, by ornithologist Ernst Mayr and entomologists Gorton Linsley and Robert Usinger (1953). In that 328-page work, discrimination of species and subspecies occupied almost all of the substantive discussion of methods. About one-third of the volume dealt with rules of nomenclature. Approximately two pages were devoted to the connection between characters and classification, and techniques for discovering relationships among groups of organisms, be they species or taxa above the species level, were barely mentioned. The book contained only eight figures intended to portray phylogenetic relationships, with no indication as to what, if any, data supported the topologies illustrated. In the largely neontological perspective of the three authors, knowledge of relationships among taxa was fundamentally bound to, and presumably thought to flow directly from, the microevolutionary diversification of populations.

A more strongly paleontological view of systematic biology, but one nonetheless closely associated with "evolutionary taxonomy," was portrayed a few years later in George Gaylord Simpson's *Principles of Animal Taxonomy* (1961). Simpson, a specialist on fossil mammals, devoted many pages to the discussion of interrelationships among groups of organisms, while also emphasizing the importance of the temporal perspective that could be gained from geology and the study of change in populations of fossils through time. He observed, "The construction of formal classifications of particular groups is an essential part and a useful outcome of the taxonomic effort but is not the whole or even the focal aim. The aim of taxonomy is to understand the grouping and interrelationships of organisms in biological terms" (p. 66). Simpson's perspective, reflecting systematists' historical incapacity to agree upon objective criteria for inferring phylogenetic patterns by means of traditional methods, was that taxonomy "is a science, but its application to classification involves a great deal of human contrivance and ingenuity, in short, of art" (p. 107).

Another fundamental aspect of the evolutionary taxonomists' perspective— the view that not only phylogenetic divergence of taxa but also phylogenetic inference by taxonomists—stems from the wellspring of Darwinism, can be observed by examining passages from *The Growth of Biological Thought* (Mayr 1982:209), where the author noted: "That Darwin was the founder of the whole field of evolutionary taxonomy is realized by few ... the theory of common descent accounts automatically for most of the degrees of similarity among organisms ... but also ... Darwin developed a well thought out theory with a detailed statement of methods and difficulties. The entire thirteenth chapter of the Origin is devoted by him to the development of his theory of classification."

A few pages later (p. 213), in what appears to be a direct contradiction of this argument, Mayr stated, "As far as the methodology of classification is concerned, the Darwinian revolution had only minor impact." However, this pronouncement is couched as a criticism of contemporary practicing systematists, whose manifest success in organizing biological diversity was tainted, in Mayr's view, by their benighted metaphysical commitment to non-Darwinian essentialism. As we will discuss in Chapter 8, metaphysical correctness is a concern that continues to inflame the passions of more than a few systematists even in the twenty-first century. Mayr summarized his opinions by noting that Darwin's decisive contributions to taxonomy were that common descent provided an explanatory theory for the Linnaean hierarchy and that it bolstered the concept of continuity among organisms, propositions with which we agree.

The most steadfastly defended tenet of Mayr and Simpson's classical evolutionary taxonomy is that biological classifications must express what they considered to represent the maximum amount of evolutionary information, incorporating both phylogenetic branching order and subsequent anagenetic change within lineages. Another proponent of this approach, Mayr's student Walter Bock, claimed, "formal classification is an attempt to maximize simultaneously the two semi-independent variables of genetic similarity and phylogenetic sequence," with the caveat that a one-to-one correspondence between classification and phylogeny is impossible (Bock 1974:391). He further opined that improvements in comprehension of systematic relationships among organisms must come through the more thorough study of organismal attributes, not through the introduction of new philosophical approaches. Given the evolutionary taxonomists' quixotic desire to meld imprecisely inferred phylogenetic patterns with the vague concept of similarity, according to unarticulated criteria, it is not surprising that not everyone agreed with Bock. In fact, contemporaneously, two alternative, quite different philosophical approaches to systematics were being advanced in the literature.

The Phenetic Point of View

Frustrated with the artful nature of evolutionary taxonomy and the authoritarian posturing of its proponents, a group of self-styled "extreme empiricists" began to propound a new, explicitly quantitative approach to taxonomy in the late 1950s. The pheneticists—including Robert Sokal, Peter Sneath, Arthur Cain, G. Ainsworth Harrison, F. James Rohlf, Paul R. Ehrlich, Donald Colless, and others—were motivated to make taxonomy objective and "*operational*," with the ultimate goal to produce general purpose classifications in which relationships among groups of organisms are formed on the basis of overall similarity. The notion that groups based on maximal correspondence of many characters provide the most general classifications was propounded by the positivist philosopher John S. L. Gilmour (1940), following the empiricist writings of John Stuart Mill (1843). According to Gilmour, a natural classification is not one that expresses phylogeny (which in his view was unknowable), but is a classification that grouped taxa to reflect the greatest overall correspondence of features among the organisms classified. To a certain extent, the philosophical perspective that phylogenetic relationships cannot be known with certainty, but can only be inferred from empirical evidence, is shared among present-day systematists of all the schools discussed in this book. However, the measure employed by pheneticists to assess the correspondence of features was overall similarity. For example, the presence of some feature would be evidence to unite the group that possesses it, and the absence of the same feature would be evidence to unite the complementary group (e.g., feathers present or absent, uniting both birds and nonbirds as complementary taxa). Although this approach was initially called "numerical taxonomy" by proponents such as Sokal and Sneath (1963), the term phenetic—as introduced by Cain and Harrison (1960) and elaborated by Mayr (1965; see also Brower 2012)—was not only shorter but seemed more apt because not all quantitative taxonomic approaches embraced the overall-similarity-based school of thought espoused by the pheneticists.

Sociologist-philosopher David Hull (1970) summarized the views of the pheneticists as including (1) the desire to exclude completely considerations of evolution from taxonomy because in the vast majority of cases phylogenies are unknown; (2) the belief that the methods of the evolutionary systematists were not sufficiently explicit and quantitative; and (3) the observation that classifications based on phylogeny are by their very nature designed for a "special purpose." The reader should keep in mind that the distinction between special purpose— or artificial—classifications and general purpose—or natural—classifications has been a subject of debate since the time of Linnaeus. The first textbook-length exposition of phenetic methods was entitled *Principles of Numerical Taxonomy* (Sokal and Sneath, 1963).

The methods of phenetics are numerous and have never been consistently codified. Some authors have claimed that phenetic methods are—or ought to be—"atheoretical," an obviously nonsensical proposition since "similarity" itself is a complex theoretical concept (see Chapter 3). In practice, the "atheoretical" aspect of phenetics is usually manifest simply in the view that, in order to avoid subjective bias, many characters should be recorded and all of them should be weighted equally, a notion first expressed by the eighteenth-century French botanist Michel Adanson. For this reason, pheneticists have also been referred to as neo-Adansonians. Another common phenetic tenet is the rejection of character polarity (discussed in depth in Chapter 4).

The phenetic approach, as expounded by Sokal and Sneath, was based on Gilmour's precept that classifications incorporating the maximum number of unweighted observations would be general purpose, rather than being disposed toward some particular scientific theory, such as organic evolution. "Operationalizing" taxonomy would, in the view of the pheneticists, make the process of data gathering unbiased, practicable by an "intelligent ignoramus" (i.e., a nonspecialist in the group under study) and possibly amenable to automation and the use of computers.

Phenetic techniques are implemented by converting the numbers of character-state similarities and differences among all characters into a summary matrix of pairwise distances of the type shown in Figure 1.1. Phenetic algorithms originally treated rates of evolutionary change across lineages as equal, as exemplified by the unweighted pairgroup method of analysis (widely abbreviated in the literature as UPGMA). More recent applications of the phenetic approach (such as the Neighbor Joining method) incorporate less restrictive assumptions and allow for variable rates of change across lineages. Nonetheless, both of these algorithms are "phenetic" because they convert the original data into "distances" between pairs of taxa and form groups on the basis of overall similarity. The exact nature of that similarity, and of the evidence supporting hypotheses of relationship in the resulting classification, cannot be specified, however, because they are derived from a summary matrix of similarities with a single value for each pair of taxa, not from the original matrix of the characters themselves.

The Phylogenetic (Cladistic) Point of View

The German entomologist and systematic theorist Willi Hennig believed that "the task of systematics is the creation of a general reference system and the investigation of the relations that extend from it to all other possible and necessary systems in biology" (Hennig 1966:7). He first propounded a set of "methods and principles" in a 1950 work entitled *Grundzüge einer Theorie der phylogenetischen Systematik*, which was later revised and published in English under the title *Phylogenetic Systematics* (Hennig 1966). Hennig viewed the hierarchic classifications long produced by systematists as the general reference system of biology, but he argued that the utility of that system could be maximized only if it reflected the phylogenetic relationships of the organisms involved. In distinct contrast to the pheneticists, Hennig advocated an approach that he believed would directly reflect the pattern of relationships resulting from the evolutionary process.

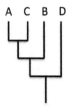

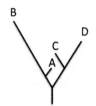

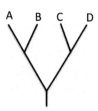

Phenogram, showing levels of clustering based on overall similarity.

Phylogram (evolutionary tree), showing phylogenetic relationships and degree of divergence

Cladogram, showing phylogenetic relationships

CHARACTER MATRIX

Character / Taxon	1	2	3	4	5	6	7	8	9	10	11	12	13	14	15	16	17	18	19	20
Group X	0	0	0	0	0	0	0	0	0	0	0	0	0	0	0	0	0	0	0	0
Group A	1	1	0	0	0	0	0	0	0	0	0	0	0	0	0	0	0	0	0	0
Group B	1	0	0	0	0	1	1	1	1	1	1	1	1	1	0	0	0	0	0	0
Group C	0	0	1	1	1	0	0	0	0	0	0	0	0	0	0	0	0	0	0	0
Group D	0	0	1	0	0	0	0	0	0	0	0	0	0	0	1	1	1	1	1	1

DISTANCE MATRIX

	A	B	C	D
A	0			
B	10	0		
C	5	13	0	
D	9	17	8	0

FIGURE 1.1. Character matrix and the corresponding phenetic distance matrix for four taxa. Character states shared with Group X, the outgroup, are considered to be ancestral states; states that differ from the outgroup are considered to be derived. The outgroup thus determines the polarity of the characters and provides a root for the evolutionary tree and cladogram (see Chapter 4). Distances are computed by counting the number of character state differences between all possible pairings of taxa. The phenogram, phylogram, and cladogram depicting relationships among the taxa are determined by the methods of phenetics, evolutionary taxonomy, and cladistics, respectively. The pattern of grouping shared by the phylogram and the cladogram is based on characters 1 and 3, which exhibit the only derived states shared by more than a single taxon. The phenogram, driven by the similarity of ancestral states in characters 6–20, has a different topology: the taxa with the smallest amount of change are clustered. This example reveals the sensitivity of phenetic methods to variability in evolutionary rates among the different groups of organisms (modified from Farris 1971).

Hennig's approach, at first labeled phylogenetic systematics—now commonly called cladistics (see Sidebar 1)—forcefully articulated the idea that phylogenetic hypotheses intended to reflect genealogical relationships should be based on special similarity (presence of shared derived character states) alone and that only those relationships should be reflected in formal hierarchic classifications. Despite the obvious evidentiary connection between shared derived character states and hypotheses of common ancestry, these points were not accepted by the pheneticists, nor were they accepted by the evolutionary taxonomists, as seen in Simpson's (1961:227) enduring belief that classification is "an art with canons of taste, of moderation, and of usefulness."

Sidebar 1
Origins of "Clade," "Cladistics," "Cladist," and Other Terms

The term *clade* (from the Greek *klados*, branch) was introduced by Lucien Cuénot (1940) and first used in English by Julian Huxley (1957), who apparently independently derived it from Bernard Rensch's (1954) concept of kladogenesis. As Huxley said, "cladogenesis results in the formation of delimitable monophyletic units, which may be called clades." Cladistic relationships (using patterns of shared derived character states to infer phylogenetic branching order) were contrasted with phenetic relationships ("arrangement by overall similarity based on all available characters without any weighting") by Cain and Harrison (1960), and as early as 1965 the term *cladistics* was applied by Joseph H. Camin and Sokal and by Mayr to phylogenetic systematic studies of the type espoused by Hennig. The graphical depictions of phylogenetic relationships produced by these methods were called *cladograms* by those same authors. The term *cladist* was also soon in use, initially often as a pejorative, to refer to one who employed the methods of Hennig.

At the same time that "cladistics" was joining the lexicon, Mayr, Sokal, and Camin popularized the term *phenetics* for the approach widely known earlier as "numerical taxonomy." Mayr made it clear that it was the numerical taxonomists' particular methods of grouping by overall similarity and the rationale for their approach that were distinctive from his preferred methodology, rather than the use of numerical techniques per se. The diagrams of relationships produced with phenetic techniques were called *phenograms* by Mayr (1965) and Camin and Sokal (1965). Those who practiced phenetics were soon called *pheneticists*.

The additional terms *syncretist* and *gradist* are also to be found in the literature. They usually refer to individuals whose approach to taxonomy reflects a combining of intermingled cladistic and phenetic methodologies with idiosyncratic, subjective criteria for classification into what is often called *evolutionary taxonomy*. We consider the preoccupation with "*branch length*" (see Sidebar 12 in Chapter 5) as a quantity potentially affecting patterns of phylogenetic relationship to fall into this category as well.

In fairness to history, methods akin to those described by Hennig had apparently been applied by earlier workers—as for example P. Chalmers Mitchell, working with birds (1901) and Walter Zimmermann (1943), working on plants. And, as was pointed out by Norman Platnick and H. Don Cameron (1977), the fields of textual criticism (stemmatics) and historical linguistics both use methods nearly identical to those propounded by Hennig for establishing historical relationships among manuscripts and languages, respectively. Thus, the approach of grouping by special similarity seems to have a general applicability to systems involving descent with modification and diversification of lineages over time (think of the "mutations" introduced to sequential manuscript copies by illiterate monks). However, within biology, the earlier applications of cladistic-like approaches—as by Mitchell—seem to have been neither sufficiently influential nor compellingly enough articulated to revolutionize systematics in the way that Hennig's work has done.

How Evolutionary Taxonomy, Phenetics, and Cladistics Differ

It may seem counterintuitive that by 1965, after more than 200 years of research effort, the field of systematics still did not have a clearly codified—and broadly accepted—set of methods. Yet, that was indeed the case.

Let us pose four questions as a way of examining the basic precepts of the three "schools" introduced above, each of which by the late 1960s was competing for the primacy of its point of view as the most efficacious approach to the study of biological systematics.

1. Can and should we attempt to group taxa in a manner that reflects their putative pattern of phylogenetic divergence? The phenetic point of view was clearly "no," whereas evolutionary taxonomists and cladists felt just the opposite.

2. Does evolutionary change proceed at the same rate in different lineages? Since they deliberately eschewed evolutionary biases, the pheneticists were not concerned—and perhaps even approved—that the methods they applied assume equal rates of change among lineages, an assumption that contributed to their ultimate undoing. The evolutionary taxonomists thought that rates varied and wished to incorporate that information in their results, particularly with regard to grouping together "primitive" taxa with relatively low rates of change. The cladists applied methods that were unaffected by variation in rates among lineages and came to conclusions distinct from those of the other two schools about how the results of differences in rates of divergence across lineages might best be portrayed in formal classifications.

3. What type of information is counted as evidence of grouping? As noted, phenetic algorithms transform character matrices into distance matrices by converting the proportion of character-state matches between each pair of taxa into an overall percent similarity. Thus, phenetic groups are those with the highest similarity scores, regardless of whether the similarities are due to shared changes or shared lack of change, as in the feathers present/absent example above. Cladists count character-state transformations, forming groups of those taxa united by the greatest number of inferred shared derived transformations. Evolutionary taxonomists also group taxa based on shared derived character-state transformations, except sometimes they choose to ignore these data in favor of phenetic similarity by treating differently groups that have many unique features. Consider the following example:

Taxon   A   B   C   D
State   x   x   y   y

A pheneticist would recognize (A + B) *and* (C + D), while a cladist would recognize (A + B) *or* (C + D), depending on whether state x or state y is inferred to be the derived condition, and would not recognize the other pair as a group (by cladistic reasoning, taxa sharing the ancestral state do not form a group that does not also include the taxa sharing the derived state). It is hard to say whether an evolutionary taxonomist would prefer the phenetic or cladistic interpretation of these data because the choice would be determined by taxonomic judgment (Simpson's "art") rather than by a rule.

4. Are all attributes of organisms useful in forming classifications? Pheneticists thought the answer was yes. Because they count state similarities regardless of whether they are ancestral or derived, their

methods explicitly used techniques that measured the degree to which groups were similar and different; to the extent that they were intended to reflect intuitively "natural" taxa, these measures were implicitly dependent upon the assumption of a constant rate of change among lineages. Cladists took the view that groups could be formed only on the basis of shared derived attributes, for to do otherwise would be to allow any possible grouping. Many evolutionary taxonomists accepted Hennig's cladistic point of view concerning the formation of groups based on shared derived characters that imply common ancestry but maintained that overall degree of difference among lineages should be recognized in assigning rank in formal classifications. The result was the formation of groups based on artful combinations of ancestral and derived character states.

The salient attributes of the three taxonomic approaches are characterized in Table 1.1. Figure 1.1 shows a data matrix, a phenetic distance matrix derived from that data matrix, and the trees implied by different analyses of those data. As can been seen, cladistic methods group taxa by the presence of shared, derived attributes alone. In contrast, phenetic methods group taxa by degree of similarity, taking into account both shared derived similarities and shared ancestral similarities, and counting as "differences" features that are unique to a single taxon. Thus, groups A + B and C + D are found in the cladogram, but neither group is seen in the phenogram. This is because B is so different from A, and D is so different from C, that neither forms a group with its "nearest relative"; rather A and C group together because they are less different from each other, due to absence of those inferred changes, than either is from the other two taxa. Thus, under phenetic methods, a large number of uniquely derived attributes (e.g., characters 6–14 in taxon B, and characters 15–20 in taxon D) will cause a group to be formed (A + C), even though the members of that group share no derived attributes in common. In contrast, cladistic methods form groups only on the basis of shared derived attributes (group A + B, character 1; group C + D,

TABLE 1.1   Attributes of the "three schools" of systematics

ATTRIBUTE	PHENETICS	EVOLUTIONARY TAXONOMY	CLADISTICS
Data type	Character data converted to matrix of distances between taxa	Discrete characters	Discrete characters
Grouping method	Overall similarity	Special similarity	Special similarity
Diagram type	Phenogram	Evolutionary tree	Cladogram
Hierarchic level determined by	Amount of difference	Amount of difference	Sharing of unique attributes
Sensitive to rate	Yes	Yes	No*

* See Statistical Inconsistency and Long-Branch Attraction section in Chapter 5.

character 3) and treat attributes unique to a single taxon (e.g., those occurring in taxa B and D) as irrelevant to the recognition of groups. In this example, an evolutionary taxonomist would recognize the same *topology* as the cladist, but might also be influenced by the degree of divergence (difference) contributed by the uniquely derived attributes.

Alternative Approaches to Classification

The pheneticists argued for forming and recognizing groups on the basis of overall similarity, an approach implemented through the acquisition and analysis of as many "objective" observations as possible and converting those data into distance matrices. Their antiphylogenetic stance and quantitative persuasion set them apart for a while, but it was soon pointed out that phenetics and classical evolutionary taxonomy actually shared a critical element in common; that is, both approaches emphasized the importance of "overall similarity" in establishing rank in formal classifications. Cladistics, on the other hand, recognized groupings based solely on "special similarity," what Hennig called *synapomorphy*— the core principle of the Hennigian revolution (see Sidebar 2). If the classical evolutionary taxonomists and cladists could have agreed on the significance of *autapomorphies*—features unique to a single taxon—for the construction of formal classifications, they might have agreed on the choice of methods. Such was not the case, however.

Sidebar 2
The Writings of Willi Hennig: From Relative Obscurity to Preeminence

Emil Hans Willi Hennig (1913–1976) was a German dipterist of humble origins, whose theoretical work revolutionized systematic biology in the 1970s. With the recent centennial of his birth and 50th anniversary of the publication of his most celebrated work, several biographies and encomia have been published shedding light on his life and professional impact (Schmitt 2013; Rieppel 2016; Williams et al. 2016). The original works by Hennig on "phylogenetic systematics" (e.g., Hennig 1950), and his publications applying that method, were not widely read and appreciated because they were written in German, published in East Germany during a time of paper shortages, and therefore were not readily available to western systematists. Two 1966 publications, in English, changed this situation: Hennig's *Phylogenetic Systematics*, which was translated from a German-language manuscript, and a monograph by the Swedish entomologist Lars Brundin (1966) entitled *Transantarctic Relationships and Their Significance, as Evidenced by Chironomid Midges.*

As Henry Walter Bates' (1862) work on mimicry in butterflies served as an elegant corroboration of Darwin's (1859) theory of natural selection, so Brundin's monograph was unique in arguing persuasively for the theoretical merits of Hennig's methods, while simultaneously applying them to a relatively complex real-world problem. The end result was a magnificent taxonomic and biogeographic example supporting the Hennigian approach.

Gareth Nelson, late a postdoc from the Swedish Natural History Museum, zealously transported the ideas of Hennig and Brundin to the offices and lecture halls of the American Museum of Natural History in New York when he joined its staff as a curator of ichthyology in 1967. Nelson further invigorated discussion of the subject through his editorship of the journal *Systematic Zoology* from 1973 to 1976. Whereas some entomologists, such as Pedro Wygodzinsky (also of the American Museum), had independently become familiar with and applied Hennig's approach in their work, it was primarily Nelson's influence that changed the thinking of vertebrate systematists.

The transmission of phylogenetic methods to botanists took place more slowly. As late as 1978, Kåre Bremer and Hans-Erik Wanntorp lamented that botanists still had not recognized the importance of Hennig's work and pointed to some of the glaring cases of *paraphyletic* groups, such as Dicoteledoneae, still recognized in all major classifications of higher plants at that time. In the twenty-first century, the cladistic revolution in botany has certainly caught up with zoology, not only in the recognition of *monophyletic* groups through the application of cladistic principles but also in the production of phylogenetic trees implied by extensive databases of DNA sequence data (Angiosperm Phylogeny Group 2016; Soltis et al. 2017) and the preparation of second-generation textbooks embracing phylogenetic classifications (Judd et al. 2015; Christenhusz et al. 2017

Owing to their commitment to the dual macroevolutionary processes of *cladogenesis* (splitting) and *anagenesis* (progression) (Rensch 1954; Simpson 1961), evolutionary taxonomists recognized the importance of synapomorphy but also wanted to accommodate degree of difference in their circumscription of named taxonomic groups. For this reason, the pheneticists' and classical evolutionary taxonomists' classifications are often more similar to one another than to cladistic classifications. If this association at first seems contradictory—in light of the philosophical gulf separating these two groups of practitioners—consider the following statements by Mayr (1982:230): Cladistic "classifications are based entirely on synapomorphies [shared special similarities], even in cases, like the evolution of birds from reptiles, where the autapomorph characters [unique special similarities] vastly outnumber the synapomorphies with their nearest reptilian relatives," and (p. 233) "the main difference between [classical evolutionary taxonomy] and cladistics is in the considerable weight given to autapomorph characters." In other words, because birds have so many attributes unique to themselves, they should not be classified as a subgroup of Reptilia, even though all of their novel features are simply modifications of the more general attributes (*symplesiomorphies*) shared among dinosaurs, crocodiles, lizards, and other "reptiles." Mayr, in his role as one of the most vocal boosters for the classical evolutionary taxonomists, made it clear in this statement that autapomorphies must be considered in the recognition and ranking of groups in classifications (the logical corollary of separating a taxon out of a group based on autapomorphies is recognizing the remaining taxa in the "group" based on their sharing of complementary symplesiomorphies). Mayr cited examples to demonstrate why excluding unique characters in assessing rank produces what he believed to be absurd results. Mayr's taxonomic abominations included the grouping of man with the great apes as proposed by Linnaeus, among others, and more poignantly for him as an ornithologist, the treatment of birds and crocodiles as each other's nearest relatives among living organisms, or birds as a subgroup of dinosaurs when extinct taxa are also considered. The rejection by cladistic taxonomy of symplesiomorphies, such as lack of feathers and presence of teeth that make Reptilia a "group" separate from birds, is what Mayr argued against, and it is exactly that methodological attribute—the counting of symplesiomorphies as evidence of grouping—that distinguishes cladistics from classical evolutionary taxonomy and phenetics. As we shall see in subsequent chapters, this key distinction remains to the present day.

To further illustrate this point, consider the following classifications of the living amniote tetrapods as they might be produced by the three schools of taxonomy:

Evolutionary taxonomic classification:
Class Mammalia
Class Reptilia
    Subclass Testudines (turtles)
    Subclass Squamata (lizards, snakes, crocodiles)
Class Aves

Phenetic classification:
Class Mammalia
Class Reptilia
    Subclass Testudines
    Subclass Squamata
Class Aves

Cladistic classification:
Class Mammalia
Class Reptilia
    Subclass Testudines
    Subclass Sauria
        Infraclass Squamata (lizards and snakes)
        Infraclass Archosauria (crocodiles, birds)
            Order Crocodilia
            Order Aves

Even though the evolutionary taxonomists were aware that birds are phylogenetically part of the Reptilia, based on synapomorphies such as the diapsid skull, their desire to express degree of difference in formal classifications led them to obfuscate that relationship. Only the cladistic arrangement of groups within groups accurately represents in the formal classification the nested pattern of shared, derived attributes that is considered to portray genealogical relatedness and the evolutionary history of the groups. While these phylogenetic patterns have been uncontroversial for more than 30 years, general biology textbooks have continued to refer to Reptilia and Aves as separate classes as recently as 2014 (Raven et al. 2014), and the scientific literature is replete with reference to "the class Aves" (over 15,000 records in Google Scholar, 2016–2020).

Our discussion of classificatory methods so far has been based largely on the work of zoologists because their works clearly express the controversies surrounding classificatory methods and because the taxa discussed will be familiar to many readers. In botany there is a tremendous amount of synthetic taxonomic work, but a lingering influence on its preparation seems to stem from tradition rather than from a more critical approach to the selection and consistent application of methods. Nonetheless, the underlying philosophy employed in many botanical classifications is clear, as it is in zoology. Consider for example the classification of seed plants at the highest levels.

Traditional classification (e.g., Lawrence 1951):
Phylum Spermatophyta
    Subphylum Gymnospermae
        Class Cycadales
        Class Ginkgoales

        Class Coniferales
        Class Gnetales
     Subphylum Angiospermae
        Class Dicotyledoneae
        Class Monocotyledoneae

Cladistic classification (Loconte and Stevenson 1990, 1991):
Phylum Spermatophyta
    Subphylum Cycadales
    Subphylum Cladospermae
        Infraphylum Ginkgoales
        Infraphylum Mesospermae
           Microphylum Coniferales
           Microphylum Anaspermae
              Class Gnetales
              Class Angiospermae
                 Subclass Calycanthales
                 Subclass [Unnamed]
                    Infraclass Magnoliales
                    Infraclass [Unnamed]
                       Microclass Laurales
                       Microclass [Unnamed]
                       Et cetera

In the traditional classification, the Gymnospermae (including fossil taxa) are a group without any characteristics distinctive to them but rather are recognized by having seeds with two cotyledons and not having flowers. Obviously all Spermatophyta have seeds, most of which have two cotyledons, and therefore this attribute is not distinctive for the Gymnospermae. The Dicotyledoneae, the classification of which is listed only in part, are defined in a similar way: plants with flowers and two cotyledons. But most seed plants other than monocots have two cotyledons; therefore the characteristics of the Dicotyledoneae are more accurately stated as plants with flowers and without seeds with one cotyledon. The cladistic approach, which distinguishes synapomorphies from symplesiomorphies, has shown that Gymnospermae and Dicotyledoneae, and other long-recognized higher groupings within the green plants, are not substantiated by derived features but rather by a combination of attributes that also occur in other groups, or by the absence of the apomorphies that define their sister taxa. Clearly, traditional, noncladistic methods, whether explicit or not, recognize taxa based on eclectic combinations of synapomorphies and symplesiomorphies.

The first great period of modern systematic controversy ended with the decline of the phenetic and narrative evolutionary taxonomic schools as a general consensus emerged in the 1980s that the cladistic approach offers compelling methodological and philosophical advantages over those alternatives. Hennig's desideratum (1966:239) that "the phylogenetic system may be regarded, for inherent reasons, as the general reference system for biology," has been almost universally accepted. Few today would disagree that groups are natural, to the extent that they are *monophyletic*, based on inferred *synapomorphies*, and that the hierarchy of biological nomenclature should reflect this pattern. We may now confidently turn around Theodosius Dobzhansky's (1973) famous dictum "nothing in biology makes sense except in the light of evolution" to say, nothing in evolutionary biology makes sense except in the light of an empirically supported phylogenetic hypothesis. It is in light of this consensus and these advantages that this book is primarily focused on the explication of cladistic methodology. We will explore further the philosophical strengths and purported weaknesses of cladistics in comparison to alternative, statistical approaches to phylogenetic inference in Chapters 3 and 5, as well as delving into the last strongholds of grouping by overall similarity when discussing the analysis of molecular data.

Some Terms and Concepts

To aid our subsequent discussion of systematics, it will be helpful to clarify the meanings of several commonly used terms and to make further observations on the place of systematics within the broader field of biology. These definitions will help us to better comprehend the types of scientific problems systematists attempt to solve and to gain perspective on the scientific contributions of systematics as a field. The glossary at the back of the book includes definitions for all italicized technical terms used in this volume.

Taxonomy and *systematics* are terms that embody the activities of systematists, but the exact meanings ascribed to them have varied widely. Referring once again to the work of Simpson (1961:7), we find the following definitions:

> Taxonomy is the theoretical study of classification, including its bases, principles, procedures, and rules.

> Systematics is the scientific study of the kinds and diversity of organisms and of any and all relationships among them.

As Simpson's definitions suggest, systematics has often been used as the more inclusive term, with taxonomy and classification subsumed within it. In contrast, others have argued that the terms are synonymous, that systematics is the term having historical precedence, and that taxonomy should therefore be supplanted. Politically artful use of the terms, stemming from eighteenth-century debates about "system" (the quest for the true Natural Order) versus "method" (development of convenient but artificial classifications), has at times suggested that taxonomy somehow represents a mundane activity most closely associated with identification of specimens (natural history) and that systematics is the more elevated form of the discipline concerned with general principles and theory (natural philosophy). Among such usages could be included "biosystematics," a term implying the integration of a broader range of biological information than would be the case within "ordinary" systematic studies and a consequent elevation of the status of the field (e.g., Ross 1974; reinvented as "integrative taxonomy" by Dayrat 2005).

Systematics is the term used throughout this work because of its wide usage and broad connotation, particularly toward the realm of theoretical methodology. Taxonomy would, nonetheless, serve equally well. The field of study covered by these terms encompasses the methods and practice of describing, naming, and ordering biological diversity, at the species level and above.

Classification represents the nomenclatural or hierarchic formalization of systematic studies. According to Simpson (1961:9), "classification is the ordering of animals into groups (or sets) on the basis of their relationships, that is, of associations by contiguity, similarity, or both." This definition is concise and clear, marred only by the evolutionary-taxonomic equivocation about the basis for determination of "relationships" discussed in the previous section. Sorting out those similarities that imply phylogenetic relationship (synapomorphies) from those that do not (symplesiomorphies), and thereby arriving at classifications that reflect the historical pattern of evolutionary divergence, as best it may be inferred, is the subject of much of this book.

Within biology, the term classification usually refers to a "natural hierarchy," a nested set of groups within groups that is intended to reflect something about the intrinsic order of biological diversity. To Linnaeus (1758) and his contemporaries, the "Systema Naturae" was God's plan of creation. Since Darwin (1859), that natural hierarchy has been explained by the "hidden bond" of genealogical relationships among organisms through time. That is not to say that other types of classifications are not used by biologists or by other scientists. For example, special purpose classifications are commonly found in the form of keys (Figure 1.2)—schemes that explicitly invoke complementary alternatives to facilitate the efficient identification of organisms, or ecological guilds, such as herbs, shrubs, and trees. Such classifications are often based on particular readily observed aspects of similarity and may bear little resemblance to a "natural hierarchy."

A classification frequently contrasted to biological classification is the periodic table of the elements, the natural arrangement of which is organized to

1. Hind femora with moderate to dense covering of appressed, scalelike setae (figs. 81, 82), or rarely with scalelike setae restricted to ventral surface of femora (fig. 91) 2
Hind femora without scalelike setae .. 10

2(1). Hemelytral membrane with scalelike setae, usually most abundant inside areolar cells and along veins (fig. 13) .. 3
Hemelytral membrane without scalelike setae 4

3(2). Antennal fossae nearly contiguous with anteroventral margin of eye; length of antennal segment II slightly greater than width of head across eyes; peritremal disk and coxae pale ....
.... *taxcoensis*, new species
Antennal fossae removed from anteroventral of eye by distance equal to or greater than diameter of antennal segment I (fig. 7); length of antennal segment II much less than width of head across eyes (ratio—0.46:1 to 0.62:1); peritremal disk and coxae dark, or disk rarely somewhat paler than adjacent thoracic sclerites .... *balli* Knight

4(2). Hind femora with scalelike setae restricted to narrow band on ventral surface (fig. 91) .... *tuthilli* (Knight)
Hind femora with more or less generally distributed scalelike setae (figs. 81, 82) 5

5(4). Tibiae uniformly dark reddish brown or black, never paler than adjoining femora; antennal segment III uniformly darkened, without pale region basally; dorsum uniformly dark brown or black, without red or yellow markings ... 6
Tibiae, at least distally, yellow or brownish yellow, rarely somewhat darker, but always paler than adjoining femora; antennal segment III uniformly pale yellow to yellowish brown, or with distinct pale region basally; dorsal coloration variable, usually with at least bases of corium and clavus, embolium, and cuneus yellowish brown or red 8

6(5). Antennal segment II strongly inflated, greatest thickness nearly twice that of segment I (fig. 41); length of gonopore sclerite in lateral view approximately 1.5 times that of the gonopore (figs. 140, 141) .... *reuteri* Knight
Antennal segment II linear or weakly clavate, not strongly inflated, greatest thickness rarely little more than that of segment I (figs. 18, 22); length of gonopore sclerite in lateral view approximately twice that of the gonopore (figs. 120, 121, 123, 124) 7

7(6). Ratio of length of antennal segment II to width of head across eyes from 0.85:1 to 0.92:1; vesica as in figures 120 and 121, spinose field on gonopore sclerite usually broad proximally ....
.... *arizonae* (Knight)
Ratio of length of antennal segment II to width of head across eyes from 0.73:1 to 0.80:1; vesica as in figures 123 and 124, spinose field on gonopore sclerite usually narrow proximally ....
.... *cercocarpi* Knight

8(5). Ratio of length of antennal segment II to width of head across eyes from 0.86:1 to 0.90:1; vesica as in figure 160, with short gonopore sclerite, and gonopore well removed from apex of vesical strap
.... *ramentum*, new species
Ratio of length of antennal segment II to width of head across eyes from 0.72:1 to 0.83:1; vesica either with long gonopore sclerite (fig. 162), or gonopore located near apex of strap (fig. 122) .. 9

9(8). Vesical strap distad of medial coil elongate, gonopore removed from apex, gonopore sclerite with elongate row of evenly distributed spines (fig. 162) ...
.... *rubidus* (Uhler)
Vesical strap distad of medial coil short, gonopore near apex, gonopore sclerite with spines mostly restricted to distal half (fig. 122) ...... *atricolor* (Knight)

10(1). Hemelytral membrane with widely distributed scalelike setae (fig. 12) ......
.... *acaciae* Knight
Hemelytral membrane without scalelike setae 11

11(10). Dorsum without scalelike setae 12
Dorsum with scalelike setae, sometimes restricted to anterior margin of pronotal disk and bases of clavus and corium 13

12(11). Head, pronotum, and base of hemelytra yellowish orange, sometimes tinged with red; remainder of hemelytra shiny black; ratio of width of vertex to width of head across eyes from 0.48:1 to 0.51:1 ....
.... *chiapas*, new species
Head, pronotum, and hemelytra uniformly reddish brown; ratio of width of vertex to width of head across eyes from 0.36:1 to 0.40:1 ....
.... *polymorphae*, new species

13(11). Hemelytra with light and dark scalelike setae .... *nicholi* Knight
Hemelytra with silvery white scalelike setae only 14

14(13). Antennal segment II yellow or brownish yellow, rarely with apex narrowly darkened 15

FIGURE 1.2. Examples of dichotomous keys. Groups are progressively subdivided using easily recognized characteristics, but those characteristics may not be those that indicate natural groups. The indented format is frequently used in botanical works. Partial key of the insect genus *Atractotomus* from Stonedahl (1990; courtesy of the American Museum of Natural History); key to the Mexican species of plant genus *Galactia* from Standley (1922).

43. GALACTIA Adans. Fam. Pl. **2**: 322. 1763.

Scandent or erect herbs or shrubs; leaves pinnately 3 or 5-foliolate, the leaflets large or small; flowers small or large, usually racemose; fruit linear, bivalvate.

Leaflets 4 to 9 cm. wide. Plants scandent------------------1. G. viridiflora.
Leaflets less than 3.5 cm. wide.
  Flowers in axillary clusters-----------------------2. G. brachystachya.
  Flowers racemose.
    Racemes stout, dense, sessile, mostly shorter than the leaves.
                                                3. G. multiflora.
    Racemes slender, interrupted, pedunculate, mostly longer than the leaves.
      Plants erect; leaflets acute or acuminate------------------4. G. incana.
      Plants scandent or trailing; leaflets often obtuse.
        Leaflets glabrous on the upper surface------------5. G. acapulcensis.
        Leaflets variously pubescent on the upper surface.
          Leaflets bright green on the upper surface, not closely sericeous on
            either surface---------------------------------------6. G. striata.
          Leaflets grayish, closely sericeous on both surfaces.
            Leaflets white beneath with a soft silky pubescence, oval or ovate.
                                                7. G. argentea.
            Leaflets grayish beneath with rather stiff pubescence, usually
              oblong---8. G. wrightii.

FIGURE 1.2. (Continued)

reflect not putative historical patterns but rather the increasing atomic numbers and electron orbitals of the various chemical elements. It is interesting to note that Dmitri Mendeleev's original conception and execution of the periodic table was realized on the basis of chemical reactivities and atomic weights, prior to the discovery of many of the subatomic properties the table is now employed to explicate (Brower 2002). This parallels the observation mentioned earlier that the "Natural System" of Linnaeus, Georges Cuvier, and other pre-Darwinian systematists was an empirically valid pattern, even though their explanation of its cause is now considered to be nonscientific. The key point shown by both of these classification systems is that the pattern of relationships may be inferred without reference to, or knowledge of, the process that produced or explains it.

As we shall see, the activity of preparing formal classifications depends in part on the adoption of a set of procedures, as discussed in Chapter 8, rather than directly representing the result of an empirical analysis. Nonetheless, biological classifications have great scientific value as indices of accumulated knowledge, although the limits of their utility and the means by which they store and transmit information have often been misunderstood. These misunderstandings have led to many erroneous criticisms concerning the flaws of taxonomic ranking. We will discuss formal classifications further in Chapters 4 and 8.

In sum, the question of whether a consensus of opinion exists as to the precise meanings of the terms systematics, taxonomy, and classification might best be made by pondering the titles of foundational works describing the methods and procedures of this general field of study. Among many possible examples, we could mention Simpson's *Principles of Animal Taxonomy* (1961), Richard Black-welder's (1967) *Taxonomy*, Mayr's *Principles of Systematic Zoology* (1969), Roy Crowson's *Classification and Biology* (1970), and Herbert Ross's *Biological Systematics* (1974). All of these authors discussed essentially the same subject and had little hesitation in using these three closely related terms to describe that subject.

Another term stemming from Hennig's work that has found its way into text-book titles (e.g., Wiley 1981) is *phylogenetics*. Originally used as an adjective, the noun form has frequently announced an author's Hennigian approach. Today, it refers mainly to the tree-building aspects of systematics (phylogenetic inference), discussed in Chapter 5, whether sympathetic to Hennig's methods or not.

Identification usually means "to place a name on." We might say that a specimen was identified as such-and-such species, or a specimen was "determined" to be such-and-such. Identification is an important day-to-day activity for most systematists and possibly for most human beings, but the activity does not form the basis for recognition of groups or for establishing relationships among them. Rather, correct identification, critical in applied fields such as medicine and agriculture, is an application of knowledge obtained from the fundamental work of systematics.

The Place of Systematics within Biology

There is a frequently held perception on the part of the general public, and some scientists, that equates science with experiment. Much of science is experimental in nature, but this is not a prerequisite for a researcher's work qualifying as "scientific" (Cleland 2002; Leonelli 2016; McIntyre, 2019). Systematists make discoveries about the natural world, but those discoveries are usually the results of observation and comparison rather than experiment. That other broad fields of science need not be experimental might be appreciated by noting that sequencing genomes, measuring seismic activity of the Earth's crust, and drawing celestial maps are not fundamentally experimental activities but nonetheless represent important and valid endeavors in science.

The field of biology has been divided in many ways. Not all of these divisions are easily compared, and indeed some are incompatible. The division can be taxonomically based, such as botany and zoology. Or it might distinguish basic research from applied fields, such as agriculture and medicine, or be based on scale of organization, such as molecular biology, cell biology, and comparative anatomy These approaches have not served well to incorporate systematics as a field of study. More constructively, we might divide academic biology into evolutionary and nonevolutionary, comparative and general, or reductionist and integrative. Even these dichotomies have their limitations: in universities, one frequently finds systematics subsumed within more general categories such as "ecology and evolution," where it may be viewed as a relatively arcane elective component of specialized graduate education. In some academic programs, systematics has been eclipsed by the more technological disciplines of genomics and bioinformatics, which focus almost entirely on computer-based manipulations of molecular data. These trends are unfortunate because a stronger emphasis on sound training in systematic principles could help forestall the uncritical proliferation of ill-conceived phylogenetic applications by geneticists, molecular biologists, and other users of the systematic "toolbox," as will be discussed further in Chapter 2.

Systematics is the most strongly comparative of all of the biological sciences, and its methods and principles transcend the differences between botany and zoology. It is also the most strongly historical field within biology, and as such provides the basis for nearly all inferences concerning historical patterns and processes. Among the earth sciences, systematics is directly comparable to historical geology, and indeed the two fields find integration in paleontology.

We might contrast systematics with "general biology," which often involves the study a single species, a single organ system, or a single biochemical pathway. Such an approach is nonhistorical and places little emphasis on comparisons among species and their organ systems. On the other hand, the reason that *Drosophila* or "the mouse" makes a generalizable model system for studying animal genetics and physiology is because many of those creatures' basic biochemical pathways and processes are homologous with those in human beings.

The Units of Systematics

A *taxon* (plural taxa) is the basic unit of systematics. This term can be used to refer to a group of organisms at any level in the systematic hierarchy, from species to kingdoms or domains. It does not refer to individual organisms, despite the fact that systematists often study individual organisms as *exemplars* of a given taxon. Because it is frequently necessary to refer to a group of organisms without reference to its hierarchic position, a number of words for making such reference have been employed. Thus, terms such as *terminal taxon* and *operational taxonomic unit* (OTU) have been proposed, the former being used primarily in cladistics, the latter arising out of phenetics. Although such terms may carry somewhat different connotations, depending on the author who is using them, their meaning is generally the same. We defined taxon above; we might define a terminal taxon as a group of organisms that for the purposes of a given study is assumed to be homogeneous with respect to variation in the characters that bear upon its relationships to other such groups. Attempting to avoid perceived evolutionary baggage of prior taxonomic hypotheses, pheneticists employed the term operational taxonomic unit (OTU) as a synonym for terminal taxon. The taxa on the tips of the *branches* of a phylogenetic tree are the terminal taxa (regardless of whether they are species or more inclusive groups), and under the cladistic approach, features that are unique to a terminal taxon in that context are *autapomorphies*. In general, cladists consider that any taxon more inclusive than an individual species should be a *monophyletic* group. We will address the varying conceptions of species in the systematic, evolutionary, and ecological literature in Chapter 7.

The Systematic Literature

The systematic literature begins with the earliest works purporting to represent biological classifications, from the mid-eighteenth century. Because of the rules of nomenclature (discussed in Chapter 8) and the tremendous biological diversity encompassed by systematic research, present day workers are confronted with an almost overwhelming number of pertinent publications. Much of the older or foreign systematic literature will not be found in physical form in local libraries, or even in many college or university libraries, but rather exists only in specialized research libraries. However, this situation has been ameliorated somewhat in the twenty-first century by digitization of many rare, classic works. A brief survey of the types of publications containing information on biological systematics will help to orient a search for relevant books, articles, and other reference materials.

Descriptive Works

Documentation of biological diversity historically has been a largely descriptive enterprise. This work includes not only the recognition of species and the higher taxa into which they are grouped but also the more detailed inquiry into the structural attributes (morphology, anatomy) of those organisms. The larger and more comprehensive publications of this type are often called monographs or revisions (Figure 1.3). They have sometimes appeared as books, particularly before about 1840, but after that time more commonly as journal articles, the periodical literature of science.

Mertila bhamo, new species
Figures 36, 38

DIAGNOSIS: Recognized by the broadly reddened bases of clavus and corium; length of antennal segment II nearly equal to width of head across eyes; prominent tylus; and the structure of the male genitalia (fig. 38), especially the large, spinelike tubercles on basodorsal margin and inner-medial surface of right lateral process of genital capsule (fig. 38b, c).

DESCRIPTION: MALE. Length 6.25; dark, metallic coloration on distal portion of clavus and corium not extending anteriorly beyond level of apex of scutellum, embolium pale to near level of apex of corium. HEAD. Width across eyes 1.40, width of vertex 0.72; tylus prominent, strongly produced basally; length of antennal segment I 0.59, basal half light reddish brown, darkening to fuscous apically, segment II 1.35, dark reddish brown, very slightly expanded distally; labium damaged distally. PRONOTUM. Posterior width 1.90.

HEMELYTRA. Dark coloration on corium and clavus more brownish black. LEGS. Femora yellowish brown, tinged with red; tibiae brown; tarsi brown or brownish yellow. GENITALIA. Figure 38. FEMALE. Length 5.50; embolium less extensively pale than for male.

ETYMOLOGY: Named for the type locality; a noun in apposition.

DISTRIBUTION: Burma (fig. 36).

HOLOTYPE ♂: BURMA: **Bhamo District:** Bhamo, Aug. 1885, Fea (Distant collection, 1911-383) (BMNH). The above specimen was incorrectly identified as *malayensis* by W. L. Distant in the early 1900s. It also seems to be the specimen upon which Poppius (1912a) based a redescription of *malayensis.*

ADDITIONAL SPECIMEN: INDIA: 1♀, [port interception with no specific locality data], July 22, 1939, "on *Vanda*" (USNM). This specimen appears conspecific with the type (same dorsal coloration, prominent tylus, and second antennal segment), but a male example is needed for positive identification.

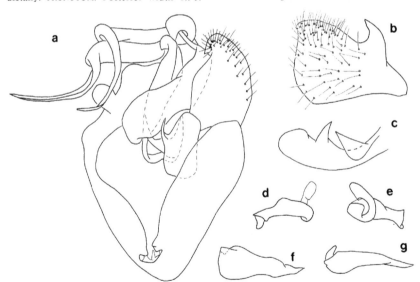

Fig. 38. Male genitalia of *Mertila bhamo.* **a.** Genital capsule, posterior view. **b, c.** Right lateral lobe of genital capsule. b. Lateral view. c. Dorsal view. **d, e.** Left paramere. d. Lateral view. e. Posterior view. **f, g.** Right paramere. f. Lateral view. g. Dorsal view.

FIGURE 1.3. Typical page from a taxonomic revision, including diagnosis, description, holotype designation, locality data, and figures of structures important in characterizing a taxon, in this case described as new (from Stonedahl 1988:37, 38; courtesy of the American Museum of Natural History).

Today there are thousands of journals that publish myriad articles on systematics. Many of these publications focus on a single group of organisms (e.g., *Lambillionea*, which focuses on Lepidoptera, and *Copeia*, which focuses on "reptiles" and amphibians); others are restricted geographically, such as the *Pan-Pacific Entomologist* or the *Journal of the Tennessee Academy of Science*; and still others may publish systematic papers among a broad spectrum of other scientific articles (if you are lucky enough to find a new species of fossil hominid or dinosaur, you may be able to publish your discovery in *Science* or *Nature*). A lot of basic taxonomy and description of new species is currently being published in the enormous online journals *Zootaxa* and *Zookeys* (animals), *Phytotaxa* and *Phytokeys* (plants), and *Mycotaxon* and *Mycokeys* (fungi). Current issues and back content of many "traditional" journals are also accessible online through digital subscriptions by research libraries.

A growing body of historical literature is being scanned to digital format and made freely available on the World Wide Web (hereafter, web) as PDF files, either by the societies that publish particular journals, by museums and libraries such as the Biodiversity Heritage Library (online) or via omnibus digitization efforts like Google Scholar.

Summaries of the primary descriptive literature often become available to the general public in the form of handbooks, field guides, faunas, floras, and similar sources that are written in a more easily understood fashion and present information in a more distilled manner. Many efforts to summarize natural history information have also migrated to the web, including such online resources as BugGuide.net, the Encyclopedia of Life, and even Wikispecies. These secondary contributions, while attractive, functional, and instantly accessible (in the case of online resources), are often neither exhaustively researched nor peer reviewed and may contain nomenclatural or identification errors. Such works generally should not contain formal descriptions of new taxa or taxonomic revisions (see further discussion of appropriate publication formats for taxonomic descriptions in Chapter 8).

Catalogs and Checklists

Comprehension of the literature dealing with animals, plants, and other organisms is a formidable task simply because of the sheer volume. This gulf is bridged by publications that are often referred to as catalogs (Figure 1.4) and checklists. These are comprehensive lists of taxonomic names, their authors, and the publications where they first appeared, sometimes also including lists of subsequent revisions and even other significant mentions of that taxon in the literature, providing a roadmap for systematists to find pertinent work on their group of

Gen. **Cyclosternum** Ausserer, 1871

♂♀ **kochi** (Ausserer, 1871)   Venezuela
   C. k. Schmidt, 1993d: 62, f. 55–56 (♂).

♀ **longipes** (Schiapelli & Gerschman, 1945)   Venezuela
   C. l. Schmidt, 1993d: 62, f. 51 (♀).

♀ **obesum** (Simon, 1892)   Brazil
   Magulla obesa Gerschman & Schiapelli, 1973b: 75, f. 14–15 (♀).
   C. o. Schmidt, 1993d: 62, f. 50 (♀).

♂ **rufohirtum** (Simon, 1889)   Venezuela
   Adranochelia rufohirta Gerschman & Schiapelli, 1973b: 62, f. 74–78 (♂).
   C. r. Schmidt, 1993d: 62, f. 57–60 (♂).

♂♀ **schmardae** Ausserer, 1871   Colombia, Ecuador
   C. s. Bücherl, Timotheo & Lucas, 1971: 124, f. 26–27 (♂).
   C. s. Gerschman & Schiapelli, 1973b: 67, f. 68–73 (♂♀).
   C. s. Schmidt, 1993d: 62, f. 52–54 (♂♀).

♂♀ **stylipum** Valerio, 1982   Costa Rica, Panama
   C. s. Nentwig, 1993: 95, f. 40a–d (♂♀).

♂♀ **symmetricum** (Bücherl, 1949)   Brazil
   Magulla symmetrica Bücherl, 1950: 1, f. 1–3 (D♂).
   Magulla symmetrica Bücherl, 1957: 391, f. 36–36a (♂).
   C. s. Schmidt, 1993d: 62, f. 61–63 (♂♀).

Gen. **Cyriocosmus** Simon, 1903

♂♀ **elegans** (Simon, 1889)   Venezuela, Brazil, Bolivia
   C. e. Schiapelli & Gerschman, 1945: 181, pl. VIII (D♂).
   C. e. Schenkel, 1953a: 3, f. 3a–c (♂).
   Pseudohomoeomma fasciatum Bücherl, 1957: 391, f. 38–38a (♂).
   Pseudohomoeomma fasciatum Bücherl, Timotheo & Lucas, 1971: 125, f. 35–37
      (♂♀).
   Chaetorrhombus semifasciatus Bücherl, Timotheo & Lucas, 1971: 126, f. 42–43 (♀).

♂♀ **sellatus** (Simon, 1889)   Brazil
   C. s. Gerschman & Schiapelli, 1973b: 67, f. 16–22 (♂♀).
   C. s. Schiapelli & Gerschman, 1973b: 65, f. 1–4, 16–18 (♂♀).
   C. s. Schmidt, 1993d: 63, f. 64–67 (♂♀).

Gen. **Cyrtopholis** Simon, 1892

Transferred to other genera:
C. cyanea Rudloff, 1994 – see **Citharacanthus.**

♂ **agilis** Pocock, 1903   Hispaniola
   C. a. Bryant, 1948b: 337, f. 6–7 (♂).

FIGURE 1.4. A page from a systematic catalog, in this case of the spider family Theraphosidae names, sexes known, figures published, distributional information, and the literature sources for these data.

interest. The exact form differs from group to group, but the general type of information contained in them is the same.

In botany, the historically most important source of this type is the *Index Kewensis*, which listed the names and place of publication of all seed-plant taxa at the family level and below. A new supplement was issued every five years. The work was started by Joseph Dalton Hooker, then director of the Royal Botanic Garden at Kew; the original four volumes were prepared with funds provided by Darwin. This database has been merged with data from the Harvard University Herbaria and the Australian National Museum to form a website called the International Plant Names Index (online). Compact, readily available sources for the names of most plant genera are Willis (1973), Wielgorskaya (1995), and Mabberley (2017). The Kew Record of Taxonomic Literature (online) is an additional valuable source for locating botanical literature.

The situation in zoology is more complicated, and no single source will serve for all groups; indeed, no definitive works exist for many taxa. Nonetheless, one will find world or regional catalogs or checklists for many groups of animals, these works usually providing information on names, distributions, and, often, including extensive lists of sources from which the information was derived. Examples include Eschmeyer (1990) for fishes, Sibley and Monroe (1990) for birds, Wilson and Reeder (2005) for mammals, and Pelham (2008) for the butterflies of North America. For many less well-known groups, it may only be possible to find a listing for Western Europe—if the group occurs there.

The type of structured information contained in catalogs lends itself to preparation in the form of computer databases, and these are becoming available at an increasing rate through major museums and international initiatives such as the Global Biodiversity Information Facility (GBIF). Creating and using digital databases (biodiversity informatics) will be discussed further in Chapter 8. Taxonomic information derived from the web should always be cross-checked for accuracy!

Theoretical Literature

Most scientific fields have technical journals that deal with the methods of the discipline. Systematics is no exception. The oldest such journal is *Systematic Biology* (*Systematic Zoology* prior to 1991), published by the Society of Systematic Biologists. Since its inception in 1952, this journal has included many important articles pertinent to both zoology and botany. *Cladistics*, published by the Willi Hennig Society, is largely a journal of papers on methods and concepts, although like *Systematic Biology*, it also publishes articles on the phylogenetic relationships of particular groups of organisms.

Taxon, published by the International Association for Plant Taxonomy, publishes articles of general interest to systematic botanists, including "official commentary" on botanical nomenclature. *Systematic Botany*, published by the American Society of Plant Taxonomists, includes articles of broad general interest to systematists as well as articles dealing more strictly with the details of plant classification.

A number of other journals, such as the *Australian Journal of Botany*, *Biological Journal of the Linnean Society*, *Journal of Zoological Systematics and Evolutionary Research* (formerly *Zeitschrift für Zoologische Systematik und Evolutionsforschung*), *Systematics and Biodiversity*, and *Zoologica Scripta*, and occasional symposium or edited volumes, also publish articles on the theory and methods of systematics. For those inclined to philosophical issues, *Biology and Philosophy* and *Philosophy of Science* contain pertinent articles.

The burgeoning study of molecular data applicable to systematics has spawned a number of journals, including *Journal of Molecular Evolution*, *Molecular Biology and Evolution*, *Molecular Phylogenetics and Evolution*, and *Molecular Ecology*. These journals deal mainly with the results of molecular studies and the analysis of molecular data (as, increasingly, do many of the traditional journals mentioned above).

Textbooks

Since the appearance of Hennig's *Phylogenetic Systematics* in 1966, a number of book-length "essays" on systematics have appeared. Each has its own strengths, but many are now somewhat dated because of advances in the field, particularly in the use of numerical methods. Others adopt a perspective on these matters that differs strongly from the views presented here. The thoughtful scholar of systematics is encouraged to investigate the diversity of views presented by these authors. A selection of their books is listed at the end of the chapter under Suggested Readings.

Literature Searches, Abstracting, and Indexing Sources

Almost all of the most useful tools for exploring the scientific literature in general, including systematics, are now to be found on the web. The web is a tremendous asset for science, since it allows instantaneous searching for, access to, and dissemination of information without the time and expense of traditional mail or trips to the library. In addition, web-based resources can be updated as new information becomes available, while traditionally published media are static. Some of the most useful online search engines for general scientific literature, now including results from as far back as 1900, include Web of Science, Scopus, and the aforementioned Google Scholar.

Most present-day systematists are actively engaged in posting the results of their work on the web either on personal or institutional websites, on taxon-oriented websites, or on more general repositories, such as ResearchGate (online) and Google Scholar. Two repositories that present information about biodiversity in a phylogenetic context are the Tree of Life Web Project (Maddison and Schultz, 2007), a now-defunct collaborative effort to graphically display the hierarchy of life in an interactive format, and OpenTree (Hinchliff et al. 2015). Tree-BASE (Piel et al. 2007) is a database of phylogenetic data matrices and results from a variety of published sources. The US government's National Center for Biotechnology Information, GenBank (NCBI 2018), is the repository for most published DNA sequence data employed in phylogenetic studies. Another major database of "DNA barcodes" (partial sequences of mitochondrial cytochrome oxidase I genes) is maintained by the (online) Barcode of Life (BoLD) initiative. Online sources that provide "authority files" of nomenclatural information include, among others, the Integrated Taxonomic Information System (ITIS) and Species 2000. Another online source—Global Names Index—provides an archival index of biological information found on the web in a nomenclatural context. Several additional websites aggregate information from a wide variety of other sites and present it in a unified context. These include the Global Biodiversity Information Facility (GBIF), which serves as a portal to a consortium of member sites, and DiscoverLife, which presents nomenclatural information, images, electronic keys, specimen data, and real-time mapping with a level of integration that none of its individual collaborating information providers is able to achieve. See further discussion and documentation in Chapter 8, Systematic Databases.

If the numbers of animal taxa seem at times almost overwhelming, so do the numbers of papers published on them. This was well recognized in the last century and was the reason for starting the Zoological Record (now online), an annual digest of the literature on animals, first published in 1864. At the time of its inception, the organization of the Zoological Record contained a bibliography whose cited articles were indexed primarily for their systematic contents. Now many additional subject areas are indexed. This basic reference is available online and most research libraries will provide access. Computer searches of the Zoological Record are now possible via the Web of Science for all records from the inception of its publication in 1864.

Literature on fossil vertebrates is summarized in the works of Hay (1902), Romer (1962), and the online Bibliography of Fossil Vertebrates published by the Society of Vertebrate Paleontology. The Paleobiology Database provides online access to a wealth of up-to-date information on fossil taxa, both plant and animal, including associated geological data. Additional abstracting sources exist for agriculture and other fields; these are often valuable sources for the more strongly applied literature.

Systematic Collections

Systematic collections serve as repositories for specimens in the same way that libraries serve as repositories for documents. The most important and broad-ranging natural history collections are in large public museums, mostly in Europe and North America. No museum has representatives of all species, but each has its strengths. Thus, it is part of the training and experience of systematists specializing on a given group to learn the repositories of specimens important to their work. Efforts to integrate collections digitally, such as iDigBio (online) help to democratize specimen data by making them globally accessible.

The Index Herbariorum (Thiers 2018) is a searchable online database of the world's herbaria, with a description of their collections, staff, and links to further information. Extensive search criteria make finding information on the deposition of plant collections and the specialists who work on them relatively straightforward. In addition, many herbaria are now scanning digital images of their collections, which may then be viewed remotely online (of course, digital imagery will never replace the need for study of actual specimens). The situation in zoology is much more complicated. Although published directories exist for workers in some groups, many such directories are badly out of date. Likewise, there are few comprehensive published listings of museum holdings. Thus, students in zoology wishing to learn who is working on a certain group or where relevant material is deposited should first consult their librarian or a specialist in the group for resource guides. If no useful sources are found, information must be acquired on a more piecemeal basis, through perusal of the primary literature, through experience, through word of mouth, and, of course, by Google or other web searches.

The internet is becoming a critical tool allowing discovery of collections' resources on a much more rapid and efficient basis than has heretofore been possible. We will continue to see dramatic changes in our ability to acquire information on collections for both botany and zoology in the coming years. Partial lists of specimen holdings are now available on the web for many collections, as well as increasing numbers of digitized images of the specimens themselves, and these numbers are increasing on a daily basis. On the other hand, as noted earlier, digitization is not a substitute for physical specimens, and the role of museums as repositories of the "real things"—including their genetic resources—will remain vital in perpetuity.

SUGGESTED READINGS

Blackwelder, R.E. 1967. Taxonomy: A text and reference book. New York: John Wiley and Sons. [An interesting counterpoint to Mayr, Linsley, and Usinger]

Crowson, R.L. 1970. Classification and biology. Chicago: Aldine Publishing. [An independently conceived exposition on the thought process and methods in biological classification]

Eldredge, N., and J. Cracraft. 1980. Phylogenetic patterns and the evolutionary process. New York: Columbia University Press. [Essay on cladistics and its relationship with the study of evolution]

McKenna, M.C., and S.K. Bell. 1997. Classification of mammals above the species level. New York: Columbia University Press. [Part I includes comments on the history and theory of classification]

Nelson, G., and N. Platnick. 1981. Systematics and biogeography: Cladistics and vicariance. New York: Columbia University Press. [Chapter 2 is a readable and stimulating review of the history of systematic thought]

Panchen, A.L. 1992. Classification, evolution, and the nature of biology. Cambridge: Cambridge University Press. [A compelling historical and philosophical review of the development of modern systematic theory]

Rieppel, O.C. 1988. Fundamentals of comparative biology. Basel: Birkhäuser Verlag. [A philosophical treatment of the cladistic approach]

Schoch, R.M. 1986. Phylogeny reconstruction in paleontology. New York: Van Nostrand Reinhold. [Excellent literature review of cladistics, with special reference to paleontology]

Wägele, J.-W. 2005. Foundations of phylogenetic systematics. Munich: Verlag Dr. Friedrich Pfeil. [A continental view of phylogenetics]

Wheeler, W.C. 2012. Systematics: A course of lectures. Hoboken, NJ: John Wiley and Sons. [Another, more quantitative take on systematics]

Wiley, E.O., and B.S. Lieberman. 2011. Phylogenetics: Theory and practice of phylogenetic systematics. Hoboken, NJ: John Wiley and Sons. [An update of Wiley's classic 1981 book, adopting a more syncretic approach than the original]

Wiley, E.O., D. Siegel-Causey, D.R. Brooks, and V.A. Funk. 1991. The compleat cladist: A primer of phylogenetic procedures. Lawrence: University of Kansas Museum of Natural History. [A workbook of basic cladistic methods; also available as a PDF on the web]

Williams, D.M., and M.C. Ebach. 2008. Foundations of systematics and biogeography. New York: Springer Science+Business Media. [A historical review of the development of systematics]

Williams, D.M., M. Schmitt, and Q.D. Wheeler, eds. 2016. The future of phylogenetic systematics: The legacy of Willi Hennig. Cambridge: Cambridge University Press. [A collection of essays marking the centennial of Hennig's birth, assessing his impact on systematics]

Winston, J.E. 1999. Describing species. New York: Columbia University Press. [A how-to manual for alpha taxonomy]

SYSTEMATICS AND THE PHILOSOPHY OF SCIENCE

> **Science—which is essentially critical—is also more conjectural and less self-certain than ordinary life because we have consciously raised to the level of a problem something which normally may have been part of our background knowledge.**
>
> —Karl R. Popper (1979:80)

At the end of our historical introduction to systematics in Chapter 1, we had explained the major differences between the three "schools" of systematics that were competing against one another in the 1960s and 1970s. By about 1985, cladistics had emerged as the victor in that competition. The reasons were straightforward: other than a few recalcitrant holdouts such as Mayr (e.g., 1997), no systematist disagreed in principle that hypotheses of relationship reflecting inferred patterns of evolutionary history were desirable as a general reference system for comparative biology. James S. Farris (1979a, 1979b, 1982) had shown that the cladistic method of grouping by synapomorphy provided a transparent connection between empirical evidence and putative patterns of historical divergence, thus deflating the pheneticists' claims that hypotheses of relationship that reflected phylogeny could not be achieved or did not adequately convey information about the Natural System. The evolutionary taxonomists' objections to strictly monophyletic classifications also subsided, as empirically supported cladograms proliferated that plainly contradicted their preferred classifications and revealed the hypocrisy of their stance, in light of their own claims that "every taxonomic category should thus, ideally, be monophyletic" (Mayr 1942:276). With the tools in hand to infer the existence of those monophyletic groups, demonstrably paraphyletic taxa were largely banished to the periphery of systematics, and "metaphysical cladism" (Brower 2018a) ruled the day.

Yet there was a fly in the ointment: statistical geneticist Joe Felsenstein (1978) observed that cladistic *parsimony* could be statistically *inconsistent*. What this means is that, under particular circumstances, the most parsimonious tree might not be the "true tree," and the more data gathered under those circumstances, the more strongly supported the wrong tree would become. Following early efforts by population geneticists Luca Cavalli-Sforza and Anthony Edwards (1967), Felsenstein proposed an alternative approach to phylogenetic inference—maximum likelihood—that he claimed was not susceptible to this defect, thereby reframing phylogenetic analysis as a problem of statistical inference. And so began the proliferation of ever more complex statistical models and ad hoc probabilistic workarounds that has led systematics to the bewilderingly complex state of affairs in which it is currently mired. Felsenstein's claim about the potential inconsistency of parsimony is demonstrably the truth, but it is not the whole truth. Before we reveal the denouement of the controversy that Felsenstein's attack on cladistics ignited, we will explain the cladists' perspective on the merits of their own approach and the philosophical arguments that have been advanced to support it. There are a number of critical issues to unpack, not the least of which are the principle of parsimony itself, what we mean when we talk about the "true tree," and whether, in the world of empirical data and scientific hypotheses, we can ever be certain, statistically or otherwise, that we know the truth.

Why Philosophy, and Why 80+-Year-Old Philosophy?

In the eighteenth and nineteenth centuries, systematics was considered to be a branch of natural philosophy, whose goal was to reveal the pattern of relationships among biological entities and, since Darwin, to elaborate a naturalistic explanation for that pattern. Today, we recognize the scope of pertinent philosophical issues arising from this project as ranging from questions about the nature and evaluation of evidence, to how the evidence supports one or more hypotheses of relationships, to how to judge the quality or plausibility of such hypotheses and compare them against one another. These theoretical/empirical problems do not have straightforward answers that can be calculated like the solution to a mathematical equation. As an account of a historically contingent singularity, the chronicle of life on Earth does not offer opportunities for statistical verification through replicated sampling from independent, identically distributed pools of evidence. We can still infer those patterns and processes, but there is a multiplicity of inferential techniques that may lead to different, incompatible solutions. Nor are there criteria for assessing the reliability or plausibility of a given hypothesis that are independently verifiable against an external reality. Each of these conundrums is laden with philosophical background assumptions, whether one chooses to acknowledge them or not.

One of us, Randall T. Schuh, was witness, participant, and "referee" (in his capacity as editor of *Systematic Zoology*, 1976–1979) to the decisive contests of the Hennigian revolution. The philosophical champion for many cladists at that time was the noted twentieth century Austrian philosopher of science, Karl Popper, whose ideas we will discuss in depth below. Much of the conceptual and semantic framework of modern cladistic research was inspired by Popper's thought (cf. Platnick and Gaffney 1978), but it is legitimate to question whether or not, 40 years later, these ideas are still pertinent to the challenges outlined above. We believe they are: Popper's writing is clear and compelling, and his critical approach meshes well with the Hennigian perspective. Perhaps, in particular, the cladistic criterion of parsimony and the asymmetry of synapomorphy and symplesiomorphy seem harmonious with Popper's falsificationism. This is not to suggest that everything Popper wrote pertains to cladistics, nor that his ideas are themselves immune from criticism (Popper would certainly not have wished that to be the case!).

Of course, Popper is not the only philosopher whose work bears upon the questions above, nor is he the only philosopher we enlist in our discussion. We consider the more recent ideas of Elliott Sober, David Hull, and Bas van Fraassen, as well as the contributions to the philosophical dialog by systematists such as Norman Platnick, Kirk Fitzhugh, Arnold Kluge, Olivier Rieppel, Maureen Kearney, and Lars Vogt. Of course, we also offer our own perspectives on these matters. The enduring intractability of philosophical problems related to describing and explaining natural phenomena perpetuates the relevance of classical contributions by the likes of David Hume and Emmanuel Kant as well. For that matter, Aristotle's works, written more than two millennia ago, still cast a long shadow upon the philosophical face of systematics today.

Much of the philosophical discussion pertaining to systematics in recent years has either disputed the relevance of Popper's framework to historical science, questioned the legitimacy of parsimony from a likelihood perspective, or drilled down on relative minutiae, such as the fit between Popper's corroboration formula and Bayes' theorem. Thus, we consider this chapter as an opportunity to review and reboot, allowing us to frame first principles and chart what we hope is a coherent philosophical landscape to provide readers unfamiliar with this background a context for understanding, even if in the end they do not agree with, the principles, methods, and applications described in the rest of the book.

Objective Knowledge versus Truth in Science

Science, including systematics, deals with the discovery of knowledge about the natural world. Yet, the path from basic observation to "knowledge" is neither direct nor obvious. Nevertheless, many scientists are content to gather and interpret evidence within the established theoretical framework of their discipline as "facts," without questioning the underlying assumptions of their methods. The work of "normal science," as described by philosopher Thomas Kuhn (1970), is not directly concerned with how we judge the truthfulness of the knowledge we acquire, that is, which ideas represent mere conjecture and which represent knowledge or whether or not we can trust our observations of the real world. Philosophy as a field of inquiry may help us to gain understanding of these issues (see, for example, Daston and Galison 2007, for an exploration of the "objective" activity of scientific illustration).

Above, we referred to "metaphysical cladism," which represents the somewhat naive belief that phylogenetic trees produced by cladistic (or other) methods, and the monophyletic groups they imply, represent accurate depictions of evolutionary history. Such are claims about the world as it truly is (or has been). Confronted with such propositions, a thoughtful person will before long begin to wonder, "how do we know that?" Within philosophy, it is *epistemology*—the inquiry into the origin, nature, and limits of knowledge—that makes science science, rather than a mystical system of beliefs. Thus, it is to this area that we now turn.

First, let us contrast alternative understandings of the concept of truth. Some scientists evidently believe that, as per the slogan from *The X-Files*, "the truth is out there." For example, Taran Grant (2002:100) said, "the ultimate aim of scientists is to achieve accurate knowledge of the world as it truly is (i.e., to attain Truth)." This is called foundationalism, or a *correspondence* theory of truth. Under such a theory, our empirical observations and the theories constructed upon them are either accurate or inaccurate, and the aim of science is to develop an accurate description of the way the world really is.

Other scientists, the authors of this book among them, do not consider that truth with a capital *T* is empirically accessible, and that instead, the aim of science is to develop a picture of the world that is internally consistent and provides a good explanation of the observable evidence. This is called a *coherence* theory of truth, and may be associated both with the hypothetico-deductivism of Popper and the constructive empiricism of van Fraassen, as we will outline below. A coherentist view is expressed in the continuation of Grant's statement quoted above: "verisimilitude cannot be defended as a scientific aim or a virtue of scientific hypotheses, because barring some sort of providential revelation, we are incapable of recognizing Truth and therefore of knowing (or estimating) how verisimilar any hypothesis is." Together, we might view the two halves of Grant's position as "the incoherent theory of truth"! Brower (2016a) discussed correspondence/coherence, concluding that Hennig was a coherentist.

The terms *fact* and *theory* are often juxtaposed in common parlance, the former connoting a kind of certainty that the latter is considered to lack. Thus, the theory of evolution is often disparaged as "only a theory" by those who would supplement or supplant it with creationist accounts in US school curricula. Ideas in science are often referred to as *theories*, or *hypotheses*, terms that will be used interchangeably in this book, even though *theory* is often used in reference to a more general or empirically substantiated concept than is *hypothesis*. A *fact* is simply a theory that is so well corroborated as to be uncontroversial. Facts, theories, and hypotheses are propositions to be critically evaluated in light of pertinent evidence and supported or rejected. The edifice of science is constructed from an interwoven network of theories of greater and lesser generality, the less general underlying the more general. Nonetheless, the terms "truth" and "fact" are frequently used in the popular press when referring to scientific statements (not to mention their widespread use in the legal profession) and are sometimes encountered in the writings of scientists and philosophers themselves.

Science produces what Popper (1979) referred to as "objective knowledge." The question of whether such knowledge is, or can ever be, absolutely truthful is a metaphysical problem beyond the scope of this work. Echoing Popper, Oreskes et al. (1994) observed that "it is impossible to demonstrate the truth of any proposition, except in a closed system," and that "we can never verify a scientific hypothesis of any kind." A commitment to absolute truth would seem to involve an act of faith, an approach to the acquisition of knowledge that, in our view, empirical science does not admit. Thus, in this book, all scientific statements will be treated as hypothetical, albeit some of them receiving stronger support from the available evidence than others.

Some examples may help put the tenuous nature of truth in science into perspective. Few people today doubt the correctness of the heliocentric theory of our solar system as propounded by Nicolaus Copernicus. Indeed, most probably consider the idea that the earth (and the other planets) revolve around the sun to be irrefutable. Nonetheless, the Ptolemaic theory, in which the earth was fixed in the center of the universe with the sun, planets, and stars rotating around it, was so widely believed—after centuries of detailed observation—and so entrenched in church dogma that the Copernican theory was not generally accepted until the eighteenth century, more than two centuries after Copernicus's death.

Within biology, it is commonly said that biological evolution has been shown to be a fact (e.g., Futuyma and Kirkpatrick, 2017). Probably few—except those who believe in a literal interpretation of divine creation—would deny that life on earth has evolved. Yet, the exact mechanisms resulting in the diversity of forms we see in the fossil record and living today are still the subject of considerable controversy. The fact of evolution may appear manifest in the parade of taxa found in the fossil record through time, and it seems to be further confirmed by the pattern of hierarchic relationships observable among taxa, both living and fossil. But even though the general pattern may appear to be factual, our understanding of neither phylogenetic nor microevolutionary processes is close to being incontrovertibly resolved in detail: evolution is a complex, high-level inductive theory.

Consider the issue of bird–dinosaur relationships. Until quite recently, it was widely held that dinosaurs went extinct at the end of the Cretaceous period, some 65 million years ago. Yet, cladistic evidence now unequivocally supports the theory that birds find their closest relatives in the Theropoda, a subgroup of saurischian dinosaurs, suggesting the extinction only of dinosaur lineages other than birds (e.g., Dingus and Rowe 1998). Thus, what was not so long ago thought to be a fact is now widely regarded as a rejected theory. While many clades of dinosaurs are extinct, one lineage from among them exists today in profusion as birds!

Knowledge of issues such as how (and whether) life has evolved, or whether all dinosaurs went extinct 65 million years ago, is not acquired simply through observation. Evolutionary transformations are not events we can observe, nor were we born with a preformed concept of dinosaur. These ideas represent scientific theories that devolved from more than two centuries of concerted observation, analysis, and synthesis on the part of biologists, geologists, and others interested in the natural world.

Prior to the early 1970s, most of the literature on systematics made no mention of the terms hypothesis or theory and gave little consideration to what a systematic theory might be or to how theories might be tested or evaluated. It has been said that systematics is something you do, not something you think about. That anti-intellectual viewpoint ignores the potential importance of inquiry into the scientific thought process itself. The philosophy of science might have remained tangential to the field of systematics but for the clash between phenetics and cladistics over divergent methodologies and their concomitant conceptual frameworks. As noted, cladists won that fight.

Regrettably, concern for philosophy has dwindled in the ensuing 40 years, replaced by the conviction that phylogenetic inference represents a statistical problem that may be solved by increasingly complex and sophisticated models. Yet as Platnick (1979:538) presciently observed:

> It may seem paradoxical that systematics (or any science) must adopt methods without being able to attest to their efficacy. But the fact is that we use our methods in an attempt to solve problems. If we already knew the correct solutions to those problems, we could easily evaluate and choose among various methodologies: those methods which provide the correct

solutions would obviously be preferred. But of course, if we already knew the correct solutions, we would have no need for the methods.

Platnick's point is that a quest for accuracy, often claimed as a goal by critics of the cladistic approach, is fundamentally misguided because we do not have access to the foundational truth of phylogenetic history. We are limited to constructing coherent pictures of the world with methods that are transparent and intelligible, to a greater or lesser degree. As the professed aim of model-based phylogenetic inference is to achieve greater realism (Swofford et al. 1996; Huelsenbeck et al. 2011, Warnow, 2018), this metaphysical limitation seems to be neither accepted nor even understood by many interested in the study of systematics, or "phylogenetics," today. For this reason, we will review some general and specific implications of this conundrum.

Scientific Theories and the Concept of Falsifiability

Approaches to science are often divided into the inductive and the deductive. The *inductive* approach holds that scientific knowledge accumulates from repeated observations of the facts. Alternatively, induction might be defined as a process in which scientific inference leads from the specific to the general. This approach, usually attributed to the sixteenth-century English philosopher Francis Bacon, has long been considered a basic method of science. As twentieth- century German philosopher Hans Reichenbach said (1930:67), "the principle of induction is unreservedly recognized by the whole of science, and . . . there is no one who seriously doubts this principle, even for daily life." Similar expressions of belief are widely scattered in the philosophical, scientific, and popular literature (e.g., Gould 1965). It is no surprise, given the structure of Darwin's (1859) argument for evolution by natural selection in *The Origin of Species*, that the two epigrams at the book's front are by Bacon and William Whewell, originator of the terms, "consilience of inductions" and "*uniformitarianism*." At a basic level, science and technology in general depend upon profound philosophical assumptions that the world is an orderly place, that the future resembles the past, and that similar causes produce similar effects. This apparent orderliness and predictability are why, for example, almost no one worries about being conveyed by modes of transport that are propelled by the explosion of petroleum distillates or poisoned by the medicine prescribed by a physician.

An alternative, more skeptical track in the history of epistemology derives from David Hume's (1748) *An Inquiry Concerning Human Understanding*. Hume argued that inductive inferences about the empirical world were impossible to confirm because no number of observations could guarantee the certainty of predictions based upon them: "Whatever *is* may *not be*" (p. 172). His view was that human convictions regarding the uniformity of nature, the notion that the future will resemble the past, and even the relationship between cause and effect, were conventional results of mere custom or instinct, of the same quality found in other animals. Recognizing this critique as a nonstarter for scientific or philosophical progress, however, Hume conceded (p. 159): "To begin with clear and self-evident principles, to advance by timorous and sure steps, to review frequently our conclusions and examine accurately all their consequences—though by all means we shall make both a slow and a short progress in our systems—are the only methods by which we can ever hope to reach truth and attain a proper stability and certainty in our determinations."

Karl Popper extended Hume's skeptical worldview, condemning the naive perspective that observations—"elementary statements of experience" and "judgments of perception"—ought to be treated as scientific facts as "positivist" and directly related to inductive approaches. The inductive nature of such a scheme becomes clear from arguments such as, "if a large number of elementary observations are judged to be true, then by extrapolation, statements that summarize general agreement among them must also be true." This sort of *positivism*—the idea that observations can be "theory-free"—was manifest in the "extreme empiricism" of the phenetic school of thought discussed in Chapter 1.

In contrast to the naive inductive approach of the pheneticists, the *deductive* empiricist approach advocated here holds that all observation should be construed as theoretical or, stated another way, all observations stem from a context of pertinent theoretical *background knowledge*. To paraphrase Rieppel (1988), observation and the search for lawful regularities can never be unbiased. Observation must be based on some expectation; research is expected to reveal something of significance, something worth recording. That significance derives from the observer's theoretical context. As van Fraassen (1980:14) said, "the way we talk, and scientists talk, is guided by the pictures provided by previously accepted theories." Science can thus never be "theory free." The theories and assumptions that make up background knowledge are themselves not certain, but they are *intersubjectively corroborated* to a degree that we accept them as uncontroversially reliable "facts." Of course, one of the key philosophical differences among current schools of systematics is the question of what and how much we are willing to accept as background knowledge. Cladists tend to be more parsimonious in this regard.

In his attempts to solve the "problem of induction," Popper reasoned that because no number of singular statements could prove the truth of a universal statement, "falsifiability" was the only suitable criterion of demarcation between science and nonscience. For example, we could not be certain that the inductive hypothesis "all insects have compound eyes" was universally true, no matter how many observations we might have made, because we can never observe all insects in all of time. However, the universal veracity of the hypothesis could be rejected easily by finding a single insect without compound eyes. Thus, even if we could all agree on the *consistent* verisimilitude of our observations—*intersubjective corroboration* of the singular statements—we would still have no rationale for believing that a more general or "universal" claim is true. On the other hand, one nonconforming observation or singular statement could suggest that the universal hypothesis was false.

Popper drew a distinction between *strictly* universal statements, which apply to all instances for all time, and *numerically* universal statements, which apply to entities restricted in space and time. Strictly universal statements are falsifiable but not verifiable, while in Popper's view, numerically universal statements are conjunctions of singular existential statements and are therefore in principle both verifiable and falsifiable. Thus, to Popper, only strictly universal statements are scientific. This is one of the aspects of Popper's philosophy that some have criticized as perhaps overly idealistic in relation to biology and historical geology, given that Earth and its inhabitants represent a finite, spatiotemporally bounded system (see Sidebar 3). Another problem is the conundrum that the certain falsification of a hypothesis relies the upon the verisimilitude of the observations and background knowledge providing the falsifying evidence, leading to a sort of skeptical infinite regress (Lakatos 1974).

Sidebar 3
Nomothetic versus Idiographic Science—Whither Systematics?

According to Karl Popper, strictly universal statements apply everywhere, at all times. Examples of universals include physical laws like gravity, the particle theory of matter, or the motion of the planets (of course, the universality of these is debatable in its own right from a philosophical perspective—as we learned from Hume, there is no necessity that the sun must come up tomorrow). Popper contrasted strictly universal statements with existential statements, which are about individual things, occurring at a particular time and place. Numerically universal statements represent a conjunction of existential statements. "All swans are white" is a strictly universal statement (all possible swans, anywhere in the universe, through all time); "all swans alive in North America today are white" is a numerically universal statement. In Popper's philosophy, only strictly universal statements are falsifiable but not verifiable, and this precept forms his demarcation between science and metaphysics (Popper 1968).

Following Wilhelm Windelband, Ernest Nagel (1961:547) described research into the immutable realm of strict universals as nomothetic sciences, "which seek to establish abstract general laws for indefinitely repeatable events and processes," and contrasted them to ideographic (sic) sciences, "which aim to understand the unique and nonrecurrent." Nevertheless, Nagel's discussion clearly shows that the distinction between his alternatives is not clear-cut: nomothetic sciences sometimes refer to unique things and events (e.g., the Big Bang), and historical sciences may refer conversely to "determinate empirical regularities" (p. 549).

Life on Earth evidently had an origin and has a history that, available evidence suggests, is unique and nonrecurrent, and so we might view the study of living things as clearly falling within the idiographic realm. Arnold Kluge (2005b, 2007; Frost and Kluge 1994) and Walter Bock (2007) have argued as much. Unfazed by self-contradiction, Kluge (1997, 2001, 2009) has also argued the opposite—that the hypothetico-deductive framework of systematics is nomothetic. Systematists who have read these essays might well be confused by their antithetical implications for the methods and results of cladistic inference.

In our view, hypotheses of homology that provide the evidence for phylogenetic inference (discussed in Chapters 3 and 4) are more productively considered as "determinate empirical regularities" with a nomothetic character. Hennig (1966:3) advocated this position, at least to some degree, and concepts such as synapomorphy clearly characterize observable features that are "the same" among different taxa—empirical regularities—despite their post-hoc interpretation as singular evolutionary events. We observe that such a philosophical perspective jibes with that implicitly held by physiologists, biomedical scientists, and molecular biologists, who treat the components of biological entities as though they have predictable attributes and stereotypical responses to experimental stimuli (Brower 2019).

Some have argued that determining what is "normal" in biological systems is an inherently statistical endeavor, but when all is said and done, quantitative treatments of any data are only a means to encapsulate and highlight patterns in the evidence the researcher wishes to emphasize. Nonbiological systems have nondeterminate components as well—consider

Brownian motion or weather. The utility of science to explain and predict empirical patterns seems dependent upon uniformitarian assumptions about the regularity of nature, regardless of the inconvenient idiosyncrasies of singular stochastic events. We will return to questions about nomothetic versus idiographic perspectives when we address the problem of individuality later in this chapter and in Chapter 7.

The issue of whether systematic statements about organisms on Earth are strictly universal or not, and whether they count as "science" under Popper's demarcation, has been much discussed (e.g., Rieppel 2003; Fitzhugh 2006a; Vogt 2014; Brower 2019), but no consensus has been achieved. Given our example of insects and compound eyes above, some might argue that since insects occur in finite, albeit large, numbers and have an origin at some point in the past and an extinction point at some time in the future, the prediction about compound eyes represents a numerically universal statement, verifiable at least in principle, and therefore outside Popper's demarcation of science. Of course, the same could be said of molecules of water in the universe (VandeWall 2007) or any other assemblage of empirically observable, "real" things. It seems that most biologists focused on empirical questions view the requirement of strict universality as largely irrelevant. Even Popper never claimed that his view of science was inappropriate for systematics. Although some of his earlier writings were critical of historical sciences (e.g., Popper 1957), as he became more aware of the application of his philosophy of science in systematics toward the end of his life, Popper specifically reversed his claim that evolution is a "metaphysical research program" and admitted comparative biology inside his demarcation of "science" (Popper 1980).

Even with a relaxed demarcation that admits testing of historical patterns of relationship, Popper's hypothetico-deductive view of science holds that the growth of knowledge derives from the formulation of hypotheses or theories that imply certain predictions (deductions) that can be judged by observation. These observations represent tests of the theory. Observations that conform to the predictions are said to *corroborate* the theory; those that do not are said to *falsify* it. The greater and more severe the number of tests passed, the greater the degree of corroboration. A single nonconforming observation may present a persuasive reason to reject the theory (but see below).

A major distinction between the inductive and deductive approaches revolves around the concept of "truth" and how we can determine what is true and what is not. You might say, "Well, let's first agree on the observational aspects before

we move on to the more elaborate theories that purport to summarize the simplest observations, and then we will know whether we are right or not." Such a
proposition would reflect Hume's "clear and self-evident principles," accepted
provisionally by what Popper called *intersubjective corroboration*. Truth, then,
is not some Holy Grail that we naively seek in the world. Rather, objectivity is
constructed from the coherence of evidence that results in hypotheses whose
corroboration is transparent to multiple observers. A clear relationship between
evidence and inference, unencumbered by adumbrating a priori assumptions
and ad hoc hypotheses, is therefore a desideratum.

The deductive method of testing, which has also been called the "falsificationist approach to science," suggests that theories can be tested in the following ways
(Popper 1968):

1. Logical consistency, whereby the conclusions are compared among
   themselves
2. Investigation of the logical form of the theory to determine whether in
   fact it is an empirical or scientific theory
3. Comparison with other theories to determine whether a given theory
   would represent a scientific advance
4. Empirical testing of conclusions (deductions) from the theory

Empirical testing is probably the most important for our purposes, although
the others are also relevant. Failure of empirical observations to support the predictions of a theory would suggest that the theory has been *falsified*. Observations
that match predictions would *corroborate* the theory.

As Popper (1968) pointed out, there is no truth in science, and one should
not equate degree of corroboration with metaphysical truthfulness. For example,
one would not say, "A theory is hardly true so far" or that "it is still false" simply because it had not been rigorously tested. Instead, one might observe that it
remains poorly supported. Popper also rejected the idea that the truthfulness of
a statement can be interpreted in the sense of the probability of that statement
being true. He claimed, "The advance of science is not due to the fact that more
and more perceptual experiences accumulate in the course of time. Nor is it due
to the fact that we are making ever better use of our senses." (Popper, 1968:279–
280). Science is rather a matter of conjecture and refutation.

In Popper's view, degree of corroboration is directly related to the degree to
which a hypothesis has survived attempts to falsify it. Any hypothesis that by virtue of its formulation conforms to any and all observations is not highly corroborated but rather immunized at the outset from critical testing. *Degree of corroboration compares the relative merits of alternative hypotheses, not the degree to which
a given hypothesis is corroborated in its own right.* The view that corroboration

tells us how well a hypothesis has stood up to its tests, and not that we are warranted to believe that it is true (Popper 1968:415), is clearly in keeping with an empiricist view of science.

In applying the deductive approach, we might adopt the following methodological rules as outlined by Popper in *The Logic of Scientific Discovery* (1968):

1. Science is in principle without end. Once you decide on the absolute truth, you retire from the game.

   > Of course, the individual scientist may wish to establish his [or her] theory rather than refute it. But from the point of view of progress in science, this wish can easily mislead him [or her]. Moreover, if he does not himself examine his favourite theory critically, others will do so for him. The only results which will be regarded by them as supporting the theory will be the failures of interesting attempts to refute it; failures to find counter-examples where such counter-examples would be most expected, in light of the best of the competing theories. (Popper 1975:78)

2. A corroborated hypothesis should not be dropped without reason, that is, without a stronger competing hypothesis. In systematics, this rule applies to the problem of *homoplasy* (conflicting evidence).

3. Parsimony should be applied in the testing of hypotheses. That is, *ad hoc assumptions* [special pleading] designed to dispose of observations that would otherwise provide evidence against a theory should be avoided.

Underlying these rules, particularly 2 and 3, is the principle of *parsimony,* which is of fundamental importance in modern systematic practice. We will therefore discuss the idea in its own right.

Popper was not committed to "absolute truth," or, for that matter, certain falsity. Rather, he approached science much as described in the quote from Hume, above, advocating the evaluation of hypotheses based on empirical evidence, the tentative acceptance of those that become well corroborated, and the maintenance of a critical attitude toward the projectability of their predictions (i.e., inductive generalizations) (Popper, 1983). Closely related to this "realist" view is the constructive empiricism of van Fraassen (1980), who argued that the aim of science is to develop theories that are empirically adequate to explain observable phenomena, not the realist desideratum that they be either inductively or deductively true. Both van Fraassen and Popper held that every observation is theory laden and that truth is elusive. Applying these perspectives to the context of biological systematics, we consider the "true tree," reflecting the actual history of the origins of biological diversity through time, to be an unattainable metaphysical goal. Our epistemological approach seeks hypotheses of relationship that are empirically adequate to explain the available evidence (synapomorphy), which is, itself, underlain by theoretical considerations (Chapters 3 and 4).

Parsimony and Ad Hoc Hypotheses

Parsimony as an epistemological criterion for evaluating hypotheses is often attributed to fourteenth-century English philosopher William of Occam (or Ockham) and referred to as Occam's Razor. As articulated by Elliott Sober (2015:2), the principle of parsimony states that, "a theory that postulates fewer entities, processes, or causes is better than a theory that postulates more, so long as the simpler theory is compatible with what we observe." Parsimony is a fundamental aspect of many foundational scientific principles, including the uniformity of nature, which underlies the notion that physical laws are universal, the idea that the future will resemble the past, the principle of common cause, and other theories that serve as background knowledge for many scientific and engineering applications.

The basic *epistemological* rule of parsimony is that the number of assumptions required to explain an observation should be minimized. Extra assumptions beyond the necessary minimum are often referred to as "ad hoc"—superfluous to the problem, or sometimes even intentionally invoked to preserve a favored, less parsimonious alternative. Although Hennig never used the term "parsimony," and did not refer to Occam's Razor, the role of parsimony is clear in his original formulation of phylogenetic methods. As pointed out by Farris (1983), parsimony as a methodological criterion is implicit in Hennig's "auxiliary principle," which states that origin of a feature by convergence should not be assumed a priori (see Hennig 1966:121). An alternative formulation might be that a single origin (or "common cause") for similar structures, genes, or behaviors in different organisms should be presumed as a null hypothesis, in the absence of evidence to the contrary. Thus, parsimony is the criterion that allows systematists to select as the preferred hypothesis of relationships the one that minimizes both implied character-state transformations and ad hoc hypotheses of homoplasy.

Hennig (1966:21) addressed the problem of disagreement among characters, which could necessitate the invocation of ad hoc hypotheses, through the "method of checking, correcting, and rechecking" in an attempt to maximize congruence among characters. This approach has been labeled "circular" by some because they say the process allows for the possibility of bringing observations into agreement simply for the sake of doing so, with the ultimate aim of salvaging a favored theory. We should note that although "checking, correcting, and rechecking" is often conflated with "*reciprocal illumination*" (perhaps even by Hennig himself), we view the latter concept to come into play at a higher level, in which the *consilience* of different types of data reveals a general pattern, such as the various sources of geological evidence that support the theory of plate tectonics, or for that matter, Darwin's "one long argument" regarding evolution itself.

In our view, Hennig's approach of checking, correcting, and rechecking has merit for the same reasons that it has been criticized. Why would you not reexamine your data? There are at least two basic reasons for such circumspection. First, there is no reason to believe that initial observations are always correctly interpreted and not in need of rechecking and corroboration. This point of view derives from the idea established above that all observation is theory laden, and in the case of systematic character data—the theories of homology and transformation discussed in Chapters 3 and 4—may be influenced by prior observations. Recall, for example, the complex structures in Figure 1.3. The parameres and other parts could be described in many different ways, as one complex character or many independent characters, and those interpretations might be influenced by the conditions of those structures in other taxa. Second, it is only the agreement among observations that corroborates initial hypotheses of homology, and rechecking observations can be defended as a test of preliminary conjectures of homology as opposed to manipulation of data (de Pinna 1991; Rieppel and Kearney 2002, 2007). Thus, if we have initially coded legs as "absent" for snakes, subsequent observation of snakes with vestigial limb bones, or the recognition, based on other synapomorphies, that snakes are diapsid amniotes for which the tetrapod condition is a symplesiomorphy, could lead us to the more parsimonious interpretation that legs ought not to be coded as "absent" but as "secondarily lost." Of course, not every homoplastic feature may be reinterpreted by reciprocal illumination, and most empirical data sets (and in particular DNA sequence data) contain at least some unexplained homoplastic character states. Post hoc explanation of homoplasy (via adaptation, for example) is a fruitful research area but is outside the focus of systematics (Brower et al. 1996).

Characters or character data have frequently been referred to in the literature as "good" or "strong" based on what can only be regarded as subjective grounds. For example, structurally complex characters have been deemed more reliable indicators of relationship than less complex characters on the assumption that it is hard to imagine them evolving more than once. Conversely, structurally "simple" characters have at times been relegated to a lesser status simply because they lack complexity of form. Functional significance has also played a role in judging the value of characters in some phylogenetic analyses (a tradition stemming from Cuvier's subordination of characters; Appel 1987). Furthermore, certain character systems have attracted the attention of some authors because these authors were trained in the techniques that reveal them, and they have therefore judged them more important than other types of characters: for example, vertebrate paleontologists have a predilection for bones and teeth, while molecular biologists may dismiss morphology altogether. The parsimonious empirical methods of phylogenetic systematics bring into question all of these arguments because of their necessarily subjective quality. Like the fabled blind monks examining the elephant, each argument places a differential value on different kinds or sources of evidence, a priori, rather than examining its agreement with other data.

If a priori character weighting is eschewed, the strength of a character in support of a phylogenetic hypothesis may be judged by two factors: *consistency* and *congruence*.

Consistency refers to the degree with which a transformation series is explained by a given phylogenetic topology. Note that this is not the same as statistical consistency, a separate property introduced above, and discussed further below, in Statistics, Probability, and Models in the Historical Sciences, as well as in Chapter 5. A frequently used measure of character consistency is the *consistency index*, which is the ratio of observed changes relative to the minimum possible number of changes required for the character(s) to fit the topology (see Chapter 6 for further discussion).

Congruence refers to the degree with which distributions of individual characters agree with one another. This property has sometimes been described using other terms such as "concordance" or "consilience." The strength of a given theory (in this case a character) derives from (1) the degree to which it is corroborated through repeated observation, and (2) its conformity with observations of other characters.

These approaches to judging character data represent direct applications of the parsimony criterion in systematics. We will encounter parsimony many more times throughout this book.

Sidebar 4
Dichotomy in Cladograms: A Case Study Contrasting Metaphysics and Epistemology

Hennig's methods, in their original exposition, strongly implied the desirability of arriving at completely dichotomous phylogenetic schemes—trees in which every internal node has just two daughter branches and all relationships are unambiguously resolved. Hennig's critics wrongly construed this to mean that the methods of phylogenetic systematics worked only if speciation actually is strictly dichotomous—a metaphysical claim about evolution. Yet, Hennig was quick to point out that we desire dichotomous trees only because they imply the most informative hypotheses of relationship. Thus, the desideratum of dichotomous branching says nothing about the manner in which evolution actually proceeded but only how to get the most information out of our data. The nodes in cladograms do not represent speciation events, but rather hypotheses of grouping based on evidence from characters. It may be the case that a given empirical data set does not provide resolution for every node. As stated by Hennig (1966:209):

> Strange to say, the view is often advanced that phylogenetic systematics presupposes a dichotomous structure of the phylogenetic tree. Because dichotomy is not the rule, it is said that a system that gives the impression of a continuously dichotomous differentiation cannot be regarded as a true presentation of actual kinship relations. If phylogenetic systematics starts out from a dichotomous differentiation of the phylogenetic tree, this is primarily no more than a methodological principle.

Subsequently, the terms *trichotomy* (three branches) and *polytomy* (more than three branches) have become widely used when referring to a nondichotomous branching pattern in a cladogram or consensus tree (see Chapter 6). Polytomies can be interpreted to imply either a lack of evidence or equivocal evidence that results in the failure to find a dichotomous resolution of the node in question—an epistemological ambiguity (Nelson and Platnick 1980) or a simultaneous multifurcation, or evolutionary radiation, of the taxa stemming from the node—a metaphysical statement about what actually happened in the course of evolution. Hennig and fellow cladists generally have viewed the latter interpretation with circumspection or agnosticism: "it is very improbable that a stem [ancestral] species actually disintegrates into several daughter [descendant] species at once, but here phylogenetic systematics is up against the limits of solubility of problems" (Hennig 1966:209).

The Empiricist Approach to Systematics
Criticisms

Like most philosophers of science of his time, Popper wrote his discussion of scientific methods primarily in regard to the physical and experimental sciences.

Substantial commentary exists in the literature suggesting that the Popperian view of science requires "lawlike" qualities of universals (Hull 1983:178; Stamos 1996; Vogt 2008, 2014; Quinn 2016) that are not a good fit to systematic biology (see Sidebar 3). There are ongoing controversies among systematists, and even among cladists, over the meaning and relevance of Popper's philosophy, with various authors elaborating criticisms, alternative schemes, and defenses of the views outlined in this chapter (e.g., de Queiroz and Poe 2001, 2003; Faith and Trueman 2001; Farris et al. 2001; Crother 2002; Grant 2002, Kluge 2003, 2009; Rieppel 2003, 2005a, 2007b; de Queiroz 2004; Franz 2005; Fitzhugh 2006a; Helfenbein and DeSalle 2005; Kluge and Grant 2006; Rieppel et al. 2006; Vogt 2007, 2008, 2014; Lienau and DeSalle 2009; Vergara-Silva 2009; Farris 2013, 2014; Velasco 2013; Quinn 2016, 2017; Brower 2019). Although we will not explore the details of most of these somewhat esoteric debates here, we encourage the interested reader to delve into the primary literature to gain an understanding of these varied points of view.

As noted earlier, the Popperian approach to systematics has been criticized as representing absolute or naive falsificationism—failure to recognize that observation itself is subject to error and interpretation, or more specifically, that in order for a hypothesis to be falsified, the falsifying evidence itself must be accepted as "true" (Lakatos 1974). In addition, critics argue that the variability inherent in most biological systems is not taken into account by systematists who apply Popper's methods. Under the absolute interpretation of falsification, a theory for which any contrary evidence is adduced would be absolutely and forever discarded without consideration of the preponderance of evidence.

For example, consider the statement "all pterygote insects have wings." Finding a pterygote insect without wings, such as an ant, would reject the theory under the absolute falsificationist perspective, even though many other attributes clearly suggest that some nonwinged organisms are pterygote insects. We infer, based on many other characters, that ants belong to a group of alate insects and even observe that some ants have dehiscent wings. Given this additional information, is it better to conclude that an ant is not a pterygote insect or that it is a pterygote insect that has secondarily lost its wings? Considering the observed prevalence of homoplasy, rejecting a hypothesis because of a single nonconforming observation (instance of homoplasy) is counterproductive—virtually all phylogenetic hypotheses would be falsified under such a criterion. Rather, systematists aim to minimize the number of ad hoc statements (such as "secondary loss of wings") necessary to identify the best available theory. Parsimony calls for minimization, not minimality (Farris 1983). The "absolute falsificationist" criticism is thus a red herring.

Neither the argument that systematics and other biological disciplines are not "science" nor the argument that the existence of homoplasy makes falsification-ism impossible strikes a substantive blow to the empirical basis of systematics or to what has actually taken place in the application of the empiricist approach. Examination of the systematics literature reveals numerous applications of the minimization of ad hoc hypotheses but few, if any, applications of absolute falsificationism.

Benefits

We can identify at least four issues beneficial to the development of systematics as a science that derive at least in part from critical attempts to apply the deduc-tive approach. *First*, whereas previously it was at best unclear what represented hypotheses to be tested in systematics, it is now clear that systematic hypotheses are the distributions of character states and the alternative hierarchic patterns of relationships (trees) they imply. Characters represent lower-level theories closely associated with the realm of observation and the formulation of hypotheses of homology. Trees represent higher-level theories that are tested by the lower-level character theories. Lack of agreement between theory and observation, formerly explained away with ad hoc hypotheses, is now minimized via the parsimony criterion in the search for theories of relationships among taxa that show great-est agreement with available data and require the fewest ad hoc hypotheses of homoplasy. This represents the intersection of Hennig's auxiliary hypothesis and Popper's "principle of parsimony in the use of hypotheses." It should be noted that "parsimony" as used here does not connote the cladistic method in particu-lar: any systematic method that applies an optimality criterion to select among alternative taxonomic hypotheses employs parsimony to favor better hypotheses over worse ones (see Brower 2002, 2016a).

Second, the meaning of prediction in systematics was long muddled, notwith-standing its central place in characterizing the "power" of science. For example, Bock (1974), Mayr (1988), and Kluge (2005b) have each expressed confusion and/or doubt about the ability of systematics to make predictions. Yet *prediction* in systematics is now widely—if not universally—understood to mean the abil-ity to predict the distributions of previously unobserved (unstudied) characters among taxa. For example, if production of bacteria-killing chemical compounds is a synapomorphy of a more inclusive clade including *Penicillium chrysogenum*, it is parsimonious to expect chemical compounds with properties similar to pen-icillin to occur in a newly discovered species of mold belonging to that clade. The capacity to make such predictions provides the basis for chemical prospecting, a valuable research tool for the pharmaceutical industry. The strength of phyloge-netic hypotheses is based not only on what predictions about character distribu-tions can be made but also on what character distributions are phylogenetically constrained. For example, we would be very surprised to discover an amniote with both forelegs and wings, (like a griffon or an angel) because their duplicated anterior appendages violate the tetrapod ground plan. In all known winged tet-rapods (bats, birds, pterosaurs), the wings are modified forelimbs. Lest there be any lingering confusion, we reemphasize that prediction as applied in systematics does *not* refer to the forecasting of future evolutionary events, as has sometimes been suggested.

Third, the issue of how we judge the explanatory power of systematic theo-ries through the application of parsimony has been clarified. Ad hoc hypotheses are minimized; congruence among character distributions is maximized. Thus, optimal agreement between theory and observation is achieved (Farris 1983).

Fourth, a core principle of empiricist systematics is emphasis on falsifiability of theories and the corollary tentative acceptance of theories that are empirically adequate to explain available observations. Engelmann and Wiley (1977) and Gaffney (1979a, 1979b) stressed the significance of the falsifiability criterion in systematics, particularly in relation to paleontology. They noted that ancestor-descendant relationships, as commonly invoked in the paleontological literature, are not testable in the Popperian sense, and that such "schemes" can only be labeled as scenarios. In this regard, Gaffney (1979b) concluded that even though it would be wonderful to have testable hypotheses of (for example) selection pres-sure and adaptive zones for Devonian tetrapod vertebrates, we are constrained to construct classifications through the formulation of hypotheses testable with available empirical evidence. Gaffney saw no data available for testing theories concerning selection pressures in the Devonian period (see further discussion in Chapter 4, Ancestors, Sister Groups, and Age of Origin).

The Assumptions of Parsimony

The application of parsimony in selecting among phylogenetic hypotheses has been directly criticized on the grounds that the process of evolution is not nec-essarily parsimonious. This metaphysical argument makes a category mistake, assuming something about the nature of what is being observed, rather than the mode of interpretation of the observations. The distinction is analogous to the navigational application on your phone. Obviously, the shortest distance between two points is a straight line, but the route the application finds is almost never straight—it is just the most parsimonious navigable alternative. In the same way, no matter how convoluted the historical process of macroevolution may have been, our only empirical access to that process is by observation and efficient interpretation of its end products. As with any scientific inquiry, parsimony is applied in systematics to minimize the number of ad hoc explanations of data or, stated in another way, to maximize the power of the hypothesis to explain those data. Parsimony is an epistemological criterion for preferring one hypothesis over another and is in no way related to the verisimilitude of models or scenarios about the evolutionary process. It is simply a methodological tool and makes no ontological claims about the way things are in the world.

On this view, the application of parsimony in cladistics has been said to be free of assumptions concerning evolutionary models (Platnick 1979; Nelson and Platnick 1981; Brady 1985). But this perspective has not been unchallenged in the literature, either. For example, Sober (1983, 1988) has stated that he believes parsimony contains presuppositions, and although he was unclear as to what they might be, it was his view that if the history of evolution violates them, then character distributions will not be usable in inferring phylogenetic trees. Revisiting Felsenstein's (1978) observations, Swofford et al. (1996:426), in a hermeneutical discussion, asserted that parsimony does make assumptions and that the violation of these assumptions can lead to problems. In their view, "The difficulty lies in stating precisely what the assumptions are." Delimiting what they believed to be the yet-unstated conditions implied in the application of parsimony, Swofford et al. stated, "At a minimum, acceptance of an optimal tree under the parsimony criterion requires one to assume that conditions that can cause parsimony to estimate an incorrect tree are unlikely to have occurred." From a coherentist, empirical perspective, this concern represents a counterfactual conditional statement: if the evidence were misleading, then our inference would be wrong. We agree that the evidence could be misleading in some absolute, metaphysical sense, and the hypothesis of relationships implied by it could be false with respect to the actual historical genealogy. But since there is no independent means available to test whether or not data in phylogenetic investigations of actual taxa are misleading, the point is moot and has no bearing on the empirical adequacy of the cladistic approach. Thus, the arguments of Sober and Swofford et al. seem to us to be no more compelling than Descartes's *genius malignus*—the notion that all of our perceptions might be figments placed in our heads by an evil demon. The solipsist stands unassailable in his fortress, but neither can he sally forth, so we will pass him by.

Statements like those of Sober and Swofford et al. might also be interpreted as suggestions that no criteria exist for judging the results of parsimony analyses, whereas in reality the criterion is exactly that the data are explained with the minimum interpretation necessary. There are no confidence limits to be calculated for judging the truth content of a given hypothesis under the parsimony criterion. The standard of judgment is rather the degree of falsification (or corroboration, or empirical support) compared against alternative hypotheses. The same is true for phylogenetic inferences based on maximum likelihood— the most likely (or, in parsimonious terms, least unlikely) tree is selected, even though its absolute likelihood is extremely low (Gatesy 2007). We will return to these issues in Chapter 6.

A final point on assumptions in phylogenetic analysis was well expressed by Farris (1983:35):

> Parsimony analysis is realistic, not because it makes just the right suppositions on the course of evolution. Rather, it consists exactly of avoiding uncorroborated suppositions whenever possible. To the devotee of supposition, to be sure, parsimony seems to presume very much indeed: that evolution is not irreversible, that rates of evolution are not constant, that all characters do not evolve according to identical stochastic processes, that one conclusion of homoplasy does not imply others. But parsimony does not suppose in advance that those possibilities are false—only that they are not already established. The use of parsimony depends just on the view that the truth of those—and any other—theories of evolution is an open question, subject to empirical investigation.

Although Farris said parsimony analysis is "realistic," the term is clearly used in an ironic sense, since all of the "realistic" suppositions he indicates are hypotheses open to test by parsimony, not a priori assumptions baked into the method of phylogenetic inference. Thus, while we embrace necessary background assumptions about the relevance of evidence to the phylogenetic problems we wish to solve, as well as the basic premise that there are coherent patterns in nature to explain, both of which might be considered "realist" perspectives, we strive to avoid ad hoc hypotheses and are skeptical about claims of "greater realism" that dilute parsimonious explanations of the evidence.

Statistics, Probability, and Models in the Historical Sciences

The foundationalist viewpoint of Swofford et al. (1996) quoted above is derivative of Felsenstein's (1978) argument that because, under particular simulated conditions generated by deterministic models of character evolution, parsimony did not provide the known "true" answer, parsimony must therefore be flawed because it has the potential to be "statistically inconsistent." As noted at the beginning of this chapter, inconsistency is the statistical property that support for an incorrect result increases with the addition of data. It is possible to contrive a scenario in which parsimonious interpretation of a given data set does indeed support an incorrect topology. Felsenstein concluded from this result that phylogenetic theories might better be judged on a statistical basis, through the use of a maximum likelihood approach (see further discussion of maximum likelihood in Chapter 5).

Farris (1983:17), in criticizing the application of a statistical approach such as maximum likelihood to phylogenetic inference, noted that it "was wrong from the start, for it rests on the idea that to study phylogeny at all, one must first know in great detail how evolution has proceeded. That cannot very well be the way in which scientific knowledge is obtained. What we know of evolution must have been obtained by other means. Those means . . . can be no other than that phylogenetic theories are chosen, just as any scientific theory is, for their ability to explain available observations."

Felsenstein (1978:409) himself admitted that "the weakness of the maximum likelihood approach is that it requires us to have a probabilistic model of character evolution which we can believe. The uncertainties of interpretation of characters in systematics is so great that this will hardly ever be the case." Yet he asserted that true tests of the robustness of phylogenetic methods can be undertaken only statistically, and that this goal must be pursued, even though "establishing that robustness . . . by examining a wider range of models is a daunting task." Beyond the practical challenges of finding believable models, we question how one would recognize the believability of a model to begin with, when the object of that model is to correct for potentially erroneous empirical evidence. You can run as many simulations as you please, but they may or may not have any relation to patterns in the "real world" they are supposed to emulate. In the words of van Fraassen (1980:60), "empirical adequacy concerns actual phenomena: what does happen, and not what would happen under different circumstances."

Parsimony was further impugned by Felsenstein (1981:194) because it "implicitly assume[s] very low rates of change"; in other words, parsimony assumes that relatively few evolutionary events have occurred in the diversification of a given group. Farris (1983:13) dispelled Felsenstein's argument, asking whether a "procedure that minimizes something must ipso facto presuppose that the quantity minimized is rare?" In refuting this notion, he used an example from statistics, pointing out that in regression analysis, just because the variance may be large, there is no reason to doubt that the regression line found is the best description of the data; one simply needs to expand the confidence limits to accommodate the variance. Subsequent empirical research (e.g., Källersjö et al. 1999) has supported Farris's theoretical claim by showing that even highly homoplastic data (implying a high rate of change) increase support for a cladistic hypothesis. Recall the mapping example above: the shortest route between two places is still the shortest route, even if it is not a straight line.

Many advocates of model-based phylogenetic inference tout it as being more "realistic" than parsimony (cf. Huelsenbeck et al. 2011). The realism-motivated statistical worldview would seem to suggest that it is possible to construct some absolute measure of the goodness of phylogenetic hypotheses with respect to the truth. We argue that such is not the case. That view of systematics, when applied as a criticism of parsimony, disregards the fact that parsimony procedures have always been applied as a way of judging the explanatory power of *alternative* hypotheses, not as a way of understanding the truthfulness of hypotheses in absolute or probabilistic terms.

Finally, we stated at the beginning of this chapter that Felsenstein's (1978) claim that parsimony can be statistically inconsistent is "not the whole truth." From Felsenstein's perspective, cladistic parsimony represents a specific model of the evolutionary process. What he chose not to emphasize is that *ANY* such model, couched in likelihood terms or otherwise, could also be statistically inconsistent. Likelihood as a general analytical framework is guaranteed to be consistent for a given empirical data set only if the specified model matches the way that evolution really happened. When conducting a phylogenetic analysis of empirical data to infer patterns of relationship among actual taxa, there is no way to tell whether a given model is consistent with the underlying evolutionary process or not (Farris 1999; Brower 2018c). Thus, the specter of statistical inconsistency in phylogenetic analyses is empty rhetoric, deceptively spun by Felsenstein to disparage cladistic parsimony and advance his alternative likelihood-based agenda.

More generally, we can characterize the statistical inconsistency argument as an instance of Descartes's *genius malignus* problem, that our perceived reality might be a delusion. As skeptical empiricists, our response to this is the same as Hume's: "whatever *is* might *not be*." We cannot know whether our empirical data are truthful or not, so we acknowledge these unfalsifiable metaphysical trepidations and shrug them off, for to obsess about them leads to epistemological paralysis with regard to scientific knowledge. Parsimony is the lighthouse that cuts through this fog.

The likelihoodists' alternative approach to this indeterminacy is to specify models they postulate to be more realistic than parsimony. However, statistical models cannot solve the problem of empirical correspondence but only encapsulate it in a Platonic bubble of a priori quantitative assumptions. The result is then a parsimonious explanation of the evidence, assuming the model—which is logically less parsimonious than a parsimonious explanation of the evidence, assuming parsimony.

Models are accessories to parsimony, not alternatives to it. The predictive power of statistical inference to infer probability distributions, reject null hypotheses and assess degrees of confidence in results from empirical data fundamentally relies upon rigid assumptions about the future resembling the past (for example, the requirement of independent samples drawn from an identical distribution). As we pointed out earlier, assumptions about the uniformity of nature are based on parsimony. Therefore, contrary to Sober's (1985, 1988, 2015) repeated efforts to articulate a "likelihood justification for parsimony," likelihood and other forms of statistical inference as applied to empirical problems appear instead to depend on underlying parsimony assumptions. We will return to these key points about statistical phylogenetics in Chapter 5.

The Basis of Systematic Knowledge

In elaborating his arguments establishing a general reference system for biology on the basis of phylogeny, Hennig concluded that we can expect to observe a hierarchic distribution of attributes in a system of descent with modification, once modifications become fixed in lineages (Hennig 1966; Davis and Nixon 1992). Regardless of whether we choose to presuppose "descent with modification" or not, the natural hierarchy of attributes of organisms has for centuries been an assumption of the method for discovering relationships (Brady 1983; Brower 2000b). It is the fixity of attributes at the minimum level (species—see Chapter 7) that admits those units as valid for analysis in such a system (Davis and Nixon 1992; but see Brower 1999).

Hennig's view of the goals of systematics was heavily grounded in biology. A more philosophically oriented statement of purpose was offered by Hull (1965, in Ereshefsky, 1992:202), who declared that, "from the beginning, taxonomists have sought two things—a definition of 'species' which would result in real species and a unifying principle which would result in a natural classification." Hull's statement points to two partially overlapping lines of argument: (1) whether systematics informs our knowledge of evolution, and particularly evolutionary history, or (2) whether knowledge of evolution is a prerequisite to understanding systematic relationships.

Hull (1965) argued that taxonomy has a history tainted with "*essentialism*"— the idea that species and higher taxa are immutable, unchanging entities—and that only when the undesirable influences of that ancient worldview are removed will the scientific status of the field be established. In his discussion of "the species problem," Hull (1965) asserted that taxa are *individuals* and not *classes*, and that the entities given taxonomic names (including species) cannot be defined as classes by sets of properties because to do so would require that the properties be distributed both universally and exclusively among the members of the taxon. If this were not the case, then essentialistic definitions would not be universally applicable.

This view was adopted by de Queiroz and Gauthier (1990), who noted that numerous taxonomic papers contain a section entitled "definition," which to those authors indicated that the referents of taxonomic names continue to be treated as if they were defined by lists of organismal traits and that such "definitions" must then be directly associated with the definition of an immutable essence, in the Aristotelian sense. As such, a definition stipulates some necessary and sufficient conditions and the presumably invariant nature of the properties composing the definition. What seems apparent is that "define" as used in most papers was never intended in any such semantic sense: for most systematists, the "*definition*" of a taxon is interchangeable with its *diagnosis* and is simply a brief statement of those features that allow it to be reliably distinguished from other taxa (in cladistics, sometimes the definition of a group, particularly for higher taxa, is the list of synapomorphies that distinguish it from its sister taxon). In contrast, the description is a more lengthy account of the habitus of different life stages, sexual dimorphism, and other variability that might be encountered in examination of multiple individuals belonging to the taxon in question

Rejecting these practical considerations, de Queiroz and Gauthier proposed "phylogenetic taxonomy" as a solution to the perceived benighted essentialism of their fellow systematists (see further discussion in Chapter 8), in which taxa are "defined" on the basis of lineage relationships rather than character attributes. We simply observe that the transfer of the term "definition" from characters to lineages seems to confound their claim that the word is inseparable from essentialism (unless de Queiroz and Gauthier's "phylogenetic definitions" are themselves essentialistic).

The purportedly pernicious influence of essentialism on systematics was elaborated in Mayr's (1982) *The Growth of Biological Thought* under the heading of "population thinking versus essentialism." Mayr's worldview emphasized the importance of variation in biological systems in general and of "species" in particular. Mayr labeled most approaches not focused on the analysis of variation at the species level and below as typological or essentialist. These labels would evidently apply to most pre-Darwinians, as well as a large proportion of modern workers interested in phylogenetic inference. Hull, Mayr, de Queiroz and Gauthier, and Ghiselin (1984) would have us believe that essentialism, in the sense that they understand it, is actually being practiced by these wrong-headed taxonomists and that it must be eradicated to set things right in systematics as a science. Several systematists and historians of science have challenged the idea that pre-Darwinian systematists were essentialists, or that systematics as practiced by the rank and file suffers from this malady (Wheeler 1995; Brower 2000b, 2019; Winsor 2003, 2006; Amundson 2005). Of course, there are also those who openly embrace a Platonic mathematical framework, based on proofs, sets, and the like, for phylogenetics (e. g., Warnow 2018).

As noted by Nelson and Platnick (1981:328), the anti-essentialist argument assumes that when the attributes of a lineage change over time, the modified versions of those attributes will no longer be recognizable as evidence of their genealogical connections. As an example, Nelson and Platnick observed that under this presumptive view of taxonomy, we could not use characters such as fins as part of the definition of the group Vertebrata because some members of that group, the tetrapods, possess fins in the modified form, as limbs. Nonetheless, as we will see, the concept of character transformation plays a central role in current systematic thought of even the anti-essentialists, and all systematists necessarily make use of that concept in forming hierarchic classifications.

Hull's second imputed aspiration of taxonomists, the "unifying principle which would result in a natural classification," also bears scrutiny in our efforts to understand the basis of systematic knowledge. The "unifying principle" to which Hull alluded is widely thought to be the theory of organic evolution (e.g., Kluge 2001). Descent with modification is certainly capable of producing the hierarchic patterns of relationships that taxonomists have observed—"from the beginning"—as Hull put it. However, descent with modification does not provide—by itself—a method for discovering that hierarchy, nor does it necessarily imply that the pattern to be discovered should be a hierarchy (Sidebar 5). Having written his philosophical tract in 1965, Hull might be forgiven for lack of foresight because the details of the method that is capable of effectively and consistently recovering information on the hierarchy of life was just at the point of becoming widely appreciated. What seems less clear is why subsequent authors—particularly biologists—have adopted Hull's anti-essentialist arguments lock, stock, and barrel, long after the distinction and relationship between a theory capable of explaining hierarchy (i.e., organic evolution) and a method capable of discovering the hierarchy of life (cladistics) had been clarified.

Sidebar 5
"Descent, with Modification": Critical Background Knowledge for Phylogenetic Inference?

Kluge (2001:325) said, "I consider Popper's concept of background knowledge b to comprise only currently accepted (well-corroborated) theories and experimental results that can be taken to be true while helping to guide the interpretation of evidence *e* on hypothesis *h*." Kluge has asserted many times (e.g., 2009:182) that "'descent, with modification' . . . is the only auxiliary assumption *required* of phylogenetic systematics." On its face, that assumption represents a foundational realist extrapolation, invoking unobserved microevolutionary processes (tokogeny) as a cause for observed patterns of character-state distribution and the phylogenetic trees they imply. While readers steeped in an ontological framework received from evolutionary biology may find the process assumption of descent with modification (DWM) intuitively plausible as background knowledge, we consider DWM instead to be more appropriately viewed as a post-hoc explanation of a pattern independently discovered on the basis of empirical observations of similarities and differences among taxa, interpreted under the cladistic paradigm. As Brower (2000a) suggested, rather than metaphysical assertions about DWM, the critical background knowledge assumptions of phylogenetic inference are that the pattern we seek is hierarchical in nature, that the character distributions we observe provide evidence of groups, and that parsimony serves as the criterion for selecting among alternative hypotheses of relationship.

In regard to the role of DWM as an explanation of character distributions, Brower observed (2002:223):

> Descent with modification can in principle explain any pattern of character similarity or difference among taxa: characters that support the preferred groups are called "homologies," and are explained by retention of similarity from a common ancestor; characters incongruent with the preferred groups are called "homoplasies," and are explained by convergent evolution from disparate ancestors or reversal. A given character could have undergone any number of evolutionary transformations during the course of history, and the theory of descent with modification can explain them all. Descent with modification can likewise explain the hypothesized irregularly bifurcating hierarchy, but can equally well explain polytomies, anastomosis, or "phyletic" character state transformation in a single lineage without splitting.

Thus, DWM does not satisfy Hempel's (1966:71–72) criterion that "the assumptions made by a scientific theory about underlying processes must be definite enough to permit the derivation of specific implications concerning the phenomena that the theory is to explain." Nor, if we view distributions of character states among taxa, and the tree that they imply, as the evidence to be explained, does DWM as a universal law have obvious "testable consequences which are different from the *explicandum*"; neither, as the above quoted paragraph describes, does DWM "logically entail" any particular result (both desiderata quoted from Popper 1979:192).

An alternative framework for approaching this problem is Popper's characterization of explanation as a syllogistic deduction (1979:351). In that schema, the conclusion (explicandum) is "definitely known to us—the fact lies before us in stark reality" (p. 350), while the premise (the *explicans*) "which is the object of our search, will as a rule not be known: it will have to be discovered. Thus, scientific explanation, whenever it is a discovery, will be *the explanation of the known by the unknown*" (p. 191, italics in original). Of course, deductive syllogisms normally work the other way around, and it is the "background knowledge" that is known a priori and the conclusion that is inferred:

All men are mortal	(major premise = covering theory or "background knowledge")
Aristotle is a man	(minor premise = evidence or observation)
Aristotle is mortal	(conclusion = deductive inference)

Nevertheless, Popper's (1979) alternative account of the logic of explanation does seem to mesh well with Brady's (1985) contention that the explanandum is independent of the explanans, to the extent that the former can be empirically observed without reference to the latter: we can see that there is a dead rat without knowing why it died. If the proclamation that "descent with modification is necessary background knowledge for systematics" can be interpreted as per Popper's (1979) schema, that DWM is an explanatory theory discovered from the empirical evidence of character-state distributions, rather than an a priori causal process assumption, then perhaps we agree with Kluge after all (see Brower 2019).

Hull's argument concerning natural classifications has been wielded against cladists—including Nelson, Platnick, and many others—who have argued forcefully that cladistics is a system of pattern recognition and that a priori assumptions about evolution are unnecessary for the success of the method (Platnick 1985; Brower 2000b; 2019). Beatty (1982) criticized what he dubbed "pattern cladistics" because the approach did not explicitly justify the search for a hier-archic pattern of relationships with the widely accepted tenets of biological evolution. Beatty's criticisms ignore the idea that the theory of descent with modification gains credence through the evidence provided from systematists' independent discovery of hierarchic patterns of relationships among taxa (Hennig 1966; Brady 1985).

In responding to the arguments of Beatty, Hull, and others, Platnick (1982) wondered what causal theory was necessary to recognize the group known as spiders, comprising 35,000 species, all of which are united by their possession of abdominal spinnerets and male pedipalps modified for sperm transfer. The same might be said for the 1 million species of winged insects, 9000 species of winged and feathered birds, 250,000 species of flowering plants, and many other groups, all of which were recognized long before the formulation of any coherent theory of organic evolution. Our ability to pick out such groups in nature is based on the structural attributes that each possesses and the congruent distributions of those attributes among taxa. The existence of hierarchically related systematic groups is explained by the overarching material causal mechanism of organic evolution. However, as in other branches of science, our ability to make such discoveries does not depend on an a priori explanatory theory (Brady 1985).

To bring this issue into perspective, Platnick (1982) invoked the analogy of planetary motion. He asked if causal theories were required to discover the exis-tence and movements of Jupiter. The answer is "certainly not." The ancient Greeks, Semites, and Chinese had detailed knowledge of planetary existence and motion, even in the absence of telescopes, and predicted the positions of celestial bodies with considerable accuracy, in part evidenced today by calendars (and horoscopes) derived from these cultures. This detailed knowledge of the heavens was developed even though the earth was erroneously thought to be the center of the universe. Also, as noted in Chapter 1, Mendeleev developed the periodic table of elements in the absence of knowledge of the atomic number, the quantity of protons in the atomic nucleus that is now considered to be its "explanation" (Brower 2002).

Platnick (1982) reckoned that authors such as Beatty, who argued for the pri-macy of evolutionary theorizing over systematics, have the cart before the horse, both historically and logically. Reflecting a similar view, Patterson (1987:4) noted: "In the decades after the [appearance of] *The Origin of Species*, comparative mor-phology became phylogenetic research; common ancestors replaced archetypes, and homology became evidence of common ancestry rather than of common plan [essence]. But these were changes in doctrine rather than of practice."

Perhaps the most elegantly developed declaration of the independence of sys-tematics is that of American philosopher of science Ronald Brady (1985:114), who argued that "the hierarchy of taxa was considered by Darwin [in *The Origin of Species*] to be one of the established facts of natural history . . . The hierarchy and other patterns of natural history were not, therefore, based upon evolutionary theory [in the mind of] the founder of that theory. They occupied a privileged position of an [*sic*] independent evidence, to which one could point for a test of the theory [of evolution]."

Brady further observed that systematics is complete in itself, generating its own standards and tests, and that the expounding of evolutionary theory follows not only temporally but logically after the discovery of the patterns. It seemed clear to him (p. 125) that most cladists discover patterns of relationship by the same method, and he therefore found it odd that those singled out for attack as "pattern cladists" were the ones who insisted that the basis for our knowledge about the world is the pattern derived from observation, rather than the explanations for that pattern. In Brady's view it is the independence of the pattern that allows systematics to retain its empirical status. Failure to distinguish the empirical problem (the pattern of relationships among taxa) from the explanatory hypothesis (the theory of organic evolution) leaves us with no independent evidence to test that hypothesis. As stated by Brady (1985:117), "by making our explanation into the definition of the condition to be explained, we express not scientific hypothesis but belief. We are so convinced that our explanation is true that we no longer see any need to distinguish it from the situation we are trying to explain."

For the moment, let us summarize by saying that *systematics* is the branch of biology capable of providing a map to navigate the genealogical history of life on earth. Just as do all empirical scientific endeavors to describe and explain patterns in the world, systematics depends upon theoretical assumptions about the uniformity and continuity of nature and the reliability of observation. The methods advocated might or might not reveal the truth, but they do allow us to interpret the history of relationships among organisms in a way that conforms as closely as possible to the data at hand. Cladistics offers a methodologically coherent and empirically transparent framework integrating observation, analysis, and synthesis of systematic evidence.

The most general conclusion to be derived from this discussion is that the pattern of phylogenetic relationships is neither directly observable nor directly knowable, and even less so are the particular historical processes responsible for that pattern. Rather, acquisition of such knowledge comes from the collection of relevant data and the analysis of those data by appropriate methods. The methods of data acquisition and analysis are described in the chapters of Section II.

SUGGESTED READINGS

We recommend the following works to the reader as representative of the historical canon of cladistics. These are the pilings upon which our epistemological edifice is constructed.

Brady, R.H. 1985. On the independence of systematics. Cladistics 1:113–126. [A classic explication of the pattern cladistic perspective—a must-read for any serious systematist]

Farris, J.S. 1983. The logical basis of phylogenetic analysis. In: Advances in Cladistics. Vol. 2, Proceedings of the Second Meeting of the Willi Hennig Society, ed. N.I. Platnick, and V.A. Funk, 7–36. New York: Columbia University Press. [Perhaps the foundational explanation of the role of parsimony in systematics]

Gaffney, E.S. 1979. An introduction to the logic of phylogeny reconstruction. In: Phylogenetic analysis and paleontology, ed. J. Cracraft and N. Eldredge, 79–111. New York: Columbia University Press. [A clear, concise discussion of the hypothetico-deductive approach to systematics]

Magee, B. 1973. Karl Popper. In: Modern masters, ed. F. Kermode, 10–49. New York: Viking. [An easily understood synthesis of the philosophy of science as propounded by Karl Popper]

Popper, K. 1968. The logic of scientific discovery, 2nd English ed. New York: Harper and Row. [Popper's classic treatise on the nature of the scientific thought process]

Rieppel, O.C. 1988. Fundamentals of comparative biology. Basel: Birkhäuser. [An insightful perspective on systematics and its underlying bases]

Wiley, E.O. 1975. Karl Popper, systematics, and classification: a reply to Walter Bock and other evolutionary systematists. Systematic Zoology 24:233–243. [An early discussion of the relationship between the views of Karl Popper and systematic methods]

Section II
CLADISTIC METHODS

CHARACTERS AND CHARACTER STATES

Having laid the groundwork for a general understanding of the history and philosophy of systematics, we will now begin the detailed examination of methods for inferring hierarchic—phylogenetic—relationships among taxa. The selection of taxa and characters provides the raw materials for all subsequent steps in phylogenetic analysis, taxon delimitation, and classification. The process of observing, coding, and rechecking character information has been referred to as "character analysis." Although central to phylogenetic analysis of both morphological and molecular data, there are divergent opinions about what this process represents and how it should proceed. In this chapter, we will explore the theory and methods that allow systematists to recognize characters, character states, and the taxa they delimit.

The Role of Homology in Character Recognition

The concepts of *synapomorphy* and *monophyly* were central to the method of phylogenetic analysis developed by Hennig and were unequivocally characterized by him. The Hennigian approach has been widely adopted, and a typical introductory biology textbook discussion of systematics states something like, "The cladistic approach to inferring trees is based on the realization that relationships among species can be reconstructed by identifying shared derived characters, or synapomorphies" (Freeman 2008:544), or, "What we need to develop hypotheses of evolutionary relationship are homologies shared by some, but not all, of the members of the group under consideration. These are called synapomorphies" (Morris et al. 2013). However, statements of synapomorphy and monophyly themselves represent theories about characters and taxa, respectively, that cannot be observed directly. It is only through the methods of phylogenetic analysis that such knowledge is inferred (Fitzhugh 2006b).

Biologists have produced substantial evidence over the past two centuries revealing the mechanisms of genetic inheritance and of ontogenetic transformation, cementing our understanding of these processes to the point where we are comfortable referring to Mendel's and Von Baer's Laws and incorporating these generalities into our biological background knowledge. Yet, there is little or no direct evidence for many of the grand-scale macroevolutionary transformations that we infer from phylogenetic patterns: the development of the arthropod body plan, irrespective of its similarity to that of annelids and onychophorans; the development of insect wings; the development of turtles from more generalized tetrapods; the origin of flowers in the angiosperms. It had long been hoped that paleontology would provide the solutions to these mysteries in the form of intermediate fossils, the missing links. Although evidence exists for the sister taxon relationship of tetrapods and actinopterygian fishes, of whales and terrestrial mammalian precursors, and of birds and dinosaurs, for most groups of organisms there are no intermediate fossils and, consequently, no direct evidence of transformation. Even the fossil "missing links" that provide evidence of transitional forms between seemingly different taxa occupying distinct stations in the history of life do not provide this hypothesized continuity—they do not "connect the dots" but merely add more dots on the branches between those representing Recent taxa. What we have, by way of explanation of the apparent discontinuity in the fossil record, is the concept of "punctuated equilibrium" (Eldredge and Gould 1972). Yet this evolutionary process model does not *explain* transformation but only suggests a mechanism compatible with evidence that it has apparently occurred relatively rapidly, at intervals, between long periods of comparative morphological stasis.

Even though we have not witnessed major structural transformations in biological lineages, natural historians have observed structural correspondences among parts of organisms since the time of the ancient Greeks. These similarities of form were termed "homologous" by the British anatomist Richard Owen (cf. Patterson 1982). Owen (1843) defined *homology* as pertaining to "the same organ . . . under every variety of form and function." The criteria used by Owen were the "principle of connections" (e.g., various components of the vertebrate skeleton attached in the same way regardless of function) and the "principle of composition" (e.g., similarity of fine structure of tissues and organs), both of which appeared earlier in the work of Geoffroy St. Hilaire (1818), according to Brady (1985), Appel (1987), and Rieppel (1988). Theories of homology in the sense of Geoffroy and Owen are theories of "anatomical philosophy" that arose prior to, and exist independent of, the theory of evolution. The "common plan" of homologies found in nature constitutes the empirical pattern of transformation that allows us to postulate connections between one group of organisms and another. It is only through our theories of homology that phylogenetic analysis can proceed. The minimal assumption of the homology concept in cladistic analysis seems to be that features of taxa can be—or have been—transformed over time (see Sidebar 6).

Sidebar 6
Transformational Homology

A source of persistent confusion and controversy in the cladistic literature has been the meaning (or meanings) of the term "*transformational homology*," which is contrasted with the more straightforward "taxic homology" (features that characterize groups, or synapomorphies, cf. Patterson 1982). Brower (2015) described how "transformational homology" has been understood in several different ways by various authors. As originally employed by Eldredge (1979), it referred to the evolution of a character within a single lineage through time (*anagenesis*). Patterson (1982) applied the term differently, to refer to the atemporal, precladistic homology concept employed by Owen, which may be viewed either metaphysically, in an idealistic morphological context, as relating to some feature of a Platonic archetype, or more agnostically as what Brower and Schawaroch (1996) referred to as topographically identical features hypothesized to be "the same thing" in different taxa, without implication of hypotheses of grouping. Another meaning of transformational homology can refer to "transformation series"—either hypothesized a priori as ordered character-state trees (see Chapter 4) or mapped on a tree after phylogenetic inference (see Chapter 10). Obviously, transformation series that have directionality implied by a root may be narrated as evolutionary scenarios only after phylogenetic inference. Thus "transformational homology" refers to either (1) *topographical identity*, or (2) specific hypotheses of evolutionary transformation as inferred from a cladogram or the fossil record.

The modern formulation of the criteria for recognizing homologous structures, frequently attributed to Remane (1952), can be stated as follows:

1. Similarity of special structure
2. Position (similarity of topographical relationships)
3. Connection by intermediate forms (transformation)

Mayr (1982:232) complained that it "was unfortunate, and quite inappropriate, that Remane raised the criteria that served as *evidence* for homology, to the *definition* for homology." According to Mayr, "After 1859 there has been only one definition of homologous that makes biological sense: A feature is homologous in two or more taxa if it can be traced back to the same feature in the presumptive common ancestor of these taxa." Brady (1985:116–117) criticized Mayr's reasoning by noting that it confused the phenomenon to be explained (similarity of structure) with the explanatory theory itself (organic evolution). In Brady's view, Remane was thinking quite clearly (and consistently with his predecessors Geoffroy and Owen) when he formulated his criteria for the recognition of homology.

Considering Hennig's viewpoint with regard to homology may help clarify Brady's position, which is the one taken in this book, while simultaneously emphasizing the fallacy of Mayr's argument. Hennig (1966:93–94) observed that, "apparently it is often forgotten that the impossibility of determining directly the essential criterion of homologous characters—their phylogenetic derivation from one and the same previous condition—is meaningless for defining the concept of homology." To make his point, Hennig quoted Boyden (1947), who observed, "Today the pendulum has swung so far from the original implication in homology that some recommend that we define homology as any similarity due to common ancestry, as though we could know the ancestry independently of the analysis of similarities!" The pendulum has not swung back: the introductory biology books quoted above state that homology "occurs when traits are similar due to shared ancestry" (Freeman 2008:545) and "characters that are similar because of descent from a common ancestor are said to be homologous" (Morris et al. 2013:23–26).

To recap, if hypotheses of ancestry, monophyly, and even synapomorphy are inferences based on phylogenetic analyses, then these theoretical results should not be presuppositions of the analytical methods that provide their bases. If we want to infer cladograms that provide us with evolutionary insights and test evolutionary hypotheses, we need to establish our methods and principles in terms that do not rely on evolutionary premises. Thus, we discuss the discovery and delimitation of characters—the building blocks for forming hypotheses of relationship—in this chapter and will return to discuss the concept and mechanics of inferring homology in greater detail in Chapter 4. Nevertheless, we will address Remane's heuristic criteria of structural similarity, topographical similarity, and transformation here. First, however, let us tackle a long-standing and fundamental philosophical problem.

What Is Similarity?

Similarity is the quality of two or more entities being the same in some respect—of sharing attributes in common. In systematics, similarity is a relative relation that exists among at least three things. For a given attribute, two things are more similar to one another than either of them is to a third thing, and when multiple attributes are assessed together, the nested degrees of similarity across the range of attributes provide evidence for hypothesizing phylogenetic relationships. Yet things can be similar in one aspect but not similar in other aspects. A circle and a square are both planar geometrical forms. A circle and an orange are both round. So, is a circle more similar to a square or an orange? This problem suggests that similarity is an interpretation framed by the observer's interest and goals for comparing the objects—it relies on theoretical assumptions. It also suggests, in a fairly basic way, the epistemological challenges of the systematist's method: How do we decide whether roundness or planarity is the more important criterion of grouping for these objects?

Philosophers—as well as some philosophically inclined systematists (e.g., Kluge 2003)—have argued that, because of the apparent arbitrariness of resolving such intractable questions, similarity is a vacuous concept. Perhaps in the grand metaphysical sense that is correct, in the same way that other abstract concepts like truth and reality are vacuous, or at least distorted by the interests of the observer. Yet if we reject similarity as a meaningless concept, we de facto reject "special similarity" (synapomorphy) as well, so to do so is clearly not a productive path forward for cladistics. That the evidence bears upon the hypothesis to be tested is one of the empirical axioms of cladistics proposed by Brower (2000b; see Sidebar 5 in Chapter 2).

Furthermore, the capacity to discriminate like from unlike seems to be a basic component of human and animal cognition. For example, a toad will not accept as prey a palatable, bumblebee-mimicking robber fly after it has been stung by the bumblebee because (the observer surmises) it mistakenly assumes that the robber fly *is* a bumblebee (Brower et al. 1960). Despite Hume's skeptical critique, the uniformity of nature is a compelling regularity, and science—and the toad—assume that this is so: what stung me before will sting me again. (In this case, the predator is fooled by superficial resemblance due to Batesian mimicry, an adaptation that itself evolves owing to expectations of the uniformity of nature).

To us, the toad's intuition is more convincing than the philosopher's fulmina-tion. The ability to discern meaningful similarities is at the root of sentience and learning in many animal species, and we are confident that, as sentient animals, systematists can usefully employ the criterion of similarity to make comparative observations among distinct entities and thereby draw inferences about more general patterns in the natural world.

When we compare the structures of insect genitalia, or sequences of the cyto-chrome b gene, we are making the implicit assumption that the objects of our comparison are fundamentally the same. The act of naming a structure or a gene de facto represents a hypothesis that the named thing is the same unit of compari-son in every entity that bears it, regardless of its degree of similarity or difference. It is for this reason that, in addition to the detailed codification of nomencla-ture for organisms, the parts of organisms also have standardized nomenclatures (e.g., Gray 1918; Tuxen 1970), and lately there is much interest in "ontologies" for computerized databases (in this context, an ontology is a standardized set of correspondences between names and parts coded in a data matrix; cf. Ramirez et al. 2007; Yoder et al. 2010).

Rather than being tools to "recognize homology," Remane's criteria—similarity of special structure, position, and connection—are actually epistemological cri-teria for the formation of character theories. In order for a feature shared among taxa to be hypothesized as a character, it must pass the tests of similarity of special structure, position, and connection. For example, the special structure of the head, with its unique sensory organs, entry point of the digestive tract, and so on, allows us to recognize "head" across a wide range of animals (even though cephalization may have been derived independently in protostomes and deu-terostomes). The relative structures, positions, and connections of body parts have been guides to recognizing their identity at least since the magnificent com-parative anatomical diagrams by Belon (1555) of human and bird, with cor-responding bones labeled on both skeletons. If features of different organisms cannot be easily recognized as similar by those criteria, then they may be linked via comparative study of taxa exhibiting intermediate conditions. Features that appear quite different, such as the ear ossicles of mammals and the operculum of actinopterygian fishes, can be related as "the same" through comparison of a series of taxa that exhibit intermediate conditions linking structural and posi-tional similarities (Geoffroy St. Hilaire 1818).

A final concept pertinent to our consideration of similarity is the notion of complementarity. One of the most compelling sources of inspiration for the hypothesis that there is order in biological diversity is the presence of complex features we might want to call "adaptations" in certain groups that are not pres-ent in any other groups. As Platnick has often pointed out, "All spiders have abdominal spinnerets." Likewise, all birds have feathers, all angiosperms have double fertilization, and all papilionid caterpillars have osmeteria. Each of these features is present in all members of its respective clade and absent in all other organisms. Paul Sereno (2007) has referred to such novelties as *neomorphic* characters. Although there is nothing remotely similar between the presence of a feature and the absence of that feature, "absent" is the logical complement to the presence of a feature that occurs only in a limited subset of living things. So, these opposites are treated as representing alternate states of the same character in the formal structure of a data matrix. As we will see when we discuss character polarity in Chapter 4, the presence of a complex feature is often considered as compelling evidence that the taxa bearing it form a clade.

Formal Tests of Similarity

Once recognized and characterized in words, a theory of similarity of a feature shared among taxa may be tested in three (often interconnected) ways: (1) conjunction, (2) similarity of structure, and (3) similarity of position (Patterson 1982). *Conjunction* dictates that multiple instances of the same structure may not exist in the same organism. Under this criterion the apparent multiple occurrence of an organ or structure would suggest the treatment of nonhomologous structures as the same. The conjunction criterion serves primarily to exclude false concepts of homology. The classic examples are imaginary hexapodous tetrapods, such as angels or Pegasus, which possess both "normal" forelimbs (arms or front legs) and wings (which are structurally homologous to forearms in the actual winged tetrapods that possess them). There are, however, situations that might be considered exceptions to the conjunction test. First is *bilateral* or *radial symmetry* in organisms that display such features. Second is *serial homology*, as represented by structures such as segmental repetition of body parts in the Annelida and Arthropoda and the similarity of the arrangement of bones in tetrapod fore and hind limbs. Third is *homonymy*, sometimes called mass homology, which is the simultaneous occurrence of many similar parts of a single organism. Familiar examples include the occurrence of multiple leaves on a single plant, scales on fishes, and hairs on mammals.

At the molecular level, gene or genome duplication can result in multiple copies of similar sequences within the same genome, a phenomenon referred to as *paralogy* (Fitch 1970). These sequences are the same in the way that tetrapod forelimbs and hindlimbs are the same. In order to infer organismal relationships, it is critical to compare the same (*orthologous*) copy of the gene among taxa. Comparison of paralogous structures or genes across taxa will not provide evidence about the relationships among the taxa but rather about the relationships among the copies.

Similarity of structure and position are the classic tests of "homology": when features of similar structure and position are found in two or more different organisms, they are judged to be homologous, or different instantiations of the same thing. Note that there is no reference to common ancestry in this judgment—the hypothesis is based on observed parts. Consider the tetrapod limb, as it exists in frogs, salamanders, mammals, and "reptiles." There would seem to be little disagreement in the scientific literature, or on the part of the well-informed nonscientific public, that the forelimbs of all of these groups represent the same structures—that they are homologous—as are the hind limbs among all organisms that possess them. A femur (or a humerus), for example, generally has a ball at its proximal end and a hinge at its distal end.

Similarities can be discovered between apparently very different entities by examination of fine structure. For example, the cell walls of yeast and a bracket fungus are both composed of chitin; the notochord of a larval tunicate reveals that it is a chordate. The relative position of constituent parts, sometimes referred to as topographical identity, can also reveal underlying structural similarity of structures that are superficially dissimilar, such as a whale's flipper and a bat's wing. These judgments are based on the observations that the limbs occupy the same positions on the body, have the same bony elements that are connected to one another in the same proximal-to-distal sequence, and in many cases have the same number of digits on each limb. A tetrapod femur is always associated proximally with a pelvis and distally with a tibia and a fibula (there are cases where these structures are lost). The transitive comparison of intermediate forms (if A = B and B = C, then A = C) often allows for the formulation of theories of structural sameness between less similar-appearing features; for example, fins and wings are connected by shared structural similarities found as intermediate forms in tetrapod limbs. Such a conclusion, in the minds of most observers, is based on similarity of position and structure, even though the bony elements have been modified, sometimes to the point of loss, as for limbs as a whole in most snakes and forelimb digits of birds.

What Is a Character?

As discussed above, a character is not a physical attribute of a particular organism but a conceptual grouping of such attributes that are judged by the systematist to be "the same" or complementary. Characters are theories. It is important to note that the quality or validity of character concepts is not tested by phylogenetic analysis but rather that the quality or validity of the latter depends upon thoughtful consideration of the former. Characters may be composed of variations on a theme, such as "small versus large" or "A, G, C, or T," or they may represent logical complements, such as "present/absent." Characters comprise any *heritable* attributes that may be compared among taxa and show group-defining variation, particularly those that show congruence with other such features. Historically, macroscopic morphology, visible to the naked eye, has formed the basis for most recognized taxonomic characters. More recently, DNA and amino-acid sequences have become standard character sources for many groups, augmenting—or even supplanting—classical morphology. Nonstructural attributes such as behavior and products of behavior (e.g., wasp nests or spider webs) also enjoy a place as legitimate sources of character data (Michener 1953; Wenzel 1992). Characters may exhibit variation that ranges from simple to complex. Note that when we refer to heritability as a criterion for character legitimacy, this does not presuppose "descent with modification"—it simply refers to observable patterns of family resemblance based upon transmission genetics that have been apparent to plant and animal breeders for millennia. (Indeed, patterns of "breeding according to kind" have long been invoked as evidence *against* evolution!)

Character information often does not present itself directly. It is not necessarily self-revealing. What, for example, do you think are the characters in the tidy illustrations in Figure 1.3? Character data represent, rather, a synthesis of comparative observation and theory (Rieppel and Kearney 2002). Thus, there is the potential for divergence of opinion among investigators concerning how to recognize the "characters" possessed by a given group of organisms and concerning the possible approaches for representing that information in a form suitable for analysis.

If, as stated by Pimentel and Riggins (1987), a character includes all homologous expressions of a feature found in the *ingroup* and *outgroup*, then the various observed conditions of the feature would be called *states*. Brower and Schawaroch (1996) observed that characters and their states become formalized as different things through the construction of data matrices, in which characters are represented by columns of cells, whereas states are represented by the values that populate the individual cells. Another way to think of this is that if the character is represented by a noun (e.g., "nose"), then the character states are adjectives modifying the noun (bulbous nose/pointed nose). The assignment of the condition of a feature exhibited by two or more taxa to a particular character state is a formal statement that the condition of the feature is the same, a property referred to by Brower and Schawaroch (1996) as *character-state identity*. The theoretical-semantic aspect of determining and describing character-state identity is evident in potentially differing opinions as to what range of variability might constitute "bulbous /pointed" and how these might relate to alternative characterizations such as "round/sharp" (or similar terms in other languages, for that matter). The linguistic diversity of descriptive terms is one of the reasons for interest in formalizing character ontologies.

Because any given state may serve as a group-defining feature, the terms *character* and *state* are often used interchangeably, depending on the hierarchic level at which they apply. Platnick (1978:366), for example, argued that "different character states are merely homologies that are synapomorphic at different levels, and there is thus a hierarchy of increasingly restricted characters rather than a division of characters into alternate states." This is a compelling perspective, but it is based on a view of characters and states in light of a cladogram, which can reveal correlations and connections among characters that are not evident prior to phylogenetic analysis. Although the distinction between characters and states may indeed be semantic, treatment of features as alternate states of the same character versus different characters is necessary for the construction of data matrices. How this is done can have important implications for character weights, and potentially the outcome of analyses, as we shall see below.

Character Analysis

Systematists stand upon the shoulders of their predecessors, with respect to both prior hypotheses of relationships and the characters that underlie those hypotheses. It is not uncommon for cladists to reanalyze published data matrices (e.g., Oosterbroek and Courtney [1995] reanalyzed the morphological data of Wood and Borkent [1989] and others), and many "combined analyses" of morphological and molecular data borrow from the published work of others (e.g., Whiting et al. [1997] borrowed morphological data from Kristensen [1981], Hennig [1981], and Boudreaux [1979]; Wahlberg et al. [2005] incorporated morphological data from de Jong et al. [1996]). In such cases, critical reassessment of prior authors' character coding is not necessarily conducted, and even more rarely are actual specimens examined to corroborate the original observations. In light of these practices, it is of fundamental importance for the working systematist to analyze characters thoroughly and to explain and illustrate that analysis clearly in the resulting publication because for many taxa the work likely will be cited and mined for characters long after its author is dead. Furthermore, in many groups of organisms, most notably insects, group specialists appear at long-separated intervals, and therefore accumulated knowledge is not passed directly from one investigator to another but only remains accessible through reading of the published literature and consulting museum collections. It is therefore essential to preserve data matrices and the descriptions of character coding that underlie them (not to mention the specimens from which they were observed) in a permanent archive.

If we had only two taxa under study, the question of their relationship to one another would be trivial—no matter the number and distribution of character similarities or differences between them—because they would be (tautologically) each other's closest relatives within the context of the study. It is only when we examine three or more taxa that alternate hypotheses of grouping among those taxa exist. These hypotheses may be assessed by examining the distribution of states of one or more characters.

Characters that show no variation among the included taxa, or with variable states unique to a single taxon, provide no evidence of relationships among the constituent taxa because they imply no grouping schemes. Character states observed only in a single terminal taxon are called *autapomorphies* and merely indicate that that taxon is different from others. Note that the status of characters as synapomorphies or autapomorphies is dependent on the taxon sampling in a given study. If we are interested in relationships among higher taxa, we might sample single exemplars for genera or families, and in such a case, the features that are synapomorphies for a given genus or family would be autapomorphies of the exemplar representing them in that data set.

As we saw in Chapter 1, whether or not to take account of autapomorphies, and their complementary symplesiomorphies, was the major source of contention between evolutionary taxonomists and cladists; the current manifestation of this controversy persists in the problem of branch length, discussed in Chapter 5. The characters of interest, from a cladistic perspective, are those that are shared by at least two terminal taxa, because they provide a hypothesis of grouping, although we do not know at this stage which state is derived and therefore represents synapomorphy.

A naive person might imagine that once the characters are observed, construction of the data matrix should follow straightforwardly and that the rigor of systematics lies in the sophistication of the model employed in the phylogenetic analysis, rather than in the care with which the characters and states have been interpreted. Felsenstein's 2004 book, *Inferring Phylogenies*, adopted this stance, viewing phylogenetic inference as the sine qua non of systematic biology, a myopic perspective punctuated by the fact that the term "homology" does not even appear in its index! Although it may not have been intended as such by its author, that book has been employed as the textbook in graduate-level systematics courses at more than a few universities. The trend toward teaching systematics (often in the guise of "statistical phylogenetics") solely through the use of tree-building algorithms and absent any attempt to address the complexity of characters and taxa strikes us as a fundamentally deficient approach to training the systematists of the future. Despite the tendency in some quarters to reduce it to a numerical or computational exercise, systematics remains first and foremost a biological discipline.

Equally unfortunate is the relinquishment of epistemological consideration of topographical identity to automated assembly pipelines for phylogenomic data matrices. Understanding the nature and relationships of characters provides the keys to understanding phylogenetic patterns (Rieppel 2006a, 2007c). That is why most of the chapters in this book are devoted to explicating the nature of systematic data, rather than celebrating computational algorithms in the absence of a deep familiarity with the data themselves. We believe that a reciprocal relationship exists between the character data and cladograms (or trees produced by alternative methodologies) and that both characters and cladograms should be considered as theories whose mode of construction is explicit, transparent, and subject to test.

It is through the process of *character coding* that specific observations of the features of organisms are converted into comparable character-state data available for use in phylogenetic inference. This is a critical stage of taxonomic/phylogenetic analysis because tree-building algorithms employ differences among character states to numerically assess the relative propinquity of the taxa. However, because the features themselves do not tell us how they should be recognized and coded, we must establish some criteria by which to convert qualitative observations into discrete characters and states. We have already alluded to the hypothetical nature of character observation and coding in Chapter 2.

Coding Criteria for Morphological Characters

Under the assumption that each expression of a feature is unique to a single group, one might well conceive a system in which all characters were coded as present in one group and absent from all other groups. Such characters are often referred to as *binary* or *two-state characters*. So coded, characters represent observations, or conjectures, the only test of which is congruence with other characters. Thus conceived, character theories are presumed by some authors to be minimally theory laden. A character theory (conjecture of homology) can then be rejected when the character appears in more than one group, as implied by the most parsimonious cladogram, or, as stated by Lipscomb (1992:55), when there is "non-congruence due to non-homology in two or more different taxa."

This pattern is called *scattering* (Figure 3.1) in the parlance of Mickevich and Lipscomb (1991), the most basic form of homoplasy. Good examples are the occurrence of endothermy in birds and mammals and the loss of wings among various groups of apterous insects, such as lice and fleas.

Because many features manifest only two conditions in the groups being studied, the question of how they should be coded is not at issue. The character states are complementary: either "this" or "that." We might use as examples "wings absent or wings present" in insects or "seeds absent or seeds present" in plants. However, when there are three or more conditions for a feature, such as "corolla with 3, 4, 5, 6, or numerous petals," "antennae filiform, moniliform, serrate, pectinate, clavate, capitate, or lamellate," or "nucleotide site 630 A, G, C, or T," then the question of how the characters should be coded—the relationship among the various alternative states—becomes more complex. Any complex character can be broken down into a series of discrete binary characters (so-called *presence–absence coding* [Meier 1994; Pleijel 1995]), or it can be viewed as a more elaborate whole with multiple parts.

The implementation of multistate coding requires an approach for specifying hypotheses of transformation from one state to another. If the characters themselves are recognized on the basis of similarity of position and structure, then it is the "connection by intermediates" that allows conjectures of state-to-state transformation. Sometimes this is intuitively obvious, in cases such as "small–medium–large," in which "medium" seems uncontroversial as the conjoining state

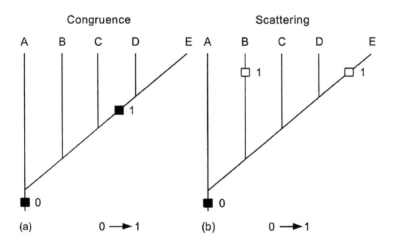

FIGURE 3.1. A two-state character, (a) parsimoniously distributed on a cladogram of five taxa, and (b) showing scattering, with the derived condition inferred to have arisen independently in more than one terminal taxon.

between the two extremes. However, in other instances, such as "red, white, blue," there is no evident transformation series. In such cases, the simplest approach to coding a multistate character allows any state of a homologous feature to transform to any other state with equal "cost" (Figure 3.2; see also Optimization in Chapter 5). This hypothesis of relationships among character states is called *Fitch transformation* (Fitch 1971), and such characters are described as *nonadditive* or *unordered*. This coding approach is most commonly used in parsimony analyses of DNA sequence data, data that would otherwise require a specified model of sequence evolution, but it is also used for many morphological characters where

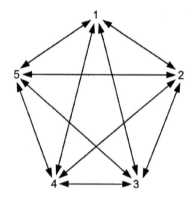

nonaddititve, unordered, Fitch transformation

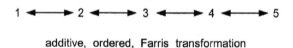

additive, ordered, Farris transformation

FIGURE 3.2.   Above: diagrammatic representation of unordered (nonadditive; Fitch) transformation, allowing any state-to-state change among five character states, as frequently used in the analysis of DNA sequence data. Below: ordered (additive; Farris) state-to-state character transformation, as ordinarily used in morphocline analysis, transformation series analysis, and homology analysis. Under the unordered approach, no state-to-state theories of transformation are implied, and any state-to-state transformation is possible without the imposition of additional steps. Under the ordered approach, transformations from one state to another are established a priori on the basis of observed similarities among the states, and additional steps are required if the implied state-to-state change on the cladogram does not conform to the postulated transformation series.

there is no apparent transformation series implied by relative similarity among states. Use of nonadditive coding achieves maximal congruence between the data and the cladogram, since only homoplasy due to scattering is taken into account. For this reason, some systematists view nonadditive coding as less assumption laden than alternative approaches—the state distributions of unordered characters are always at least as parsimoniously explained on a given cladogram as are ordered transformation series. However, in the case of morphological characters, this agreement is achieved by ignoring hypothesized information on transformation (see example in Figure 5.2). All other approaches specify state-to-state transformational order.

Mary Mickevich (1982) argued strenuously against treating all character data in a two-state (presence–absence) format and also against unordered multistate characters, urging instead the use of multistate coding when three or more conditions of a homologous feature exist. Under this view, multistate coding is said to represent a more meaningful and bolder hypothesis of character transformation. The degree to which multistate-character theories disagree with the cladogram they imply is termed *hierarchical discordance* (Figure 3.3; Mickevich and Lipscomb 1991), which is the "non-congruence due to non-homologous similarity used to hypothesize order of states in multistate characters" (Lipscomb 1992:56). At least three techniques have been proposed for coding ordered character-state transformations.

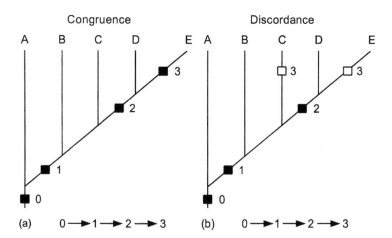

FIGURE 3.3. A multistate character (a) parsimoniously distributed on the cladogram, and (b) showing hierarchical discordance, the character not fitting the cladogram perfectly.

Morphocline analysis (Maslin 1952), as here defined, orders states on the basis of similarity alone, without regard for congruence, and may therefore produce transformation series in which homoplastic similarity is confused with homology (Lipscomb 1992). Under this precladistic approach, rules other than congruence would be invoked to adjudicate character conflicts.

Transformation series analysis (TSA) (Mickevich 1982) also orders the states of a character on the basis of similarity. The hypothesized transformation is then tested by whether or not the character is concordant (congruent) with the cladogram it defines (i.e., whether the character-state transformations are distributed parsimoniously on the cladogram). If not, the character is recoded, and the data are reanalyzed. This process is repeated iteratively, until global agreement between all multistate characters and the cladogram is maximized. This coding method might be envisioned as an extension of Hennig's idea of "checking, correcting, and rechecking." Transformation series analysis was criticized by Diana Lipscomb (1992:61) for discarding information on similarity among states in favor of codings that define a more resolved cladogram.

Homology analysis uses both similarity and congruence to evaluate character codings. Lipscomb (1992) argued that the flaws of the above-mentioned approaches to multistate coding can be avoided by employing this method. When using homology analysis, information on observed similarity is not discarded simply because the character has an imperfect fit to the cladogram. Rather, Lipscomb recommended checking all similarity-based transformation theories for congruence with other data, accepting those showing maximal congruence. However, she rejected recoding similarity-based transformations simply to achieve congruence.

Example and Discussion

We suspect few would doubt the hypothesis that the forelimbs of mammals are homologous with the wings of birds (cf. Belon 1555), and it might even seem evident without recourse to other character information. On the other hand, the hypothesis of limb loss from an ancestrally limbed condition in snakes might be viewed as stronger via assessment of congruence with other characters because most snakes possess no structural remnants of the forelimbs whatsoever, and only a few snakes show hind limb remnants. All of these conditions might also be judged homologous with the pectoral fins of fishes. In *presence–absence* format, the characters could be coded as follows, where "0" represents absent and "1" represents present, the unique or special condition:

1. (0) anterior appendages absent, (1) anterior appendages present as fins
2. (0) anterior appendages not in the form of limbs, (1) anterior appendages in the form of limbs
3. (0) anterior appendages not present as wings, (1) anterior appendages present as wings
4. (0) anterior appendages present, (1) anterior appendages absent (lost)

The same information could be coded in an ordered, multistate format where the numbering sequence indicates the adjacencies of the states:

(0) anterior appendages absent, (1) anterior appendages present as fins, (2) anterior appendages present as limbs, (3) anterior appendages present as wings, (4) anterior appendages absent (lost)

Graphically, this multistate coding can be presented as follows:

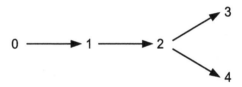

Note that in this example, state 0 of character 1 and state 1 of character 4 in the presence–absence coded character, or states 0 and 4 in the ordered character, are ostensibly "the same." Determining that a feature is lost, rather than having never been there in the first place, often depends on assessment of congruence with other features.

In a review of character coding approaches, Wilkinson (1995a) coined the terms "reductionist coding" and "composite coding." In the above example, the presence–absence coding could be termed *reductionist coding*, treating as many conditions as possible as unrelated. This type of coding was referred to by Pimentel and Riggins (1987) as the "stepwise decisions approach," which they said offered the utility of a pencil and paper solution but which could introduce errors into the phylogenetic analysis. Pleijel (1995) defended the presence–absence approach: he believed it to be simpler and more straightforward than multistate coding because it removes the necessity for making decisions about relationships of states to one another, allowing that information to emerge as a result of cladistic analysis. In spite of the benefits envisioned by Pleijel, hypotheses of transformational similarity are ignored when a feature with three or more states is coded in presence–absence format. Under presence–absence coding logic, for example, wings and pectoral fins would not be coded as alternative states of the character "forelimb" but as the presence or absence of unrelated structures.

The multistate coding shown above represents the *composite coding* of Wilkinson. This approach may combine conditions that some observers would say are not homologous and that therefore should be coded via a reductionist approach. For example, leg loss in some amphibians, some lizards, and snakes might be grouped—incorrectly—as a single character "anterior appendages absent." Such a hypothesis would, of course, be interpreted as homoplasy in light of a phylogenetic analysis including numerous other characters, and if the observer had evidence that the absence of legs was somehow not homologous among these taxa, it could be recoded specifically as such. To its advantage, multistate coding treats all "homologous" expressions of a feature as parts of a single character, thereby preserving information on similarity and transformation.

Because a hypothesized character-state identity is tested by its congruence with patterns implied by other characters, we might then wish to code all character data to maximize congruence. Whether this is best achieved via two-state or multistate codings is not always clear.

In summary, presence–absence coding would seem to have the following limitations:

1. Some states will not be coded as the result of observations. This phenomenon will probably occur most frequently for certain "absence" states in characters that would have otherwise been coded as multistate.
2. Absence connotes a non-group-forming condition of a feature. Therefore, presence–absence coding may imply knowledge of polarity a priori. For many attributes, such knowledge may appear virtually self-evident; for example, the relative recency of origin of tetrapod limbs versus fins, or the presence versus absence (notwithstanding secondary loss) of wings in insects. Yet, for many homologous features, polarity is far from obvious, especially in those cases in which detailed cladistic analysis ultimately suggests character reversal.
3. Presence–absence coding may discard evidence for transformation in complex features.
4. During optimization of hypothetical ancestral states (see Chapter 11), internal nodes may be inferred to have possessed the "absent" condition for all characters (Meier 1994).

5. For some types of data, such as DNA sequences, presence–absence coding is onerous and intuitively nonsensical. For instance, so coded, a single nucleotide site would constitute five "characters": nucleotide present/absent; adenine present/absent; guanine present/absent; cytosine present/absent; thymine present/absent.

Multistate coding, on the other hand, simply hypothesizes possible transformations among multiple states (i.e., ordering) on the basis of similarity. This approach to coding does not necessarily imply polarity, however. The optimal rooting, and therefore polarity of a character, will be determined by addition of the outgroup (Chapter 4).

The assumptions of the coding methods described above can be summarized as follows:

1. Presence–absence coding does not assume that alternative states are manifestations of the same character concept.
2. Nonadditive coding assumes a common character concept but does not assume a particular transformation order (see Figure 3.2).
3. The remaining methods (morphocline analysis, transformation series analysis, homology analysis), all of which represent different approaches to additive coding, assume ordered transformation series among character states (see Figure 3.2).

On the basis of the above discussion, it can be argued that features with three or more conditions should be coded as multistate. The value of such an approach is that (1) most coded states will be based on observations, and (2) information on observed similarity that implies a particular transformational order can be preserved in the coding. In practice, systematists often employ different coding strategies for different characters within a single matrix.

Character-State Transformation Weighting

In the above discussion, the implicit assumption was that every character-state transformation has an equal *weight*, or *cost*. In addition to the imposition of fixed transformation series on character states, it is also possible to differentially weight the costs of different types of state transformations. There are various approaches to this, and the approach chosen has potential to exert a fundamental impact on the resultant tree topology. Note that character-state transformation weighting is different from differential character weighting: The former is differential weighting of various transformations among the states of a single character; the latter is differential weighting among separate characters. Both of these types of differential weighting will be revisited in Chapter 5, where we discuss their impact on inference of tree topologies.

As discussed above, nonadditive coding allows any character state to transform to any other character state at equal cost, which is the most general hypothesis of character-state transformation. If characters are coded additively, this imposes differential costs on different types of transformation, depending on the relative *adjacency* of the character states on the character-state tree. Thus, in the ordered character in Figure 3.2, a change from state 1 to state 5 requires four steps.

Unless otherwise specified, these transformations are normally assumed to be symmetrical (change from state 1 to state 5 costs the same as change from state 5 to state 1). This need not be the case, however. Camin and Sokal (1965) described a scheme in which character change is irreversible, such that a reversal has a weight of infinity. Such a scheme tends to hypothesize independent gains of features on a tree. Dollo characters (Farris 1977) are weighted such that a derived state can arise only once but can subsequently be lost multiple times. Such an assumption might be appropriate for complex structures such as vertebrate eyes.

Some authors, such as Swofford and Olsen (1990), have labeled different types of character-state transformation as different "kinds of parsimony," which we view as an obfuscating misuse of the concept of parsimony. As discussed in Chapter 2, parsimony is a basic and venerable methodological rule that underlies all empirical science and, more specifically in cladistics, provides the rationale for selecting the tree with the smallest number of implied character-state transformations, or the lowest cost (sum of weighted state-change values) if characters and/or transformations are differentially weighted. Neither the general nor the more specific meanings of parsimony should be confused with state-transformation weighting schemes per se, which represent only one of the factors that can affect the length and topology of the shortest tree implied by a given data matrix. Nevertheless, so-called generalized parsimony, a framework often employed in models of molecular evolution, is an unordered transformation matrix with assignable costs for each of the various types of character-state transformations. A common example of such an assignment is the differential (usually down-) weighting of transitions with respect to transversions in DNA sequence, known as the Kimura two-parameter "model" (which in its most extreme form is "transversion parsimony," in which transitions [A $\leftrightarrow$ G, C $\leftrightarrow$ T] are weighted zero). The oft-employed generalized time-reversible "model" in maximum likelihood analyses is a symmetrical matrix of specified differential transformation costs for each of the six nucleotide transformation types, A $\leftrightarrow$ G, A $\leftrightarrow$ C, A $\leftrightarrow$ T,

G ↔ C, G ↔ T, C ↔ T. Note that "generalized parsimony" is not a model any more than a grid of empty cells is a data matrix.

Data Matrices and Coding Methods

When being prepared for processing by a computer algorithm, character data are normally assembled in the form of a matrix (Figure 3.4). By convention, the taxon names are listed as the rows on the left, and the characters are listed as the columns across the top. Some data sets could be coded adequately and accurately based only on the discussion up to this point. Other data will contain complexities that require knowledge of additional techniques, however.

Linear versus Branching Characters

Multiple states of a character can be coded in a linear format, up to a total of 10 states, for most computer programs. Such coding may take on a branching form

```
Character        0     5     10    15    20    25    30    35    40    45    50    55    60    65
Hypochilus       00001 11000 00000 00000 00000 00000 00000 00000 00000 00100 00001 -0000 00000 00
Ectatosticta     00001 11000 00000 00000 00000 00000 00000 00010 00000 00100 00000 -0000 00000 00
Cradungula       00010 00111 11020 00100 00000 00000 00000 00001 00001 00000 0000? -?000 00000 00
Pianoa           00010 00111 11020 00100 00000 00000 00000 00001 00001 00000 0000? -?000 00000 00
Hickmania        00010 00111 10110 00100 00000 00000 00020 00001 00001 00000 0000? -1000 00000 00
Austrochilus     00010 00111 10110 01100 00000 00000 00020 00000 00001 00000 0000? -1000 00000 00
Thaida           00010 00111 10110 01100 00000 00000 00020 00000 00001 00000 00001 -1000 00000 00
Filistata        00110 00101 10001 12000 00000 00000 10011 10001 00001 10000 00010 -1000 00010 11
Kukulcania       00110 00101 10001 12000 00000 00000 10011 10001 00001 10000 00010 -1000 00010 11
Scytodes         11110 00111 10001 12000 00000 00000 -0141 10101 0000- -0000 00000 -0010 01010 00
Sicarius         11110 00111 10001 12000 00000 00000 -0141 11101 1000- -0000 00000 -0000 00010 00
Drymusa          11110 00111 10001 12000 00000 00000 -0131 10101 0000- -0000 00000 -0010 01010 00
Loxosceles       11110 00111 10001 12000 00000 00000 -0141 11101 1000- -0000 0000- -0100 01010 00
Diguetia         11110 00111 10001 12000 00000 10001 -0121 10101 1000- -0000 0000- -200- 11010 10
Segestrioides    11110 00111 10001 12000 00000 10001 -0121 10101 1000- -0000 00000 -2001 1101? 00
Plectreurys      01110 00111 10001 12000 00001 10000 -0121 10111 100-- -0000 00000 -210- 1111? 00
Kibramoa         01110 00111 10001 12000 00001 10000 -0121 10111 100-- -0000 00000 -2100 1111? 00
Pholcus          01110 00111 10001 12000 00000 00000 -0141 10101 110-- -0000 00000 -1001 1111? 00
Caraimatta       11110 00111 10001 12010 00000 00000 -0020 10111 100-- -0000 0000- -1000 00110 00
Nops             01110 00111 10001 12010 00000 00000 -0041 10101 1000- -0000 00000 -1000 00010 00
Ochyrocera       11110 00111 10001 12010 00000 00000 -0130 10101 1101- -0000 00000 -1000 01010 00
Segestria        11010 00111 10201 12010 00000 00000 -0040 10011 0000- -0000 00000 -0000 0001? 00
Dysdera          11010 00111 10201 12010 00000 00000 -0040 10011 1000- -0000 00000 -?000 00010 00
Mallecolobus     11010 00111 10201 12010 00000 00000 -0040 11011 0000- -0000 00000 -?000 00010 00
Dysderina        11010 00111 10201 12010 00000 00000 -0040 11111 1000- -0000 00000 -1000 00010 00
Appaleptoneta    11010 00111 10001 12000 00010 00000 -1140 10101 1001- -0000 00000 00000 00000 00
Usofila          11010 00111 10001 12000 00010 00000 -1040 10101 1001- -0000 00000 00000 01010 00
Archaea          01010 00111 10001 12001 11110 00000 -0030 00010 1010- -0001 10000 01000 00001 10
Mecysmauchenius  01010 00111 10001 12001 111?0 00000 -0130 00110 1010- -0001 10000 ??000 00001 00
Tricellina       01010 00111 10001 12001 11010 00000 -0140 00110 1010- -0000 00000 00000 00000 10
Huttonia         01010 00111 10001 12001 10010 01000 -0140 00010 1010- -0000 00000 11000 00001 10
Otiothops        01010 00111 10001 12001 100?0 01000 -0140 00110 1010- -0000 00000 ??000 00001 10
Waitkera         00010 00-11 10001 12000 00010 00010 00130 00010 00-01 00000 01002 21000 00000 ?0
Tetragnatha      01010 00111 10001 12000 00010 00110 -0130 00010 10-0- -0100 00103 01000 0000? ?0
Crassanapis      01010 00111 10001 12000 00010 00110 -0130 00100 1010- -0000 00103 01000 00000 10
Oecobius         00010 00111 10001 12000 00010 00000 00100 00010 10000 11000 00010 01000 00001 10
Stegodyphus      00010 00111 10001 12000 00010 00000 00100 00010 00-00 11000 00012 01000 00001 10
Deinopis         00010 00111 10001 12000 00010 00010 00100 00010 00-01 01000 01002 21000 0000? ?0
Dictyna          00010 00111 10001 12000 00010 00000 00140 00010 10101 01010 0000? 21000 00000 ?0
Callobius        00010 00111 10001 12000 00010 00000 00120 00010 10101 11010 00002 11000 00001 10
Araneus          01010 00111 10001 12000 00010 00110 -0130 00010 0010- -1100 00003 01000 00000 10
Mimetus          01010 00111 10001 12001 10010 00000 -0130 00010 0010- -1100 00000 01000 00000 10
Pararchaea       01010 00111 10001 12001 111?0 00000 -0130 00100 1010- -1100 0000? ??000 00000 10
```

FIGURE 3.4. A typical data matrix, with the first taxon (*Hypochilus*) representing the outgroup. Missing information is represented by a dash. (From Platnick et al. 1991: table 2. Courtesy of the American Museum of Natural History.)

if the condition found in the outgroup is placed somewhere in the middle of the sequence of states, rather than at either end, as shown below.

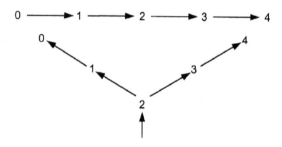

The branching approach was labeled *internal rooting* by O'Grady and Deets (1987), who observed that although it minimizes the number of variables (columns in the data matrix) required to code a simple branching multistate character, it has the disadvantage of showing the outgroup condition (primitive condition) as something other than the conventional zero. In the example, 2 becomes the outgroup condition in the internally rooted character, whereas it would be 0 in the linear character.

Not all computer phylogenetics programs currently in use support complex branching in character-state trees. Thus, if you wish to code character data that take on the form of more complex branching character-state trees, you must do so with techniques such as *additive binary coding* or *nonredundant linear coding* (see Farris 1970; O'Grady and Deets 1987; Pimentel and Riggins 1987; O'Grady et al. 1989). These methods will allow for any character-state tree to be described accurately. However, such recoded characters will occupy more than one column in the matrix, complicating comprehension of the relationship of the various states of the character in synapomorphy lists and other diagnostic output from the programs. Both types of coding maintain the state-to-state order assigned at the time of coding. Characters need not be coded in nonadditive formats, even if one wishes to process the data under that approach to optimization, because all computer phylogenetics programs allow, as an option, the processing of additively coded characters on a nonadditive basis. The converse is not true, however, and it is therefore always preferable to code characters in a format that includes hypothesized state-to-state transformations based on observation, if such hypotheses are warranted. This caveat does not apply to DNA sequence data.

Additive binary coding requires variables (characters, columns) equal in number to the number of terminals plus the number of internal nodes for the hypothesized character-state tree if the feature has a distinctive condition for each of the taxa being coded. If the feature does not possess a distinctive condition for every taxon being coded, then the number of variables necessary to code the information is smaller. The matrix values are determined by assigning a "1" (derived) state to each terminal contained in the most inclusive group on the character-state tree. The process is then repeated for the second most inclusive group on the tree, and so on, until all inclusive groups have been coded. As with

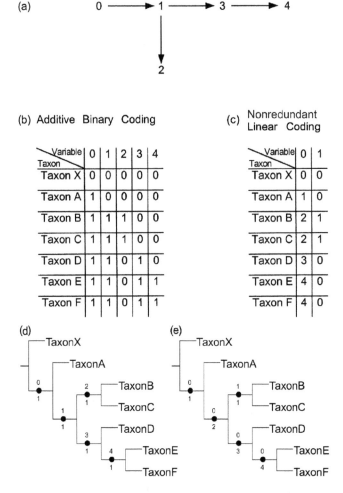

FIGURE 3.5. A (a) hypothetical branching character-state tree coded in (b) additive binary format and (c) nonredundant linear format. The relationships of the taxa derived from additive binary coding are shown in cladogram (d) and for nonredundant linear coding in cladogram (e). The character numbers for each taxon are shown above the branches in the cladogram; the states present in each taxon are shown below the branches. The nonredundant linear coding assumes that the characters are treated as additive; variable 0 represents characters 0, 1, 3, 4 in the character-state tree; variable 1 represents character 2. See text for further explanation.

any coding technique, autapomorphies are unique to the terminals that possess them, and therefore a "1" (derived) state is assigned as a unique value for each terminal, if autapomorphies are included in the matrix.

An example of additive binary coding for a complex hypothetical character is shown in Figure 3.5. Six hypothetical taxa exhibit states corresponding to the character-state transformation tree shown in Figure 3.5a. The matrix shown in Figure 3.5b contains the five binary variables required to recover the branching pattern shown in Figure 3.5d (note that taxa B and C, and taxa E and F, are coded identically). The groups formed by each binary-coded variable, numbered 0–4 in Figure 3b, are shown on the cladogram in Figure 3.5d above the corresponding node. In the matrix shown in Figure 3.5b, no autapomorphic states of the character exist in any of the taxa, and therefore none are coded in the matrix.

Nonredundant linear coding requires fewer variables than additive binary coding because multistate coding is used to code the states for the longest ordinal subset (longest sequence of states) on the character-state tree. Additional binary variables are then used to code the branches that remain. For this reason, the method has also been referred to as "ordinal-additive binary coding" and "mixed coding" by Pimentel and Riggins (1987). Returning to the example character-state tree shown in Figure 3.5a, we see one possible approach to nonredundant linear coding for this hypothesized transformation in Figure 3.5c. Variable 0 (the 0 → 1 → 3 → 4 transformation series), now in multistate form, codes for the nodes on the cladogram using four conditions. The branch of the character-state tree comprising the transformation 1 → 2 is coded in variable 1. No autapomorphic states of this character exist in any of the taxa, and therefore none are coded in the matrix (note that although taxon A is the only one exhibiting state 1, and taxon D uniquely possesses state 3, these are considered plesiomorphies and not autapomorphies in this scheme because of the nested nature of the coding scheme).

In this example, two variables code the same data that required five variables using additive binary coding.

Sidebar 7
Character Independence

The literature contains many statements arguing that character independence is a prerequisite to conducting phylogenetic analysis. As such, independence of observations as separate items of evidence is simply a methodological requirement of quantitative science: one does not wish to skew a result by counting the same information multiple times. Cladistics identifies groups supported by evidence in the form of synapomorphies, a point made by Hennig when he observed that the more characters defining a group, the greater our confidence in the monophyly of the group. If characters supporting the same branch were not independent, we could place no more assurance in groups supported by multiple characters than in groups supported by a single character.

Structures existing on different parts of the body would appear, de facto, to represent independent characters. Yet the potential for correlation among morphological features presents the systematist with a difficult problem of interpretation. One would be disinclined to view multiple measurements of size (e.g., length of a tibia and length of a fibula) or of bilaterally symmetrical parts as independent observations, yet characteristics of serially homologous features like teeth in mammals and wing venation in insects represent some of the pillars of our understanding of relationships in those groups. For characters that are structurally associated, there may be no direct evidence for independence; rather, independence is assumed if the structurally allied characters serve to define groupings of taxa at different levels in the cladogram. Problems such as these highlight the theoretical nature of observation discussed in Chapter 2.

Arguments for independence among different categories of characters or "data partitions" have been advanced for DNA versus gross morphological data, for DNA data derived from different parts of the genome and known to have different functions, and even for different parts of the same gene that may be evolving under different selective pressures or constraints (e.g., introns versus exons, different nucleotide positions in a protein coding sequence, or stems versus loops in the secondary structure of rRNA molecules, as argued, for example, by Bull et al. [1993]). However, this perspective has the undesirable corollary that if various different partitions are considered to be "independent" as sources of phylogenetic evidence with respect to one another, then the characters (individual nucleotides) amalgamated within each partition must therefore be nonindependent of one another, implying that they should be down-weighted or otherwise discounted (DeSalle and Brower 1997). This is clearly counterproductive (see further discussion in Total Evidence, Chapter 6).

In practice, under most circumstances, character independence is assumed by necessity, in order for characters to have the additivity necessary for them to serve as evidence in the calculus of quantitative systematic analysis. Such an assumption may be viewed as a corollary of Hennig's auxiliary principle (see Chapter 4). Systematists acknowledge that this convention underlies our analyses and that we therefore are obliged to remove redundant characters when we see evidence of nonindependence due to structural, functional, developmental, or adaptive constraints.

Autapomorphies

One might assume that a matrix of character data would always include as much information as possible. Consider, however, *autapomorphies*—characters unique to a terminal taxon. As we have indicated previously, autapomorphies do not imply relationship and therefore provide no evidence of grouping in cladistic analyses. We might therefore wonder whether autapomorphies should be included in data matrices used for phylogenetic analyses. On the one hand, it has been argued that they should not, because autapomorphies inflate the value of "fit statistics" such as the *consistency index* (discussed in Chapter 6), which provides an elevated but artificial sense of character congruence for a given matrix and therefore makes comparisons of such statistics across matrices less meaningful. On the other hand, if the taxa being analyzed are not species but rather collections of species, the autapomorphies add information by documenting the (assumed) monophyly of the terminals, even though they offer no information on grouping of those terminals in relation to one another. Another role of autapomorphies may be their contribution to the inference of *branch length* (see Chapter 5).

We conclude that if the goal of a study is to make comparisons of the amount of homoplasy among matrices, then removal of autapomorphic characters is a desirable approach. If, however, the primary goal of assembling data matrices is part of a continuing effort to document the distinctive attributes of the taxa under study and document the monophyly of supraspecific terminal taxa, then autapomorphies should be included in the matrix. Furthermore, today's autapomorphy may become tomorrow's synapomorphy if additional species belonging to the clade are discovered. Phylogenetic computer programs make it possible to compute trees and fit statistics using only informative characters, even if autapomorphies or invariant characters are present (as often found, for example, in DNA sequences). Thus, we recommend that all observed variable attributes be included in a data matrix. The uninformative characters can be excluded easily for the purposes of computation without physically modifying the data matrix.

Missing and Inapplicable Data

There are two types of absences for which character states are truly not observable and thus cannot be meaningfully coded. These have been referred to as missing (unavailable) and inapplicable data.

Missing data, in the sense used here, are those characters that cannot be coded for a given taxon because the information is not available at the moment. This does not mean that the data might not be available in the future with the acquisition of more specimens, or specimens in a proper state of preservation. *Inapplicable data* are those that can never be coded for a given taxon because the taxon does not possess the attribute on which the character is based.

An example may help to illustrate these points. Most insects have wings, and the wings themselves possess many distinctive attributes useful for inferring groups of the taxa in which they occur. However, secondarily wingless insect taxa cannot be coded for features of the wings themselves, such as, for example, patterns of venation. Therefore, attributes of wings are inapplicable in secondarily wingless forms. On the other hand, some species of insects exist in both winged and wingless morphs (e.g., ants, aphids). If a matrix were coded using the wingless morph because only that form was available at the time, the relevant states should be coded as though the data were unavailable, in lieu of acquiring winged specimens.

Another example is the absence of nucleotides in DNA sequences. Nucleotides could be absent from a data matrix because the researcher failed to generate complete reads of the sequence for one or more taxa. Such data would be characterized as *missing.* Alternatively, many gene regions vary in length from one taxon to another, and when these are aligned to imply homology of nucleotide sites across taxa, shorter sequences will possess implied gaps or deletions where nucleotides that are present in other taxa are inferred to be absent. These empty cells in the aligned data matrix may be either equated with missing data or coded as a fifth character state to indicate their inapplicability (see Alignment of DNA Sequence Data, below).

In light of the above discussion, we conclude that it is desirable to distinguish between missing and inapplicable data in preparing a data matrix. Such a distinction will facilitate future character analysis and additions of data to matrices. Also, unless missing and inapplicable data are specifically coded as distinct, phylogenetic computer programs do not distinguish between the two types of absences.

There may be many cells coded as missing in a matrix of data sets combined for "simultaneous analysis," as, for example, in the case of morphological and DNA sequence data collected by different investigators for the same higher taxon but not representing identical terminal taxa. Missing observations sometimes cause lack of resolution or other undesirable perturbations to the analysis, particularly if they are complementary, such that, for example, characters 1–50 are missing for taxon A and characters 51–100 are missing for taxon B. Such distributions of missing data can result in thousands of spurious equally parsimonious trees and a complete lack of resolution (Wilkinson 1995b; Wiens 2003; Brower 2018b).

What approaches might we take to minimize the numbers of empty cells in the matrix? We might choose to *fuse terminals* or to use *exemplar terminals* (Nixon and Carpenter 1996). If it is acceptable to combine DNA sequences from one terminal taxon with morphology from another because each partial data set is a fair representation of variation for the combined taxa as a more inclusive entity (e.g., data from two different species to encode a genus), terminal fusion is an option. The methods section of a publication should explicitly indicate such fusions in the description of data coding and analysis to facilitate future testing of results and to allow subsequent incorporation of new data that eliminates the requirement for fusing terminals. In using the exemplar approach, we might choose to remove from the analysis some terminals represented by incomplete data, on the assumption that other included terminals with complete data will fairly represent the higher taxon to which all of the terminals in question belong.

Strictly speaking, all systematic investigations are de facto exemplar studies, since they rely by necessity on nonexhaustive sampling of species or individual organisms that could in principle be sampled (Vrana and Wheeler 1992). In practice, however, an alternative to fused terminals and exemplar terminals is to code higher taxa using a "*ground plan*" (a set of inferred plesiomorphic conditions for the taxon in question) that itself is derived from a rigorous analysis of the data. This was an approach favored by classical taxonomists such as Hennig, but it relies upon a priori assumptions about character polarity that can influence the outcome of an analysis. It was only with the introduction of computer-based analytical approaches and molecular data that most authors abandoned the ground-plan approach (Prendini 2001).

Common cases in which particular taxa are missing data in empirical data sets include larval stages of holometabolous insects, which are often not collected or not associated with the corresponding adults for some of the taxa, missing DNA sequences from rare or extinct taxa represented solely by historical specimens in museum collections, and missing soft parts and DNA sequences from fossil representatives of both invertebrate and vertebrate taxa. Such taxa can nevertheless provide critical evidence that may change interpretations of homology for individual characters and alter tree topologies (Kearney 2002).

With the growth of large molecular data sets, the prevalence of missing data has become a significant issue. The questions of how many missing data are too many, and the relative costs and benefits of including taxa with a lot of missing data in an analysis, have been the subject of much discussion (Gauthier et al. 1988; Wheeler 1992; Wilkinson 1995b; Wiens 1998, 2003; Kearney 2002; Philippe et al. 2004; Lemmon et al. 2009; Simmons 2012; Wiens and Tiu 2012; Simmons and Goloboff 2014; Streicher et al. 2016; Xi et al. 2016; Brower and Garzón-Orduña 2018). The principle of total evidence (see Chapter 6) suggests that all pertinent data should be included, but it may also be argued that taxa with too much missing data burden an analysis by lowering resolution and support. This is an empirical problem that should be explored in data sets with substantial missing data.

Polymorphisms

Some authors have coded as "missing" characters that are polymorphic in terminal taxa, the so-called X-coding of Doyle and Donoghue (1986). Nixon and Davis (1991) criticized that approach because it does not correctly represent the specific pattern of variation among the terminals during phylogenetic analysis; consequently, the resulting cladograms may be inferred incorrectly. Therefore, polymorphic terminal taxa should be explicitly coded with all possible combinations of character states that are known to occur within such lineages. This approach will allow for the discovery of parsimonious solutions in which polymorphic lineages may be found to be polyphyletic. Some phylogenetic software allows the alternate interpretation of polymorphic characters as "or" (uncertainty, assigned to the inferred state of the sister taxon) or "and" (each polymorphic terminal has both, or all, states). "And" trees will be as long as or longer than "or" trees, and selection of one or the other option needs to be considered when various descriptive metrics are being calculated. In DNA sequence data, which often exhibit intra-individual polymorphism due to the diploid nature of the nuclear genome, there is a standardized ambiguity code for all combinations of nucleotides that may occur at a given site.

Intrinsic versus Extrinsic Data

A number of authors have drawn a distinction between intrinsic and extrinsic data. *Intrinsic* characters are those obviously subject to the observable mechanisms of parent-offspring inheritance and/or ontogeny, such as DNA sequences and morphology. *Extrinsic* characters are those not governed directly by the rules of heredity, such as biogeographical distributions and host associations. The former type of character data has traditionally been used in the construction of cladograms. The latter type has usually been examined to determine the degree to which it is congruent with intrinsic character data (see further discussion in Chapters 9 and 11).

Kluge and Wolf (1993) argued for the inclusion of "extrinsic" as well as "intrinsic" data in phylogenetic inference. They suggested that such an approach involves a "consilience of inductions." What Kluge and Wolf failed to acknowledge, in their advocacy for inclusion of extrinsic characters, is that there is no method for hypothesizing homology or transformation series for extrinsic data other than the results of intrinsic-character-based phylogenetic results themselves. Therefore, there is no independent rationale for treating such observations as part of a parsimony analysis. Kluge and Wolf (1993) quoted Mickevich (1982) as supporting their view, but they misrepresented the meaning of her statements to include character data beyond the types she intended.

Take, for example, host-parasite relationships, one of the most commonly treated cases in the coevolution literature and a subject discussed in detail by Hennig. The sequence of host changes might be looked on as a problem in cospeciation, whereby parasite speciation parallels host speciation. Brooks (1981) made that assumption in his proposal for a "solution to Hennig's parasitological method." However, the assumption would seem to be risky and potentially error prone because monophyletic groups of parasites are often observed to occupy distantly related hosts, such as bed bugs on bats and swallows, with the clear implication that cospeciation was not involved. Another example is that restricted monophyletic groups of phytophagous insects may occur on conifers and angiosperms, or legumes and composites, strongly suggesting that host shifts within the group are not solely the result of host-related cospeciation events. Coding such information as character data for phylogenetic inference could result in either false inferences about the relationships of the hosts or the needless introduction of homoplastic data into the character matrix.

Although geographic occurrences, host relationships, and other ecological associations may all represent aspects of the evolutionary history for a group of organisms, there are strong arguments for the view that the methodologies for assessing agreement among these data are different than those used to analyze data subject directly to the "laws" of inheritance, contrary to the views of Brooks (1981) and Kluge and Wolf (1993). Kluge (1997) remained undeterred in his view, asserting that host-parasite (coevolutionary) and biogeographic relationships will serve as more critical tests of phylogenetic theories than synapomorphies themselves, a notion whose justification he did not explain further. We suggest that if agreement between extrinsic and intrinsic data occurs, then it represents an example of what Hennig referred to as "reciprocal illumination." In such cases, the implications for hypotheses of relationship are benign, as patterns are not altered but corroborated by the additional evidence. When there is disagreement, then ad hoc explanations for the extrinsic characters, such as dispersal or host shifts, must be invoked. Methods for evaluating extrinsic data are discussed further in Section 3, Application of Cladistic Results.

Alignment of DNA Sequence Data

DNA sequences have in recent decades become dominant as a vast source of characters for systematics. However, DNA sequences still have not proved to be the phylogenetic Rosetta Stone heralded by molecular systematic boosters in the 1980s and 1990s (e.g., Gould 1985; Hedges and Maxson 1996). Just as morphology is not self-evident as a source of characters, DNA requires processing and interpretation to transform its raw information into a form amenable to comparative analysis. Much of this work is now performed by computerized "pipelines" that assemble and align sequences automatically, but such algorithms entail numerous underlying assumptions about the raw data, of which the perspicacious systematist should be aware.

A fundamental issue in molecular systematics is determining the correspondence of individual nucleotide sites among comparable ("homologous") DNA (or RNA) sequences sampled from a range of taxa, a process referred to as *alignment*. A sequence alignment identifies the characters that are compared among the taxa, a foundational assumption which can have a profound impact on the resultant phylogenetic hypothesis.

There are three main functional types of sequences that are compared to address phylogenetic or population-genetic problems. These are protein-coding sequences, ribosomal RNA gene sequences, and noncoding regions such as introns, flanking regions, and the mitochondrial control region. Each of these presents different alignment challenges. For taxa that are relatively closely related, such as species within a single genus, the sequences usually exhibit little or no length variation, alignments are usually straightforward, and relationships can be examined with sequences of almost any DNA region. The most common problem at these low levels of divergence is not inferring the correspondence of homologous nucleotides across taxa, which is often manifest, but finding a gene region that provides enough variation to distinguish among terminals. (It should be noted that in protein-coding genes, most third positions are two- or fourfold redundant, and so all of these sites are more or less silent and evolving at the genome's neutral mutation rate. Therefore, in principle, any protein-coding gene should provide evidence for relationships among closely related taxa at one-third of its nucleotide sites). As the taxa being compared span deeper branches of the Tree of Life, variation among their DNA sequences become more substantial, rendering the alignment process more important, more complex, and more problematic.

Alignment of protein-coding gene regions is often relatively easy because the sequence is organized into codons of three nucleotides that encode a corresponding amino acid sequence. Length variation in protein-coding genes, if it occurs at all, often takes the form of three-nucleotide (or multiples of three) additions or deletions, reflecting the addition or deletion of one (or more) amino acids from the polypeptide chain encoded by the DNA. The congruence criterion can be used to test theories of nucleotide site homology in protein-coding genes because if amino acid translations of sequences correspond to one another across taxa, this offers a measure of corroboration of the underlying nucleotide alignment. Likewise, it is relatively easy to determine if a protein-coding sequence is not in the correct "reading frame" by translating it and looking for stop codons, which occur on average approximately 1 per 20 amino acids in random sequence but not at all in a contiguous stretch of protein-coding DNA. An apparent out-of-frame sequence from such a gene region is almost always the result of a misread by the sequencing machine, and the source of the problem can usually be pinpointed by visual inspection of the raw output. Next Generation sequence alignment is believed to overcome such errors by sequencing multiple overlapping copies of the same gene region from the sample, aligning the individual sequences, and accepting a majority rule consensus sequence as "the sequence" on the assumption that errors are relatively infrequent and occur at random. Paralogs and other repetitive elements of the genome can cause problems in such instances.

Straightforward alignments, such as those found in protein-coding genes, have been traditionally done "by eye." Indeed, there is the view—although certainly not universally held—that if alignment is not obvious, then the homology of characters being compared is suspect, and any attempt at analysis of such data may lead to erroneous conclusions.

Nonprotein-coding regions of the genome often present distinct challenges to alignment because they vary in length from taxon to taxon, a condition brought about by hypothesized insertions and deletions (*indels*). Note that while the mechanisms responsible for sequence length variation remain poorly understood, it is apparent that sequences can change in length by multiple bases in single events. This has important implications both for the methodology of aligning sequences and for the theory of character independence (see Sidebar 7). Thus, with only four possible nucleic acid character states at any given site, variation in the number of sites often makes determination of site correspondence across taxa less than clear-cut. For example, if we observed two sequences, GGACAT and GAACAT, we would probably infer that the second nucleotide in one of the sequences has undergone a G ↔ A transition—a parsimonious explanation of the pattern observed. But if we observe a third sequence, GAGACAT, which is one nucleotide longer, the determination of site correspondence is less straightforward. Depending on whether we believe it is more plausible for two "gaps" to appear at different sites in the two shorter sequences or for there to be a single "homologous" gap and a mismatch at the other site (which we hypothesized to be parsimonious earlier), there are at least three different potential arrangements of the nucleotide sites in these sequences with respect to one another:

Alignment	1	2	3
Sequence 1	GAGACAT	GAGACAT	GAGACAT
Sequence 2	G-GACAT	G-GACAT	GG-ACAT
Sequence 3	GA-ACAT	G-AACAT	GA-ACAT

This example shows that even a minimal amount of length variation among observed sequences can lead to interpretational ambiguity regarding which nucleotides are topographically identical with which others, potentially leading to different homology statements and implied hypotheses of relationship. Nonetheless, the length-variable mitochondrial 12S and 16S, and the nuclear 18S and 28S ribosomal RNA genes, show variation deemed informative at higher levels of relationships among eukaryotes and therefore have been extensively used in phylogenetic analysis over a broad range of taxa. Homologous ribosomal genes with differing lengths and names (12S, 16S) are also important in prokaryotic (and mitochondrial) molecular systematics. For these gene regions, corroboration of sequence alignment via comparison with amino acid translations is not possible. Furthermore, visual alignment is untenable for any data set with a large number of taxa and substantial sequence variation because there is no way to evaluate anywhere near the number of actual possible alignments without the use of a computer (Wheeler 1994; Slowinski 1998), and even then the result is at best a heuristic estimate of the optimal hypothesis.

The crucial nature of determining positional correspondence among nucleotides, especially its determination in nonprotein-coding genes and for data matrices representing widely divergent taxa, would seem to mandate that alignment procedures be treated as an integral part of the phylogenetic analysis of such data. Nonetheless, Hillis et al. (1996), in a text dealing with molecular systematic methods, admitted that alignment was difficult and poorly understood, and they relegated discussion of alignment to a chapter dealing with sequencing and cloning under the subheading "Interpretation and Troubleshooting." More recently, Löytynoja and Goldman (2008:1634) warned, "[S]equence alignment remains a challenging task, and alignments generated with methods based on the traditional progressive algorithm may lead to seriously incorrect conclusions in evolutionary and comparative studies."

Today, with large genomic data sets, many of these fundamental decisions are determined by assembly algorithms and, alarmingly, are never examined by human eyes. As noted, the voluminous assemblies from Next Generation Sequencing protocols often rely on automated majority-rule consensus of multiple reads from what is assumed by the algorithm to be "the same" stretch of DNA. Note that assembly, just like alignment, represents a hypothesis of homology. Faulty inference of character correspondence is fatal to a phylogenetic analysis. For example, in a meticulous dissection of a large phylogenomic data set for mammals assembled by Liu et al. (2017), Gatesy and Springer (2017) found alignment errors among the sequences of 50 out of 50 gene regions they examined, including alignment of paralogous gene copies, alignment of exons with introns, and other confounding blunders. As the saying goes, "garbage in, garbage out." The abrogation of homology assessment by the researcher therefore seems to us to be, at best, a risky proposition. At worst, one might argue that the intrinsic inscrutability of the inner workings of large genomic data sets renders them unscientific, in the sense that they are not amenable to critical peer review, other than heroic efforts such as those of John Gatesy and Mark Springer.

Proposed alignment methods for large, complex data matrices are of several types. Some authors have proposed using prior conclusions about relationships as a template for the sequential alignment of sequences (e.g., Mindell 1991), but this approach is undesirable because it does not treat the sequence data as an independent source of evidence. The remaining methods can be divided into *similarity methods* and *parsimony methods*.

Alignment based on similarity has the same drawbacks found in grouping by overall similarity in nonmolecular data (discussed in Chapter 1). Nevertheless, many investigators working with molecular data (perhaps with no formal training in systematics) still calculate distance trees based on overall similarity, and it is therefore not overly surprising to encounter the use of similarity-based alignment procedures, such as CLUSTAL (Higgins and Sharp 1988) and BLAST (Altshul et al. 1990). These have the expedient of being fast and free. For example, BLAST is supported by the National Center for Biotechnology Information, and alignments of new sequences to sequences deposited in GenBank may be performed via the web. This does not, however, obviate the fact that such approaches do not adhere to the basic criteria demanded of any rigorous phylogenetic analysis (see discussion in Optimality Criteria, Chapter 5). Note that BLAST is more often used to match a query sequence to highly similar sequences based on pairwise similarity than for alignment of multitaxon matrices.

The remaining approaches to alignment attempt to maximize the number of site correspondences or minimize the number of implied character-state transformations through the application of the parsimony criterion. There are, nonetheless, many ways that this problem might be solved. Owing to the *NP-complete* nature of alignment problems, confidence in having found the solution(s) optimal for the criteria employed requires that the method be computer implemented, as noted above. The first attempt at parsimony-based multiple sequence alignment was that of Wheeler and Gladstein (1992) with the program MALIGN. This program used multiple alignment addition sequences to overcome the problem of bias introduced by pairwise alignment based on decreasing similarity. Such an approach requires substantial computing power; it attempts to produce a globally parsimonious alignment for all taxa under consideration by using multiple taxon addition orders rather than by using sequential pairwise comparisons based on decreasing similarity, as is done by CLUSTAL and other phenetic alignment programs.

Traditional alignment methods rely on the insertion of gaps to accommodate the differences in numbers of nucleotide positions among the sequences being aligned (Figure 3.6), based on a gap/change cost ratio. The calculus of alignment algorithms is such that the cost of inserting a gap must be higher than the cost of a nucleotide mismatch at a given site; otherwise the program will simply insert a gap every time there is a difference among the sequences. The higher the gap/change cost ratio, the more "expensive" the gaps are, and the fewer of them will be inserted in the alignment. In addition, there can be differential costs for gap initiation (the first implied gap site, usually a higher cost) and extension (subsequent inserted gap sites after the first one). Variation of these parameters changes the relative cost of different alignment patterns, and alignments generated under different gap/change costs cannot be directly compared with one another. This problem is discussed further in Chapter 5.

It should be noted that alignment algorithms are often designed to conform to some model of molecular evolution—such as the increased weighting of transversions relative to transitions—or can be tailored to do so. The effects of invoking one or more such models should be examined empirically with regard to maximizing character congruence and parsimonious interpretation of the data, as has been done, for example, by W. C. Wheeler (1995), rather than being applied as a priori assumptions.

Because gaps do not represent observations, but are rather a complementary absent state inserted to accommodate inferred topographical correspondence of adjacent nucleotide sites in length-variable sequences, they are often scored as "missing," thereby avoiding their interpretation as potential synapomorphies. An alternative approach to this problem is "optimization alignment," a parsimony-based technique proposed by Wheeler (1996). Under this method, insertion/deletion events are treated not as states but rather as transformations linking hypothesized ancestral and descendent nucleotide sequences (see Chapter 5 for

```
Heliconius numata     GATACACGAGCTTATTTTACTTCAGCTACTATAATTATTGCAGTACCTACACAGGAATTAAAATTTTT
Heliconius melopmene  GATACTCGAGCATATTTTACTTCAGCTACTATAATCATTGCAGTTCCAACTGGAATTAAAATTTTT
Heliconius erato      GATACTCGAGCCTATTTTACATCAGCCACTATAATTATTGCTGTTCCTACAGGAATTAAAATTTTT
Heliconius clysonymus GATACTCGAGCTTACTTTACATCAGCTACTATAATTATTGCAGTTCCTACAGGTATTAAAATTTTT
Heliconius eleuchia   GATACCCGAGCTTATTTCACATCAGCAACCATAATCATTGCAGTACCTACAGGAATTAAAATTTTT
Heliconius demeter    GATACCCGAGCATATTTTACATCAGCAACTATAATTATTGCAGTACCTACTGGAATTAAAATTTTT
Heliconius burneyi    GATACTCGAGCTTATTTTACATCAGCTACTATAATTATTGCAGTTCCTACTGGGATTAAAATTTTT
Heliconius hecuba     GATACTCGAGCATATTTTACATCTGCCACAATAATTATTGCAGTTCCTACCGGAATTAAAATCTTT
Heliconius doris      GATACTCGAGCTTATTTTACATCTGCTACAATAATTATTGCAGTTCCTACAGGAATTAAAATTTTT
Neruda aoede          GATACTCGAGCTTATTTTACATCCGCAACTATAATTATTGCAGTTCCAACAGGAATTAAAATTTTT
Eueudes vibilia       GATACTCGAGCTTATTTTACATCCGCAACTATAATTATTGCAGTACCTACAGGAATTAAAATTTTT
Dryas iulia           GATACTCGAGCATATTTTACTTCAGCAACTATAATTATTGCAGTCCCTACTGGAATTAAAATTTTT


Heliconius numata     GGGGTGATAGAAAAATTAAAATAACTTTTTTTAAAAAAAAAATAAATTACATGAATAAATGAATAA
Heliconius melopmene  GGGGTGATAGAAAAAWTAAAATAACTTTTTT--TA-AAAATAA---ACATAAATAATTGAATAA
Heliconius erato      GGGGTGATAGAAAAATTAAAATAACTTTTTTTTTAAAAAA------TTACATTAATAAATGAAATT
Heliconius clysonymus GGGGTGATAGAAAAATTAAAATAACTTTTTTTTTTAATAAAAAAAA----ACATTAATAAATGAAATT
Heliconius eleuchia   GGGGTGATAGAAAAATTTAAATAACTTTTTT------AAAATTTTACATTAATAAATGATTTT
Heliconius demeter    GGGGTGACAGAAAAATTAAAATAACTTTTTT------AAAATTTTACATTAATAAATGATTTT
Heliconius burneyi    GGGGTGATAGAAAAATTAAAATAACTTTTTTTTAAAAAAAA------TTACATGAATAAATGAATTT
Heliconius hecuba     GGGGTGATAGAAAAATTAAAATAACTTTTTTTTTAAAAAAA------TTACATTAATAAATGAATAT
Heliconius doris      GGGGTGATAGAAAAATTAATAAACTTTTTTT--CAAAAAAAA-----TACATTGATAAGTGAGTTT
Neruda aoede          GGGGTGATAGAAAAATTTAATAAACTTTTTT-----GAAAAA--TTTACATAAATAAATGAATTA
Eueudes vibilia       GGGGTGATAGAAAAATTAAAATAACTTTTTTT--ATAGA------CTAACATAAATAAGTGAATAG
Dryas iulia           GGGGTGATAAAAAAATTAAAATAACTTTTTTTTTTTTTTTT-----ATTTTAACATAAATAAATGAATAA
```

FIGURE 3.6. Alignment of sequence data. Two sets of sequences from the same group of 12 taxa. Above, a fragment of sequences from the mitochondrial protein-coding cytochrome oxidase I gene, showing the constant number of nucleotides and apparently perfect site correspondence within the fragment. Below, aligned mitochondrial 16S ribosomal DNA sequences, showing the insertion of gaps represented by dashes to accommodate the variation in numbers of nucleotides across taxa (from Beltrán et al. 2007).

further discussion of "optimization"). This insight led to the realization that calculating the shortest alignment and calculating the shortest tree based on that alignment could be collapsed into a single operation, referred to as *direct optimization*. As implemented in the program POY (Wheeler et al. 1996–2003, 2006a; Varón 2007), direct optimization uses a specified gap/change cost to produce a tree with the smallest number of implied character-state transformations. An analogous procedure using maximum likelihood was proposed by Liu et al. (2009). Patterson (1988) presaged direct optimization with his interpretation of entire DNA sequences as hierarchical characters with their variations representing nested character states. While the direct optimization approach seems in principle to represent an important advance in the globally parsimonious treatment of DNA sequence data, it has the disadvantage of obscuring the character concept as outlined above, such that there no longer is an "alignment" or the familiar homologous nucleotide sites that are traditionally represented as columns in the data matrix. In fact, Wheeler et al. (2006a:341) argued that the concept of "primary homology"—the positional correspondence of parts that allows the recognition of characters—is "conceptually misguided and operationally unnecessary" as far as nucleotide sequences are concerned. That may be so, but it should be noted that even in POY the sequences that are analyzed are assumed to be "homologous," and the nucleotides are maintained in their "sequential" front-to-back, linear order with respect to one another. These constraints obviously represent some sort of assumption or hypothesis of primary homology (see also Special Considerations When Dealing with Molecular Data in Chapter 5,). As noted before, the degree to which these assumptions are relegated to "background knowledge" in direct optimization approaches, as well as automated assembly pipelines, is disconcerting.

Other Types of Character Information

There is an extensive literature on types of information potentially bearing upon phylogenetic relationships other than those discussed so far. Many of the techniques for collecting and dealing with such data were conceived at a time before cladistic theory became preeminent in systematics. Some approaches have lost their appeal because of the methodological problems they present during analysis or because the data types have been largely supplanted by nucleotide sequences. Three of the now less frequently encountered data types deserve brief mention.

Frequency Data

The most commonly encountered frequency data come from *allozymes*, differentially charged proteins that are separated across an electrical gradient by electrophoresis.

The technique was enormously popular in the 1970s and early 1980s, and allozyme data still are commonly used in the study of populations. They have also been used to some extent in a phylogenetic context: much of the impetus for *phylogeography* stems from early studies of allozyme frequencies (see Chapter 7).

Allozyme data have three attributes that impinge on the way they might be analyzed: (1) a given allozyme locus may be polymorphic across populations or even among taxa, resulting in the lack of a fixed character-state difference; (2) when there are fixed differences among taxa, they are often difficult to interpret as shared character states, with nearly all polymorphism unique (autapomorphic) to individual taxa (or populations); and (3) because the property measured by protein electrophoresis is the net charge of the entire protein (the sum of the charges of its constituent amino acids), as measured by relative distance it migrates in an electrical field, "identical" electromorphs frequently represent heterogeneous groups of alleles at the level of DNA sequences (Barbadilla et al. 1996).

Allozyme data generally have been treated through the use of distances (see next section) instead of as discrete characters. Mickevich and Mitter (1981) discussed three possible discrete-character approaches to analyzing such data: (1) the *independent allele model*, in which each allele becomes a separate character; (2) the *shared-allele model*, in which each locus becomes a single character whose states are combinations of alleles; and (3) a *"systematic" approach*, in which separate enzymes are assumed to be characters, the states of which are defined to be the allelic combinations. It is the last approach that would seem to be most relevant if such data are to be used in cladistic analyses. While allozymes are rarely used in the twenty-first century, the same coding considerations also apply to microsatellite loci, which remain prevalent in population genetic studies.

Distance Data

The intentional conversion of discrete character data, such as morphological structures or DNA sequences, into distances deliberately discards evidence pertaining to the nature of similarity, and consequently to kinship, and should therefore be avoided (Farris 1981:21). As discussed in Chapter 1, the distance methods employed by pheneticists, while an advance over the charismatic authoritarian approach used before the advent of objective methods, were philosophically flawed and rendered obsolete by the advent of quantitative cladistic analysis.

Some data types are intrinsically based on distances. Data gathered directly in a distance format include immunological data, DNA hybridization (annealing) data, and some others (see discussion of distances in Chapter 1). Of all data types gathered directly as distances, DNA–DNA hybridization data, which were touted (most flamboyantly by the laboratory of Charles Sibley) as providing a measure of overall divergence across an entire genome, historically attracted the most attention.

The allure of molecular-level distance data is rooted in two ideas. First, that data derived from the genome are thought by some to speak more directly to the issue of evolutionary propinquity and divergence than do data derived from study of the phenotype. Second, if the rate of divergence were constant and could be quantified, we would then have a measure of evolutionary rate, a so-called molecular clock (see Chapter 11).

On the less sanguine side, at least three arguments have been put forward against the use of data collected directly in the form of distances. First, distance data contain no information on homology, despite assertions to the contrary. Gould (1985), for example, claimed that DNA–DNA hybridization techniques as pioneered by Sibley had solved the problem of distinguishing homologies from analogies. What they do, in fact, is imply a single data point (a melting temperature of heteroduplexed DNA from a given pair of taxa compared), but they offer no direct indication of what that measurement might imply about degree of relatedness and certainly do not correspond to hypotheses of homology.

Second, DNA–DNA hybridization was technically difficult. The experiments were hard to control, and reciprocal measurements of the same pair of taxa often resulted in widely different estimates of degree of divergence between them (Sarich et al. 1989; see further discussion in Chapter 11).

Third, there are methodological issues that bring into question our ability to analyze most data collected directly as distances. As noted above, the greatest attraction of distance data is their presumed ability to depict degree of divergence in a clocklike manner. In order to analyze data under the rate-constancy assumption, the data and the method of analysis must be *metric*, that is, satisfy the triangle inequality (see Sidebar 11 in Chapter 5 for a general discussion of "metrics"). The distance data must also fit the more stringent requirement of being metric. Farris (1981) compiled considerable evidence contradicting the metricity of immunological data, showing that there was no basis for an appropriate model with which to analyze DNA hybridization data. He further noted that the measures often used in the analysis of such distance data (e.g., Nei's and Rogers' distances) were not metric in their properties, and therefore any results derived from their use were meaningless when the data were analyzed under the assumption that they were metric.

Measurement Data

Some authors, particularly those advocating phenetic and statistically based techniques, have argued for the use of measurement data of sizes and shapes of features, often under the heading of "morphometrics" (e.g., Thiele 1993; Zelditch et al. 1995; Klingenberg and Gidaszewski 2010.). The seeming merit of such an approach would be the "relatively objective" nature of the data themselves. Yet measurements are not necessarily so easily interpreted, and it would be naive to assume that they are not potentially susceptible to errors of observation and character conceptualization. Possibly the most important limitation of measurement data for phylogenetic analysis, however, is that measurements by themselves do not imply homology.

Normally, character states encoded in phylogenetic data matrices are discrete, which is to say they have specific values that differ qualitatively or discontinuously from one another (unless ordered in a step matrix). By contrast, measurements by their nature occur as continuous variables and, depending on the precision with which measurements are taken, have a potentially infinite number of states that are not easily broken into the discrete character states required by most phylogenetic software. Even if we concede the potential utility of continuous variables, there exists little agreement on how to code them in a discrete-character format. One approach to converting quantitative variation into discrete character states is called *gap coding* (Thorpe 1984; Archie 1985). In this method, character states are identified by natural "gaps" in measurement or *meristic* data. For example, 2–5 mm and 10–12 mm might seem like reasonable alternate states (that could also be characterized qualitatively as "small" and "large"). But what if the ranges were 2–7 mm and 8–12 mm? In the latter instance, the range of variation within each of the character states is larger than the distinction between them. Measurement of such characters in additional taxa often results in the amalgamation of seemingly "discrete" character states. Furthermore, even if our initial observations do not overlap, measurement of additional specimens or taxa might yield results that obscure that distinction. It is easy to see that gap coding can lead to arbitrary divisions of what may actually be a continuum. The program TNT allows coding of continuous variables "as such" for cladistic analysis, and Goloboff and coauthors (Goloboff et al. 2006; Goloboff and Catalano 2010) suggested that characters so coded can add support to topologies inferred from discrete characters.

Measurements in the form of ratios can be useful for describing the general attributes of shape or form (e.g., whether a given structure is longer than wide), or the relative dimensions of structures within or among taxa, and in eliminating correlations due to overall size from a series of measurements. Morphometric techniques are also potentially capable of offering descriptions of complex shapes. Nonetheless, in spite of the nearly total quantification of cladistics, the contribution of morphometrics in the analysis of phylogenetic relationships remains marginal.

Description of Characters for Publication

The advantages of careful character analysis and explicit coding of observations into a formalized data matrix discussed above are manifest, and it is thus quite surprising to find that numerous end products of systematic research do not contain any simple way to reconstruct the methods or examine the data that were employed to produce the tree. This is a particularly pernicious problem in journals devoted to "molecular phylogenetics" and high-profile, compressed-format publications that shunt critical detail to online supplementary material or abridge it entirely. For morphological features, as we have seen, a great deal of evaluative judgment goes into the identification of characters and states, and the results should be clearly represented in the publication, both by individual descriptions of each character and its states and by a data matrix that shows the state scored for each taxon.

For molecular data, while character coding is relatively unambiguous, the alignment is deterministic of the resultant topology. Therefore, description of how homology among nucleotide sites has been determined, as well as the alignment (or the raw data and the parameters used to directly optimize them), should be accessible to the scientific community. Even if the alignment is unambiguous, it is not sufficient to simply provide GenBank accession numbers, because GenBank sequences excise gaps included in alignments and because sequences of varying length in GenBank are frequently truncated to make a tidy block for analysis. Thus, the data matrix as analyzed in the publication may be impossible to retrieve. Large data matrices are more amenable to reanalysis if they are made available in an electronic format, either on the researcher's website or (preferably) on a public database such as www.treebase.org or hosted on the website of the journal where the work is published. The long-term preservation of such resources is far from certain, which poses a threat for the accessibility of the data in the future if they are not included in the body of the publication itself.

Needless to say, with the rise of phylogenomics, the problems of data transparency have been compounded by the sheer scale of the data sets being analyzed. It is not necessarily clear that the authors of phylogenomic studies themselves are able to examine critically the characters upon which their trees are based, let alone able to provide the data in an accessible format to peer reviewers, editors, or their readers. The potential impact of these problems on the fundamental scientific desiderata of repeatability and testability of such data is worrisome.

Selection of Taxa and Specimens

Up to this point in the chapter, characters have been the primary focus. Yet, the selection of taxa and specimens for systematic studies has implications of equal importance. Yeates (1995) recognized what he called "exemplar" and "ground-plan" (intuitive) methods for representing taxa in coded character data. The entities that are studied and represent the tips of the branches in the cladogram are called *terminal taxa*, and these always involve a degree of abstraction and synthesis. For instance, characters from different developmental stages, or male and female specimens, are usually viewed as attributes of the same composite taxon (these theoretical entities were referred to by Hennig as *semaphoronts* and the *holomorph*, respectively), even though they may not have been observed from, or are not actually present in, single individuals.

Under the *exemplar method*, individual taxa serve as the terminals representing a more inclusive taxon in the analysis, and the data are coded directly from the exemplar. If the exemplars represent species, then coding should be straightforward. If the exemplars represent higher taxa, it is important that they exhibit only a single state for each character being coded. If two or more states are present in a terminal, then two or more exemplars are necessary to represent the diversity of states accurately, as was discussed in the Polymorphisms section. The greater the variation in a terminal taxon, the greater the number of exemplars necessary to allow for correct optimization of the hypothetical ancestral states at the base of the clade (Yeates 1995; Prendini 2001).

If terminal taxa in a phylogenetic study are not species, then we are faced with the question of how to diagnose them—to identify the character states that set them apart from other groups. The *ground-plan* (intuitive) method assigns character states to a terminal taxon on the basis of examination of the literature or a range of specimens. The ground plan might be based on optimization of characters from a prior analysis. It must be "deduced" (by educated guess) if no such analysis exists. Character-state assignments for a deduced ground plan are unproblematic if there is no variation for the character among the terminals. If two or more states exist for a given character in a terminal taxon, then that character must be coded as polymorphic. Otherwise, "deduced ground-plan" assignments may introduce subjectivity and can easily produce errors of optimization—that is, assignment of character-state sets for hypothetical common ancestors—for the analysis as a whole (Nixon and Davis 1991).

The above considerations apply primarily to higher-level analyses in which the terminal taxa are not species or where species were chosen to represent supraspecific taxa. When conducting species-level studies, such as revisions, the most important consideration is to acquire the broadest possible sample of taxa and specimens representing them. This will allow for corroboration or refutation of past observations, as well as for the discovery of new information.

The search for specimens is most likely to be guided by using the types of sources described in Chapter 1. In addition to corroborating previously studied characters and discovering new ones, examination of the broadest possible sample of specimens, including type material, will also allow the discovery of previously unrecognized (and therefore undescribed) taxa, as well as the creation of new synonymies where necessary. The credibility of the results will in no small measure depend upon the thoroughness of the search for relevant collections. Of course, in many instances, going to the field and collecting fresh specimens may fill in gaps in existing collection holdings and also provide tissues amenable to DNA extraction.

A final practical point worthy of emphasis is the need for proper documentation and preservation of voucher material. For issues related to taxonomic nomenclature (Chapter 8), this is accomplished by designation and deposition of type specimens, but all systematic research should include the retention of voucher specimens (or parts thereof) so that the exemplars under study can be compared with the taxa they are purported to represent, should the need arise later on. For example, specimens considered to belong to a single species may later be hypothesized to belong to a complex of multiple species. If the specimens examined are not retained, then the data from them is relegated to eternal ambiguity.

Vouchers usually take the form of museum specimens but may also be photographs or other documentary evidence if, for example, the specimen has been destructively sampled for DNA work. Publications should include documentation of voucher data, such as their unique identifying codes, collection localities, and where they physically reside as preserved specimens. The failure to preserve or adequately document vouchers, like the failure to include explicit descriptions of characters and states, reduces the objective quality of empirical systematic research.

SUGGESTED READINGS

Brower, A.V. Z., and V. Schawaroch. 1996. Three steps of homology assessment. Cladistics 12:265–272. [A discussion of characters and character states in morphological and DNA sequence data]

Deans, A.R., M.J. Yoder, and J.P. Balhoff. 2012. Time to change how we describe biodiversity. Trends in Ecology and Evolution 27:78–84. [An introduction to phenotypic ontologies]

Farris, J.S. 1985. Distance data revisited. Cladistics 1:67–85. [An extended discussion of the properties of distance data and the difficulties involved in analyzing them]

Humphries, C.J. 2002. Homology, characters, and continuous variables. In: Morphology, shape and phylogeny, ed. N. Macleod and P.L. Forey, 8–26. London: Taylor and Francis.

Lipscomb, D.L. 1992. Parsimony, homology and the analysis of multistate characters. Cladistics 8:45–65. [A concise explanation of morphological character analysis]

Mickevich, M.F., and D.L. Lipscomb. 1991. Parsimony and the choice between different transformations of the same character set. Cladistics 7:111–139. [A useful discussion of character analysis and the relationship between characters and cladograms]

Morrison, D.A., M.J. Morgan, and S.A. Kelchner. 2015. Molecular homology and multiple sequence alignment: an analysis of concepts and practice. Australian Systematic Botany 28:46–62. [A critical review of sequence alignment challenges]

Rieppel, O., and M. Kearney. 2002. Similarity. Biological Journal of the Linnean Society 75:59–82. [A thought-provoking discussion on character construction, vehemently opposed by Kluge, A. G. 2003. The repugnant and the mature in phylogenetic inference: atemporal similarity and historical identity. Cladistics 19:356–368]

Sereno, P.C. 2007. Logical basis for morphological characters in phylogenetics. Cladistics 23:565–587. [An important conceptual review of character coding]

CHARACTER POLARITY AND INFERRING HOMOLOGY

At the beginning of Chapter 3, we argued that despite the traditional notion that cladistics groups taxa "only on the basis of shared derived characters," the discovery and delimitation of characters and character states should take place without imposition of a priori hypotheses concerning which states are ancestral and which are derived. This is for both philosophical and practical reasons. Philosophically, systematists do not wish to bias their hypotheses with intuitive a priori decisions about character polarity that may turn out to be incorrect. From a practical perspective, as we shall see in Chapter 5, polarization of characters is usually an automatic outcome of the rooting of a cladogram, so designation of polarity beforehand is not necessary. However, this was not always the case. Prior to the advent of computer-based analyses, Hennig and other cladistic pioneers routinely used prepolarized characters to construct their phylogenetic hypotheses and developed a specialized terminology to describe their practices. We have introduced some of these Hennigian terms in our historical synopsis of the development of systematic theory in Chapter 1, and because of their ongoing use in systematic training and research, we will explain them here in greater detail.

Terminology

Consider the rooted cladogram in Figure 4.1. We begin with a tree we assume to be "correct"; that is, one we postulate to be so well supported by numerous characters that we believe it provides a robust and stable empirical hypothesis of

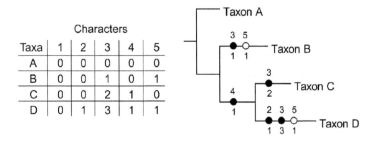

FIGURE 4.1. Four taxa and five characters analyzed as unordered with characters plotted under delayed transformation (DELTRAN) optimization (see text). The alternative, equally parsimonious accelerated transformation (ACCTRAN) optimization for character 5 is a synapomorphy at (BCD) and a reversal in C.

the relationships among the taxa represented. The base of the cladogram (to the left) is the *root*, or the position of the hypothetical ancestor. All character-state transformations on the tree occur on the branches leading from the hypothetical ancestor to the terminal taxa at the tips of the branches on the right (recall from Chapter 1 that whatever empirical knowledge regarding ancestors and their features we may possess is a result of phylogenetic inference).

Observation of similarity of a structure across a set of taxa—a hypothesis of homology—leads to five possible outcomes, which are illustrated by characters 1–5 on the tree (recall that the structure of this tree is assumed, not determined, by these exemplar characters). The implied character-state transformations are indicated by the circles on the tree (character number above the branch, state number below the branch).

Character 1 shows a feature thought to be homologous that is invariant among the taxa being examined, and neither supports nor contradicts any alternative hypotheses of grouping among them. (Because there are no implied character-state transformations, the character is not indicated on the tree).

Character 2 shows a feature thought to be homologous that has two states, but all the taxa but one exhibit the hypothetically ancestral state. The derived state present in a single terminal taxon (D) is termed an *autapomorphy* and provides no evidence of grouping.

Character 3 shows a feature thought to be homologous that exhibits a different state in each taxon. Without additional hypotheses of transformation among these states (such as an ordered transformation series, see Chapter 3), these states are all parsimoniously viewed to be autapomorphies, and this character also provides no evidence of grouping among the taxa. (Note that if the character were viewed as an ordered transformation series, $0 \to 1 \to 2 \to 3$, then state 1 would be considered a synapomorphy of taxa B, C, and D, and state 2 would be a synapomorphy of taxa C and D—because in that coding scheme state 2 is considered a derived condition of state 1, and state 3 is considered a derived condition of state 2).

Character 4 shows a feature thought to be homologous that varies in such a way as to provide evidence supporting the nested grouping of the taxa. In this case we see that state 1 is present in a subset of the taxa (C and D). State 1 is a *synapomorphy* for taxa C and D. Its complement, state 0, is a *symplesiomorphy*, a state shared by taxa A, B, and the hypothetical ancestor at the root of the clado-gram. It is important to recognize that synapomorphy and symplesiomorphy are relative concepts. A feature may be a synapomorphy for a clade with respect to other taxa that lack that feature but a symplesiomorphy at a less inclusive level. Hennig (1966) referred to this property as the *heterobathmy* of synapomor-phy. For example, since all mammals possess mammary glands, the presence of that feature does not provide any further evidence of grouping for less inclusive mammalian taxa (Brower and de Pinna 2012). It is a synapomorphy for mam-mals with respect to other taxa, but a symplesiomorphy within Mammalia.

Up to this point, the distribution of states for all four characters can be explained parsimoniously on the topology—none of them requires any more implied character-state transformations than the minimum required by the number of states (there must be at least $n - 1$ character-state transformations for n observed character states). Character 5 shows a feature thought to be homolo-gous, but that cannot be viewed as a synapomorphy on this cladogram unless the character state has undergone a $1 \rightarrow 0$ reversal in taxon C. The distribution of states of this character is referred to as an instance of *homoplasy*. Note that the alternatives of two independent gains in B and D or a gain in the common ancestor of B + C + D and a loss in C are equally parsimonious scenarios for the transformations of this character but imply different evolutionary events. The former scenario is referred to as delayed transformation (*DELTRAN*) because the character-state changes take place late, as far toward the tips of the tree as possible, while the latter scenario is referred to as accelerated transformation (*ACCTRAN*) because the changes occur early, as close to the base of the tree as possible. These alternatives do not affect the tree length or topology but can have interesting implications for the inferred evolutionary history of that trait, as well as for mapping of biogeographical and ecological data (Agnarsson and Miller 2008; see further discussion in Chapter 10).

In Hennig's view (1966:120), all the states of a character—be they plesiomorphic, apomorphic, or homoplastic—are homologous. He argued that synapomorphies—shared derived characters—provide the only evidence for the existence of natural groups. This is the fundamental aspect of his arguments for the phylogenetic sys-tem; all of Hennig's other principles are subsidiary to it. Thus, in the Hennigian view, synapomorphy is the only "kind" of homology that bears upon patterns of relationship, a distinction that has led many cladists to equate the two terms (reviewed in Brower and de Pinna 2012).

Nearly all systematic methodologies, even prior to Owen's (1843) enunciation of the concept of homology, have treated the recognition of homologues as central to the success of the method. Many of these approaches have failed, however, because they did not recognize that even though all members of a group may share some homologous feature in common, if the feature occurs among organisms at a more inclusive level of generality it is not sufficient for group definition. Some familiar examples of characters that also exist outside the groups they have been used to define are poikilothermy (cold-bloodedness) and scales in the Reptilia—because these also occur in most bony fishes—and the presence of two cotyledons in the seeds of dicotyledonous angiosperms—which also occurs in the Cycadales, Ginkgoales, and Gnetales.

Another instructive example we have mentioned several times before is tetrapod forelimbs (Figure 4.2). The limbs of terrestrial vertebrates represent novel attributes—synapomorphies—relative to the pelvic and pectoral fins of fishes. Among tetrapods, these limbs *as limbs* do not represent distinctive attributes for bats, birds, and pterosaurs. Yet, wings in those groups are distinctive—apomorphic—for each of them. The concept of transformation (see Sidebar 6, Chapter 3) is exemplified by the modification of pectoral fins into forelimbs and forelimbs into wings, the latter on three separate occasions, for birds, bats, and pterosaurs independently. Whereas the vertebrate pectoral fin–forelimb in all of its forms can be referred to as homologous, it is the nesting of the structural modifications into which this organ has been transformed that allows us to infer a hierarchy of relationships. The specialized morphological features represent synapomorphies for the groups that possess them. At the most general level within vertebrates, fins are synapomorphies. At a more restricted level, limbs are synapomorphic. And at the most restricted levels, wings are synapomorphic for each of the three separate groups. The postulated transformation series is graphically depicted in Figure 4.2. Additional examples are shown in Figure 4.3. What these examples show is that homologous features—as synapomorphies—imply a nested set of relationships among the taxa that bear them, and it is only when the hierarchic relationships of groups are supported by the nesting of synapomorphies that a "natural" classification results.

Tests of Synapomorphy

In Chapter 3, we discussed similarity and tests of similarity. But similarity by itself does not pick out nested sets of taxa. Indeed, in the counterintuitive calculus of

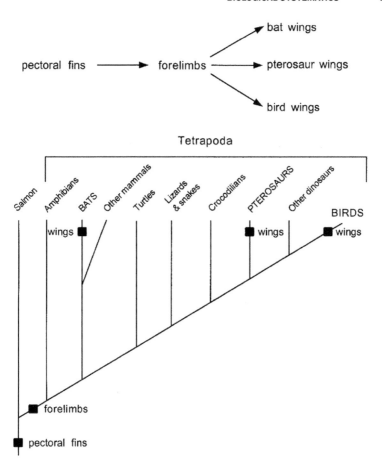

FIGURE 4.2. Nesting of structural transformation of the vertebrate forelimb. Note that the modification of fins into limbs represents a synapomorphic character state for the tetrapod vertebrates. The three distinctive modifications of forelimbs into wings represent independent apomorphic character states for restricted groups within the Tetrapoda.

data matrix construction, binary opposites (such as presence and absence) are viewed as being complementary instances of "the same" character. While there is no method to test the hypothesis that different states have been correctly grouped together as alternative instantiations of a character, it is possible to test whether two taxa share the same state due to symplesiomorphy, synapomorphy, convergence, or character-state reversal.

As mentioned previously, Hennig referred to the observation that characters can support groups at different levels as the "heterobathmy of synapomorphy." The existence of homologous features, and the systematic hierarchy they supported, had long been recognized. Nonetheless, the failure to consistently grasp

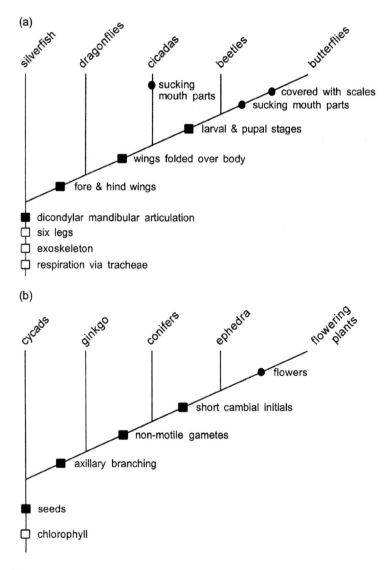

FIGURE 4.3. The distribution of plesiomorphic and apomorphic character states in (a) insects and (b) seed plants. Open boxes = plesiomorphic states; closed boxes = synapomorphic states; and closed circles = autapomorphic states (given that "cicadas," "butterflies," and "flowering plants" represent single terminals in this scheme).

the relationship between transformation and hierarchy—that only synapomorphies support monophyletic groups—had caused many groups to be formed on the basis of features that were widespread and uninformative (synapomorphies that support a more inclusive group) or on the complementary absence of characters. The resultant classifications contained unnatural groups united by combinations of synapomorphies and symplesiomorphies (e.g., "reptiles," "protists"). The observation that characters were distributed in an unorderly way in these groups (and therefore presumed to evolve at different rates) was sometimes called "mosaic evolution" by Mayr and others and cited (incorrectly) as a confounding limitation of cladistic methods (see also Farris 1971). In fact, "mosaic evolution" is only confounding to phenetic methods that assume equal rates of evolutionary change among lineages.

Congruence represents the test for theories of homology as synapomorphy. Congruence exists when two or more homologous features are observed to support the same group of organisms. Character congruence as a powerful tool for judging the weight of evidence was already well understood by Charles Darwin, who said: "We may err . . . in regard to single points of structure, but when several characters, let them be ever so trifling, occur together throughout a large group of beings having different habits, we may feel almost sure, on the theory of descent, that these characters have been inherited from a common ancestor. And we know that such correlated or aggregated characters have an especial value in classification" (Darwin, 1859:426).

Note that by "correlated," Darwin did not mean characters covarying because of a common intrinsic mechanism such as development (e.g., the length of the femur and the length of the tibia are strongly correlated with one another owing to genetic and environmental influences on growth). Rather, he was referring to characters with no evident functional relation to one another but supporting the existence of the same taxon, such as the presence of hair and mammary glands in Mammalia. Note also that Darwin was invoking shared character distributions as evidence to support his theory, not as syllogistic deductions therefrom. We may view Darwin's quote as a proto-Popperian description of a test of his theory: if the theory is correct, then we would expect to observe character congruence. As noted in Chapter 2, to provide a legitimate test in a hypothetico-deductive framework, the evidence must be independent of, and not entailed by, the theory (Brady 1985).

Congruence is also the tool that helps to resolve the question of primary absence versus secondary loss of features that are observed to be not present in some taxa. An excellent example of this occurs in snakes (and some other terrestrial vertebrates), in which we may find little or no evidence of limbs, fore or hind. In such cases we have two choices: to hypothesize that the organisms never had limbs or to hypothesize that the transformation has been from an ancestor with limbs toward complete secondary loss of all structures pertaining to the limbs. Obviously, the observed absence of limbs offers no evidence on the nature of limbs themselves—except that they are absent. Corroboration of the primitive absence theory or the secondary loss theory therefore cannot come directly from the structures themselves. Snakes, on the basis of attributes other than limb structure, such as morphology of jaws and reproductive structures, show relationships with lizards. Thus, hypothesizing secondary loss of limbs in snakes is more parsimoniously interpreted as congruent with the distribution of those other attributes. Asserting that snakes have never had legs or arose from legless ancestors would place them outside the Tetrapoda and require that most of their osteology and anatomy be independently derived, as for example from members of the Agnatha (lampreys and hagfish). Thus, in this case, congruence as understood through the application of the parsimony criterion offers the strongest evidence that the ancestor of snakes possessed limbs, even though the evidence might be construed as indirect. Further evidence comes from fossil snakes with legs (e.g., Tchernov et al. 2000; Martill et al. 2015) (if they are indeed snakes—the interpretation of those fossils as snakes is itself made possible by the congruence of other osteological characters and is the subject of some controversy). Similar arguments could be made for the absence of chlorophyll in mycoheterotrophic angiosperms such as Indian pipe (*Monotropa*) and the absence of wings in lice and fleas.

Interpreting the Literature on Homology and Synapomorphy

Most authors agree that the explanatory power of phylogenetic analysis rests on theories of homology. However, comparing even the most thorough and cogent available discussions of homology and its bearing on phylogenetic inference (e.g., Patterson 1982; Rieppel 1988; de Pinna 1991; Nixon and Carpenter 2011; Brower and de Pinna 2012) may leave the reader confused. Terminology is one source of confusion. Different authors may use the same terms to describe related but distinct concepts; for example, there are multiple definitions for the words homology, synapomorphy, and homoplasy. In a similar fashion, the nature of tests for "lower-level" (similarity) theories and "higher-level" (synapomorphy) theories are also in dispute. We shall discuss these two aspects—terminology and tests—in turn.

First, we have provided a semantic correspondence of terminologies among different authors in Table 4.1. We can see that some use the term *homology* to refer solely to "topographical and structural relations of similarity" in the original sense of Owen, in which no character polarity or directionality of evolution is implied. This is also sometimes referred to as transformational homology (cf. Sidebar 6, Chapter 3). Others, however, use the term as a synonym of synapomorphy, leaving "homology as observation" without an applicable term. The term "primary homology," coined by de Pinna (1991), has been used widely

TABLE 4.1 Homology: differing terminology and concepts

AUTHOR	TOPOGRAPHICAL AND STRUCTURAL SIMILARITY	CHARACTER CONGRUENCE	CHARACTER INCONGRUENCE
Owen (1843)	Homology	—	Analogy
Lankester (1870)	Homology	Homogeny	Homoplasy
Hennig (1966)	Homology	Synapomorphy	Convergence
Mayr (1982)	Homology	Homology	Convergence/ parallelism
Patterson (1982)	Homology– synapomorphy	Synapomorphy	Homoplasy
de Pinna (1991)	Primary homology	Secondary homology	Homoplasy
Lipscomb (1992)	Homology	Synapomorphy	Homoplasy
Brower and Schwaroch (1996)	Observation	Homology–synapomorphy	Homoplasy
Ball (1975)	Homology	Synapotypy	Convergence/ parallelism

Source: Adapted and expanded from Rieppel (1988).

as a substitute in such cases. *Synapomorphy,* proposed by Hennig, would seem to be synonymous with "homogeny" of Lankester (1870). Whereas the former term has now achieved widespread usage, the significance of the latter was apparently not appreciated at the time of its proposal (despite approbation by Darwin 1872) and consequently was never widely adopted.

Second, with regard to tests, on the one hand we might consider the viewpoint of Lipscomb (1992), who considered similarity a valid test of homology, such that among all possible transformation series for the set of possible conditions of a homologous feature, those that entail transformations between states that are least similar will be rejected. Lipscomb (1992:52), unlike de Pinna (1991), treated theories of homology as based on "the meticulous examination of all details" and concluded unequivocally that theories of homology are available for test at the level of observation and simple comparison across taxa. Concurring with de Pinna's view of primary homology as conjecture and homology as synapomorphy, Brower and Schawaroch (1996:268) argued that "homology cannot be determined prior to cladistic analysis." From this perspective, similarity of structure across a group of taxa is not viewed as a test of homology but rather as a factor that compels us to hypothesize homology. These disagreements exemplify the semantic overdetermination of the terminology: it is likely that if the word "homology" were replaced by other terms, cladists, at least, would all agree on principles and procedures of character recognition, delimitation, and polarization (cf. Wiley 1975).

In agreement with the point originally emphasized by Patterson (1982), nearly all authors—including Mario de Pinna, Diana Lipscomb, and Olivier

Rieppel—agree that hypotheses of synapomorphy are tested by character congruence. The differences of opinion among these authors rest on whether homology and synapomorphy are treated as identical concepts. Note again, however, that none of these concepts depend on metaphysical claims about common ancestry—in contrast to the "standard" evolutionary definition of Haas and Simpson (1946; see also Nixon and Carpenter 2011)

Homoplasy (Convergence and Parallelism)

Hennig's "auxiliary principle" clearly articulates the epistemological necessity that character-state identity should be accepted as homology, rather than convergence, in the absence of evidence to the contrary. This approach was introduced in Chapter 2 (see Parsimony and Ad Hoc Hypotheses). As explained by de Pinna (1991:371), "All similarities are deemed homologous initially, and non-homology is disclosed by a pattern-detecting procedure [parsimony] . . . If the analysis supports a single position for a putative synapomorphy, then the condition shared by the various taxa with that derived condition are [sic] corroborated as homologous. If a shared derived condition turns out to require independent origins in the overall scheme of relationships, then an event of non-homology [homoplasy] has been discovered."

Thus, as shown in Figure 4.1, independent hypotheses of homology may imply conflicting patterns when testing theories of group relationships. The observer is then forced to decide if similar structures arose more than once through parallel evolution or convergent evolution as analogs, or if the feature has been lost in some members of the group that share it (such as snakes and legs), or if using existing approaches we are unable to distinguish between things that are actually different. In the last case we are left with the conclusion that some of our observations are most likely mistaken. The apparent multiple evolution, reduction, or re-evolution of structures was first referred to as *homoplasy* by Lankester (1870) (see table 4.1). The resolution of such conflicts is adjudicated via application of the parsimony criterion, as discussed in Chapter 2. We will see in Chapters 5 and 6, under the discussion of character weighting, that additional logically consistent criteria have been advanced to refine further our attempts to unravel apparently conflicting hypotheses of homology. Note that although parallelism and convergence might be explained as resulting from different evolutionary or developmental processes by students of adaptation, from the perspective of systematics, they both manifest themselves simply as extra, redundant steps in the implied transformation series for a given character on the most parsimonious cladogram—as homoplasy.

There are many instances in the precladistic literature of the a priori invocation of parallelism or convergence—that similar-appearing structures have developed from different precursor types. The arguments vary from case to case, but invariably they are not based on a rigorous evaluation of the character data.

Consider the following example. Within the insect order Hemiptera, members of the family Miridae were grouped into subfamilies by O. M. Reuter (1910) primarily on the basis of morphological variation in the pretarsus—the claws and associated structures. In Reuter's system, and subsequent works of Harry H. Knight, the pretarsus was the single most important structural feature for subfamily recognition. The works of Reuter and Knight gained wide influence and were generally considered valid until the 1930s. At that time the respected British worker W. E. China (1933) observed that the structural importance of the pretarsus had been exaggerated and that it was "far too plastic to serve as a fundamental group character." The argument, as couched by subsequent workers, was that similarity of structure was the result of adaptation to similar life habits—*convergence*. The source of structural variation in the pretarsus of the Miridae is at best poorly understood. But, in the view of China, similarity of form did not necessarily imply common origin. To this day, however, pretarsal structures are used to recognize systematic groupings within the more than 11,000 described species of Miridae, not only because of the discrete variation they manifest but also because that variation is congruent with the distribution of other features in the group.

The rise of cladistics and the interpretation of putative adaptations in a phylogenetic context (see Chapter 11) have diminished the ad hoc discussion of parallelism and convergence. While in the past the existence of these phenomena was often hypothesized a priori for particular features in isolation, the interpretation of homoplastic features is now conducted through the results of cladistic analyses of matrices composed of numerous characters.

Homology Concepts as Applied to Nucleotide Sequence Data

Before we conclude our discussion of homology, the bearing of the concept on DNA and RNA sequence data should be addressed. Whereas homology statements have traditionally involved morphological structures of at least moderate complexity on a varied morphological landscape, nucleotide sequence data do not readily satisfy the traditional requirements of structural or associational complexity (cf. Remane 1952). Nonetheless, analysis of sequence data depends on theories of site homology and nucleotide transformation.

DNA and RNA are composed of only four nucleotides: the purines adenine and guanine and the pyrimidines cytosine and thymine (replaced by uracil in RNA). Therefore, the question of whether two or more sites that occur in apparently similar topographical positions are actually homologous—simply because they are occupied by the same nucleotide—may be less confidently judged than whether the wings of a bird are homologous with the forelimbs of a nonvolant tetrapod. Indeed, the equivalence of nucleotide (sequence) position in the genomes of different organisms is at times far from self-evident. Some parts of the genome, such as protein encoding regions, contain similar numbers of nucleotides across a broad range of taxa. Other parts, such as ribosomal RNA genes, may have variable numbers of nucleotides across a range of taxa, with some areas deleted, some areas duplicated, or anomalous segments inserted. These types of variation complicate the task of comparing the nucleotide sequences among taxa. Possible approaches to dealing with this issue were discussed in Chapter 3, Alignment of DNA Sequence Data.

The problem of determining homology in DNA and amino-acid sequence data was erroneously described as a statistical problem by Patterson (1987) and others. Under this characterization, if "identical" portions of the genome from two different organisms are written out side by side, about 25 percent of the nucleotide positions will match by chance alone, if it is assumed that each of the four nucleotides occurs with equal frequency. Matching frequencies of greater than 25 percent could, then, be explained as either the result of homology or convergence.

The "statistical" viewpoint has been explicitly rejected by Mindell (1991), Brower and Schawaroch (1996), and others because it treats determination of sequence homology as one of comparing the numbers of sites in common, an approach that is strictly phenetic (of course, many are unfazed by phenetic similarity: BLAST searches use exactly this criterion). If the method of homology determination in DNA sequences is to correspond to that used for morphology, then it is not the number of sites in common that is of interest but to what degree we can find sites (or groups of them) that are unique across groupings of taxa. Because there are only four possibilities for change at any nucleotide site, the problem of homology determination may be more difficult than in the case of gross morphology. Nevertheless, we still observe inferred changes at corresponding sites as they occur among taxa, not the sum of changes (differences) for a nucleotide string across taxa. In the end, correct matching (or direct optimization, sensu Wheeler et al. 2006a) should minimize the number of inferred nucleotide transformations across the taxa being compared.

At the level of the locus, as opposed to individual nucleotides, Fitch (1970) proposed the now widely adopted term *paralogy* for use in reference to duplicated gene regions, those that exist as multiple similar, if not identical, copies within the genome of a single organism. De Pinna (1991) noted that this usage corresponds to the terms *serial homology*, *mass homology*, and *homonymy* in morphology, that is, for structures that occur in multiple copies in the same organisms. Paralogous genes, according to Fitch, stand in contrast to *orthologous* genes in which "the history of the gene reflects the history of the species." De Pinna concluded that there appears to be nothing special about homology in RNA or DNA sequence data in comparison to morphology, only that new terms have been created to describe existing concepts. It is, however, increasingly evident that many genes have gained their current forms and functions by duplication and rearrangement of pieces of other genes (Lynch and Conery 2003), and even duplication of entire genomes plays a major evolutionary role in some lineages (particularly in plants: Soltis et al. 2016), rendering the genome a patchwork of DNA sequences with multiple hierarchical levels of homology both across taxa and within individual organisms.

Finally, with the growth of multigene molecular data sets, there has been rising concern that different gene regions may imply different patterns of relationship among the taxa sampled, and the argument has been advanced that parsimonious interpretation of phylogenetic evidence may obscure these separate patterns. The issues of "gene trees" versus "species trees" will be addressed in detail in Chapter 7.

Natural Groups

For centuries, natural philosophers have sought to discover and explain the pattern of relationships among living things. Today, most systematists recognize the pattern to be "natural" if groups are monophyletic, as recognized on the basis of synapomorphy. Hennig grasped the unequivocal importance of monophyly, and his advocacy of the concept is perhaps his most influential contribution to systematics. Under his conception (see also Nelson 1971; Farris 1974), such a group is characterized as follows:

> A *monophyletic* group contains the common ancestor and all of its descendants. Note that this is a metaphysical, evolutionary definition. Empirically, such groups are characterized by synapomorphies. It is now widely held that one of the central aims of systematics is to develop a natural classification in which only monophyletic groups are named as "groups." One might conclude that if all groups recognized in a classification must be monophyletic, then there should be no need to distinguish among nonmonophyletic assemblages of taxa. Hennig, nonetheless, recognized that traditional classifications often contained two other types of "groups."
>
> A *paraphyletic* group contains the common ancestor and some—but not all—of its descendants. Hennig defined this type of assemblage as

being characterized by a mixture of synapomorphies and at least some plesiomorphies.

A *polyphyletic* group contains some of the descendants of a common ancestor but not the common ancestor itself. Hennig defined this category as being characterized by the possession of convergent character states.

These definitions can be visualized in Figure 4.4.

As noted, Hennig's definitions, although generally workable, make reference to ancestors, metaphysical concepts that are neither evident from inspection of a cladogram nor testable because, as was discussed in Chapter 2, ancestors are not empirically discoverable. The internal nodes of cladograms do not correspond to any actual observable entities, which in empirical systematics reside at the terminals of the tree. Thus, we suggest that a useful heuristic means to distinguish between monophyletic, paraphyletic, and polyphyletic taxa is as follows: Imagine the Tree of Life as an irregularly bifurcating hierarchy that is actually made of wood. Any single branch or twig that can be broken off the tree at single point constitutes a monophyletic group. A paraphyletic group is such a branch with at least one subordinate branch or twig broken off of it. A polyphyletic group is one composed of two or more branches broken off the tree. Alternative character-based definitions for the groups recognized by Hennig are portrayed in Figure 4.5. Note that it is not the cladogram itself that has the property of monophyly, paraphyly, or polyphyly, but the groups we choose to recognize given the hypothesis of relationships implied by that cladogram.

Hennig's monophyly concept thus meshes well with the theory of descent with modification, grouping together all members belonging to an evolving lineage and excluding none of them (Sidebar 8). Only synapomorphies can be explained parsimoniously as unique evolutionary events that convey information about patterns of relationship. As discussed elsewhere in this book, if classifications

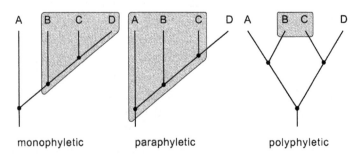

monophyletic   paraphyletic   polyphyletic

FIGURE 4.4. Monophyletic, paraphyletic, and polyphyletic groups. In the sense of Hennig's definitions given in the text, hypothetical ancestors are represented by the interior nodes of the cladogram.

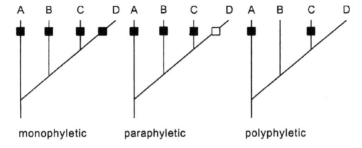

monophyletic   paraphyletic   polyphyletic

FIGURE 4.5. Character distributions for groups under the definitions of Farris (1974). Closed boxes indicate the distribution of the uniquely derived states for a character, open boxes indicate reversals. In the cladogram on the left, all taxa possess the group-defining feature in its uniquely derived and unreversed condition, thus composing a monophyletic group. That same group becomes paraphyletic in the middle cladogram, when the condition of the feature is reversed in taxon D, and D is therefore not included as a member. In the cladogram on the right, the feature uniting taxa A + C is not uniquely derived, and therefore the taxon defined by that character is polyphyletic.

contain paraphyletic and polyphyletic groups, the descriptive capacity of the characters used to indicate groups is compromised. Thus, the recognition of monophyletic groups not only brings classifications into agreement with a genealogical explanatory narrative but also serves the broader scientific purpose of recognizing groups for which there is maximal evidentiary support.

Sidebar 8
The Varied Conceptions of Monophyly

Monophyly has been defined by some authors as "a group, all of whose members are descended from a single ancestor." So phrased, the definition allows for virtually any grouping to be monophyletic. For example, a whale, an amoeba, and an oak tree are a group with a presumed eukaryotic common ancestor. Willi Hennig and others used monophyly in a more restricted sense to mean "*all* of the descendants of a single (common) ancestor." The merits of both definitions have been argued on the basis of historical precedence. Monophyly in the sense used by Hennig is thought by most authors to be the same as that of Haeckel (1866), and because Haeckel first used the term, it is this usage that should have precedence. Note that the concept applies to groups of taxa, not to groups of individual organisms—a distinction that has resulted in confusion in certain quarters (see Schmidt-Lebuhn 2012).

North American authors including Simpson, Mayr, and Ashlock (1971) each argued that the term monophyly has long been applied to groups that in Hennig's sense are paraphyletic but that those groups nonetheless merit formal recognition. These authors invoked precedence on the basis of long-standing usage but less for reasons of methodological consistency than for retaining cherished groups, such as Reptilia, in existing classifications.

Paraphyletic groups have cultural inertia in systematics, manifested in the names of divisions of natural history museums (such as "Invertebrate Zoology") and professional societies ("Ichthyology and Herpetology"), but very few active biologists would advocate the formal recognition of taxa that are manifestly not monophyletic. If any paraphyletic groups persist in classifications, it is likely that they do so only because they have not been the subject of a cladistic revision.

Finally, we note that any definition dependent upon the existence of an ancestor is a metaphysical assertion. As Hennig (1965:114) observed: "The supposition that two or more species are more closely related to one another than to any other species and that, together they form a monophyletic group, can only be confirmed by demonstrating their common possession of derivative characters ('synapomorphy')." Reemphasizing Hennig's position, Gareth Nelson (1970:378) pointed out, "Homology has often been defined in an 'evolutionary' sense, and 'homologous' features have been said to be 'traceable' back to the 'same' (i.e., 'homologous') feature of some common ancestor. Operationally, common ancestors are at best only hypothetical constructs. Thus, 'tracing homologous features back to some common ancestor' amounts only to erecting an hypothesis of ancestral conditions."

Determining Character Polarity

In Chapter 3, we discussed characters and their states and how states may be ordered into transformation series. However, the quality that makes a synapomorphy a "derived" feature is the directionality of character-state transformation from the more to the less general condition. The determination of character polarity has traditionally been viewed as the sine qua non of the Hennigian method, whereby taxa are grouped by special similarity as opposed to overall similarity. Based on the traditional Hennigian approach, polarity was usually determined on a character-by-character basis during the data-gathering phase, and controversy over interpretation of individual character polarities led to extended discussion in papers from the 1970s and early 1980s. Computer algorithms, such as the Wagner algorithm originally described by Farris (1970), minimize the number of character-state changes among taxa without regard to the polarity of the characters (as long as the characters have symmetrical transformation costs). This means that the tree is the same length regardless of where the root is placed. The orientation of the network created by the algorithm is then determined by specifying one taxon to identify the root (Figure 4.6; see further discussion in Chapter 5). This automatically establishes the directionality of character-state transformations implied by a given topology and identifies which states are apomorphic. Under this formalization of cladistics, all the older literature describing methods to determine individual character polarity really addresses a nonproblem. It is only the choice of the outgroup and the position of the root that matter. Nevertheless, a clear understanding of these issues remains heuristically valuable, and we conclude our discussion of character analysis with a consideration of this historically and heuristically important concept.

Rooting by Outgroup Comparison

An *outgroup* is one or more taxa assumed to be related to but not part of the *ingroup*—the taxa among which we wish to infer a hypothesis of relationships. Character-state polarity is determined by the choice of an outgroup, which in turn provides a *root* for the members of the ingroup. This approach was implied in the works of Hennig, although not using this terminology. The outgroup method has been discussed by a number of authors (e.g., Farris 1972; Watrous and Wheeler 1981; Maddison et al. 1984; Nelson 1985) and critically reviewed by Nixon and Carpenter (1993). Outgroup comparison, as modified from the description of Nixon and Carpenter (1993), proceeds as follows:

1. Define or circumscribe ingroup taxa on the basis of conjectured synapomorphies. *Neomorphic* (uniquely derived) character states (Sereno 2007) may offer compelling initial hypotheses of monophyly.
2. Select outgroup(s) that appear to be related to, but not part of, the ingroup on the basis of synapomorphies at a more inclusive level (higher level of generality), usually as implied by a higher-level cladistic analysis.
3. Perform unrooted parsimony analysis.
4. Root cladogram using one of the outgroups.
5. Read character polarities from cladogram.

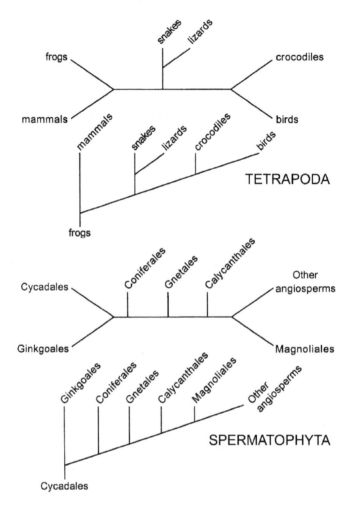

FIGURE 4.6. Tetrapoda and Spermatophyta, (above) unrooted network and (below) a rooted cladogram. Polarities of characters used to form the networks are determined by choice of an outgroup. Frogs were chosen to root the tetrapod cladogram because, among other attributes, they lack the distinctive embryonic membranes found in all other members of the group. Cycads were chosen as the basal seed plants because, among other attributes, they lack the axillary branching found in all other seed plants.

The efficacy of outgroup comparison is related to the attributes of the taxa chosen to serve as outgroups. In practice, prior classifications often serve as a null hypothesis for ingroup membership and outgroup selection. For example, if we are interested in relationships among members of a genus, we might select a member of a different, presumably closely related, genus as the outgroup. It is not necessary to select a member of the sister taxon, or even to know precisely what the sister taxon is, as long as the outgroup taxon shares enough characters in common with the ingroup to allow meaningful comparison of homologous features. A distant or otherwise poorly chosen outgroup for rooting analyses of DNA sequence data will possess few character states in common with the ingroup taxa and therefore serve its intended function poorly (an oak tree would not be a good outgroup for a genus of lizards). Ward C. Wheeler (1990) showed that an essentially random sequence of nucleotides will usually root a tree on the longest branch or internode; he concluded that such an outcome is unreliable or even meaningless. A rule of thumb might be that if an outgroup could not root a tree effectively on the basis of morphology, there should be no reason to believe that the same taxon would be any more effective in providing a root for a tree based on DNA sequence data.

Some analyses have used a single exemplar species as the outgroup; others have used a composite taxon that combines characteristics found across a range of species. The use of actual species makes character coding more straightforward than is often the case with a composite hypothetical outgroup. In the latter case, coding of character states—of necessity—will be "deduced" in the outgroup rather than observed, running the risk of unwitting subjective polarity interpretations that may bias results.

The outgroup need not be limited to a single taxon, and, indeed, analyses are probably best conducted using multiple outgroups. Multiple taxa serving as outgroups need not form a monophyletic group (Nixon and Carpenter 1993). It is advisable to include several outgroup taxa in the analysis because inclusion of a single outgroup taxon does not provide a test of ingroup monophyly—the taxon used to root the tree is assumed a priori to not belong to the ingroup. If multiple outgroups are included, one or more of them (other than the one chosen to root the tree) may appear within the ingroup, rejecting the hypothesis of ingroup monophyly. Multiple outgroups may also aid in correctly determining polarity of homoplastic characters that could otherwise result in spurious group formation. In general, the more outgroup taxa included the better because such an approach represents a more severe test of ingroup monophyly.

In some cases, no known group may serve as an obvious candidate for the outgroup. Then the root for the ingroup should be determined as occurring along the *internode* that provides the most parsimonious tree length when a hypothetical ancestor is joined to the ingroup (Lundberg 1972; Nixon and Carpenter 1993). Of course, generating character states for a hypothetical ancestor requires additional assumptions.

One might imagine that states for all characters would be derived for the ingroup, or at least some of its members, and that in the outgroup all states would be primitive. In many cases, some members of the ingroup share the ancestral state with the outgroup, and the taxa exhibiting the derived state may represent a subordinate clade within the ingroup. In fact, if this were not the case, the outgroup would provide no character polarity information for taxa in the ingroup, which would result in a lack of resolution. Although it may less commonly occur, the condition for a given character in the outgroup may also be derived (an autapomorphy). In such an instance, the character will add no information to the phylogenetic analysis unless it shows variation within the ingroup.

The breadth of distribution of a character state under study in a taxon has been referred to as the "level of generality" of that character state. Those character states that occur in all members of a group would be described as occurring at a high level of generality, whereas those that have a restricted distribution within the group would be referred to as having a lesser level of generality.

Ontogenetic Data and Character Polarity

Outgroup comparison, as described above, has sometimes been referred to as the "indirect method" of determining character polarity because one must have an a priori hypothesis of relationships to choose confidently some taxon as an outgroup that is not part of the ingroup. The venerable notion that developmental patterns directly reflect phylogenetic relationships is nearly 200 years old (von Baer 1828; Barry 1837) and was popularized as the biogenetic law ("ontogeny recapitulates phylogeny") by Ernst Haeckel (1866). Hennig (1966:95) explicitly recognized the potential value of ontogenetic data in determining character polarity, and Nelson (1973, 1978b) subsequently proposed that comparative evidence from character-state transformations that occur during the development of individual organisms offers a "direct method" for determining character polarity. Nelson (1978b:327) stated the case for using ontogenetic data in determining character polarity as follows: "Given an ontogenetic character transformation, from a character observed to be more general to a character observed to be less general, the more general character is primitive and the less general advanced."

As an example, gill slits occur in the embryos of all vertebrates and persist in the adults of relatively basal vertebrate groups (fishes) but are transformed into other structures in the adults of relatively more derived vertebrates (tetrapods). One may therefore assume that the more general condition (that of possessing gills) represents the ancestral condition, and the modification of gills into other structures in adults represents the less general—derived—condition.

To understand the nature of ontogenetic data in more depth, let us examine a specific example: the antennal trichobothria in the assassin bugs of the families Pachynomidae and Reduviidae (Insecta: Hemiptera: Heteroptera: Reduvioidea).

Virtually all Heteroptera have five larval stages, after which they become adults. It is known that some morphological details differ between the first instar and the later larval instars, and that they may differ again in the adult, thus offering evidence for a hypothesis of an ontogenetic transformation series. Trichobothria are specialized setae that occur widely in the Heteroptera (Figure 4.7). They occur on the antennae in only a few Gerridae and in all Pachynomidae and Reduviidae in all postembryonic life stages.

The antennal trichobothria of Pachynomidae and Reduviidae were surveyed by Wygodzinsky and Lodhi (1989). The patterns of ontogenetic and between-group variation demonstrated in their work are summarized in Table 4.2. Although the data are not complete for all taxa listed, we might draw the following conclusions: (1) the most general condition is 1 trichobothrium, this situation occurring in every nymphal instar and adults of at least some taxa; (2) the general condition does not occur in all first-instar nymphs, nor in all nymphs; (3) number of trichobothria increases during progression through the life stages in some taxa; and (4) trichobothrial numbers apparently never decrease during ontogeny.

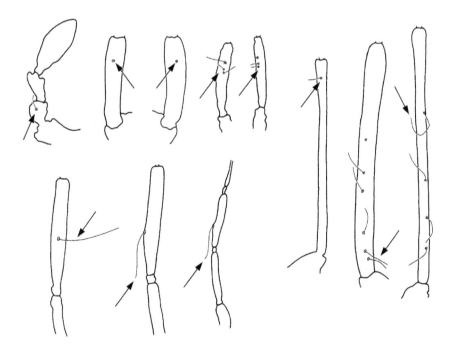

FIGURE 4.7. Distribution of antennal trichobothria on antennal segment 2 (indicated by arrows) in the larval stages and adults of some species of Pachynomidae and Reduviidae (from Wygodzinsky and Lodhi, 1989).

TABLE 4.2. Numbers of antennal trichobothria on antennal segment 2 in the Pachynomidae and Reduviidae

TAXON	INSTAR 1	INSTARS 2–5	ADULT
Pachynomidae (*Aphelonotus*)	—	—	1
Reduviidae			
Cetherinae (*Eupheno*)	—	1	5
Harpactorinae			
Amphibolus	3	4	4
Arilus	—	11	12–15
Castolus	3	—	9
Heniartes	—	7	12
Notocyrtus	6	—	8
Phymatinae	1	1	1
Reduviinae (*Leogorus*)	—	1	8
Salyavatinae (*Salyavata*)	1	1	10
Triatominae (*Triatoma*)	—	1	10

Source: Data from Wygodzinsky and Lodhi (1989).

Nelson (1978b) stated that he knew of no examples that would falsify his restatement of Haeckel's "biogenetic law" quoted above. Nelson's proposal has been viewed in three different ways: true, partly true, and false.

Weston (1988) agreed with Nelson's conclusion that the direct method occupies the logically fundamental position in cladistic analysis because it provides information on synapomorphy without recourse to previous cladistic analyses. Weston interpreted some criticisms leveled at Nelson's approach as little more than redefining terms in order to obviate the argument. He argued that ontogenetic information provides the most basic synapomorphy hypotheses in cladistics but, in an apparent contradiction of Nelson's stance, allowed that the information may not be infallible. This point had been made six years earlier by Rosen (1982:78), who noted: "As a practical matter the problem of encountering incongruence among ontogenetic character transformations within a given taxon must be addressed. It matters little whether the cladistic disagreement posed by such incongruence is due to real reversals . . . or to analytic failure; the decision as to which transformations specify the true hierarchy (of character states or organisms) must be decided, as in all cladistic disagreement, by the parsimony criterion."

Quentin D. Wheeler (1990), using an empirical example from beetle ontogeny, reasoned that *character adjacencies*—the positions of character states relative to one another in a transformation series—can be directly observed in ontogeny but polarities cannot, and that patterns of character distributions can be observed but their causal processes cannot. Wheeler viewed Nelson's rule as describing an indirect method for determining character polarity because it relies on a second taxon to estimate polarity. Furthermore, in Wheeler's view, successful polarization is dependent on the second taxon that is chosen for comparison. Wheeler concluded that Nelson's rule is a special case of parsimony and has the advantage of allowing for conclusions based on fewer comparisons than would be necessary using character data derived from only a single life stage. He also noted, as had Weston, that there are some situations in which the method does not work, as is the case with simple outgroup comparison.

Nelson's assertion that there are no falsifiers to the biogenetic law was attacked by Kluge (1985) with the observation that contradictory ontogenies (e.g., neotenic salamanders) are known. Kluge further criticized Nelson's view because he interpreted "more general" to mean commonality, with the suggestion that the argument could be reduced to "common equals primitive." However, as pointed out by Weston (1988), the more general character state is possessed by all taxa that possess the less general character state as well as some that do not; thus, Kluge's equating generality with commonness is a misinterpretation (see also Sidebar 9).

Sidebar 9
Discredited Methods of Determining Character Polarity

Several methods purporting to be useful for determining character polarity have been proposed in addition to outgroup comparison and the ontogenetic criterion. These approaches have been largely abandoned by modern-day systematists in favor of logically based criteria related to minimization of homoplasy, given that a rooted tree automatically implies the direction of character-state transformations. These discredited methods include the following:

1. *Common equals primitive.* This criterion suggests that characters more widespread among taxa are primitive relative to characters of more restricted taxonomic distribution. There are many instances, such as winged versus primitively wingless insects or viviparous versus oviparous mammals, in which taxa with derived features are more diverse than their sister taxa with complementary symplesiomorphies, and this maxim is simply not a reliable guide for inferring character polarity.
2. *More complex characters are derived relative to less complex characters.* The subjectivity of this criterion would seem self-evident. The basis

for complexity is seldom defined, let alone understood from a developmental or genetic point of view. Indeed, reduction and loss seem frequently to represent derived conditions.

3. *Characters found in fossils are primitive relative to those found in living taxa.* The rejection of the "paleontological method" in cladistics early on forced the abandonment of this approach, going hand in hand with the recognition of fossils simply as additional taxa subject to analysis and at most indicating a minimum age for a taxon (see excellent discussion in Schoch [1986] and also Chapter 11).

4. *Chorological progression.* In this view, the more advanced characters (and taxa) are to be found further from the geographic center of origin for a group. This "progression rule," which was advocated by Hennig and some later authors (e.g., Brundin 1981), rests on the unwarranted assumptions that "primitive" and "advanced" taxa can be recognized a priori, and that what might be true of one taxon in this regard should also be true of others (see Platnick 1981a).

Let us examine these arguments in light of the example given in Table 4.2 for reduvioid antennal trichobothria.

Can polarity be determined directly? Our example suggests that ontogenetic data cannot determine polarity on the basis of a single taxon. If we had only two taxa, one with a single antennal trichobothrium in all life stages (Phymatinae) and one with multiple trichobothria in all life stages (*Amphibolus*), the question of what condition was general—and therefore primitive—could not be answered. Lacking information on the Pachynomidae, the sister group of the Reduviidae, one still might not come to an unequivocal conclusion concerning the general condition in the Reduviidae, depending on the sample of taxa available. Thus, Wheeler's insight that ontogenetic evidence represents a special case of outgroup comparison, because polarity determination requires more than one taxon, would seem to be correct.

Are there any falsifiers? The answer to this question would seem to have two parts. The patterns of trichobothrial distribution, as far as they are known, do not contradict the conclusion that a single trichobothrium is the general condition. That is, there are no known instances in which there are multiple trichobothria in the larval stages and a single trichobothrium in the adult. However, the general condition does not occur universally in larvae; as seen in Table 4.2 first-instar larvae of *Amphibolus*, *Castolus*, and *Notocyrtus* have multiple trichobothria. What does appear to be regular is that if the numbers of trichobothria change during ontogeny, they always increase—an observation concordant with the generally observed terminal addition of traits during ontogeny. These observations concerning variation in ontogenetic data are similar to those of Wenzel (1993) for the ontogeny of nest-building behaviors in the paper-wasp subfamily Polistinae (see Chapter 10).

In sum, ontogenetic data appear to represent an important source of characters for phylogenetic analysis. They may contain ambiguities, as do other forms of data. Ontogenetic data, where available, should be coded like any other characters. We conclude that such data do not offer a direct method of polarity determination, but they may nonetheless offer a more clear-cut indication of state ordering than other forms of data, a subject discussed in Chapter 3 (see also Sidebar 9).

Brower (2019) has suggested an even more basic (perhaps trivial) criterion for hypothesizing polarity of certain characters—the very simple idea that presence of a novel (neomorphic) feature suggests a group with respect to absence of that feature. This "existential criterion" seems to have been used to formulate intuitive hypotheses of major groups (e.g., nuclei in eukaryotes, multicellularity in animals, backbones in vertebrates, spinnerets in spiders) from antiquity and provides a framework for the testing of other character generalities and polarities by reciprocal illumination. Of course, some such features, such as scales in "fish" and "reptiles," are symplesiomorphic, as revealed by their incongruent distributions with respect to other characters.

Ancestors, Sister Groups, and Age of Origin

The taxa studied, including fossils, in Hennig's phylogenetic schemes (and all subsequent cladistic work) have always been placed as terminals at the ends of branches, whereas many preceding phylogenetic schemes embedded taxa, particularly fossils, along the internal branches of the tree (see Figure 4.8). You might ask: What are ancestors in the sense of the Hennigian definition of monophyletic? Can we actually find them in nature? Are they represented by fossils? And, if they cannot be found in nature, does this not discredit the entire system?

Cladistic methods treat ancestors as *hypotheses* (sometimes called *hypothetical taxonomic units*, or HTUs), which have an optimized "ground plan" of attributes inferred from the parsimonious pattern of character-state transformations distributed on the cladogram. Each internal node of a cladogram represents the hypothetical ground-plan condition for the clade subtending it, based on the distribution of observed character states in the terminal taxa. For example, if taxa A–D in Figure 4.8 all share the same state for a character, it is parsimonious to infer that state was also present in hypothetical common ancestors 3, 2, and 1.

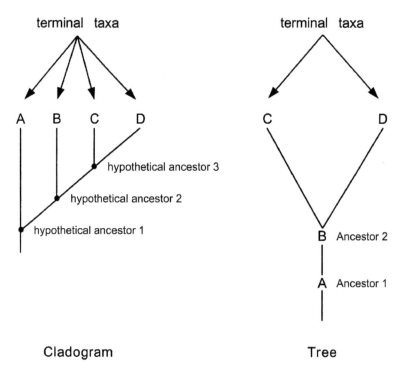

FIGURE 4.8. Comparison of cladograms and trees in the sense of Nelson and Platnick (1981), showing the nature of ancestors in the two types of diagrams. Ancestors in cladograms are hypothetical, represented by the nodes that connect groupings of terminal taxa. The characters attributed to those nodes are optimized—as described in Chapter 5—from the characters of the terminals. These deduced attributes form the "ground plan" for the hypothetical ancestor, or hypothetical taxonomic unit (HTU). Ancestors in the tree on the right are taxa represented by actual specimens, usually fossils (there is no way to know that a given fossil is the ancestor of any particular taxon).

It is critical to appreciate that ancestors, like all unobservable entities, are inferences based on the results of systematics and not things that can be recognized directly. For example, there are six or seven specimens of *Archaeopteryx lithographica*, which exhibit impressions of shafted feathers (a synapomorphy for Aves) and a variety of plesiomorphic dinosaurian features, such as forelimb digits, teeth, and a long bony tail. These specimens are pieces of limestone: we know nothing of their status as breeding members of a population of proto-birds other than what can be inferred from their static morphology. Nor do we know whether they represent part of the direct ancestral lineage that led to modern birds or merely an offshoot side branch that went extinct. What we can observe, when we interpret them as exemplars of an extinct taxon, is that they exhibit the combination of character states that places the species as sister taxon to the clade of modern birds and adjacent to a paraphyletic assemblage of small cursorial dinosaurs.

Because of these empirical limitations, the cladistic approach treats fossil taxa simply as observable terminals, like all other taxa in phylogenetic analysis (see, e.g., Hennig 1969). This view of fossil taxa stands in stark contrast to the "pale-ontological approach," where a fossil would de facto be ancestral to other recognized taxa, either those from more recent geologic strata or Recent forms.

References to fossil taxa as ancestors in the literature are usually not made in the context of an explicit scheme of relationships derived from cladistic analysis. Whether some fossilized organism actually represents an ancestor is a question that resides in the realm of the unknowable. The conventions of cladistics demand that all groups be monophyletic. Thus, if one group of taxa arises from another, the "ancestral" group is, by definition, paraphyletic (see also Chapter 8, Stem Groups and Crown Groups). From the point of view of character distributions, ancestor-descendant relationships are untestable in a cladistic framework because, as pointed out by Engelmann and Wiley (1977), corroboration of such hypotheses would be based on plesiomorphy. Ancestors necessarily would be taxa recognized by the absence of the apomorphies of their descendants.

Nelson and Platnick (1981) applied the term *cladogram* only to diagrams depicting hierarchical patterns of relationship among nested sets of sister taxa (sometimes referred to as Steiner trees [Foulds et al. 1979]), whereas they restricted the term "tree" to diagrams illustrating ancestor-descendent relationships (Figure 4.8). The distinction made by Nelson and Platnick is rarely adhered to (Brower 2016b), and in the present work, the term *tree* is sometimes used interchangeably with cladogram, in conformity with much of the literature. This usage is not meant to suggest, however, that such "trees" are thought to transmit information on ancestor-descendant relationships.

Another type of *dendrogram* that is sometimes encountered, particularly in intraspecific "phylogeographic" studies, is the *minimum spanning tree*. Frequently represented as an unrooted network with different-sized circles at the nodes, indicating the numbers of individuals genetically identical for the marker assessed, and branch lengths proportional to number of inferred mutational steps between such clusters, minimum spanning trees are often used to illustrate narrative scenarios of ancestral and derived patterns of variation in mitochondrial haplotypes or other molecular markers (e.g., Templeton 2004). It appears that the interpretation of character polarity in these networks, nominally couched in terms of a process model dependent on clocklike accumulation and spread of neutral mutations, ultimately is fallaciously based on the discredited "common = primitive" assumption described in Sidebar 9 (Chapter 4). See Brower (1999) and Posada and Crandall (2001) for alternative interpretations of minimum spanning networks, and Knowles (2008) and Templeton (2008) for controversy over "nested clade analysis." (Note that there are no "clades" in an unrooted network.)

Hennig referred to two (or more) groups arising from a common branch point in a cladogram as *sister groups* (Figure 4.9). Because sister groups share the same hypothetical common ancestor, they are—by logical necessity—of the same age. Phylogenetic hypotheses in the form of cladograms, then, offer us information on the relative age of origin of taxa. When fossil taxa are among those being analyzed, there may be an empirical source of evidence to infer a minimum absolute age of origin for a clade on the basis of the estimated age of the fossil. (We will discuss molecular clocks and dating in more detail in Chapter 11.) Cladistic methods do not, however, provide information on absolute age of origin, nor do they allow for the identification of taxa as ancestors. Neither do cladograms (or any other type of tree other than speculative and unfalsifiable narrative scenarios) represent *direct* depictions of the history of speciation. Their limitations in this regard are both practical and methodological.

The practical limitation is that our sample of taxa will always be incomplete. We have no way of knowing how many species have gone extinct and will never be sampled, or how many species from the Recent fauna have not yet been sampled. Therefore, although a cladogram based on character information may represent

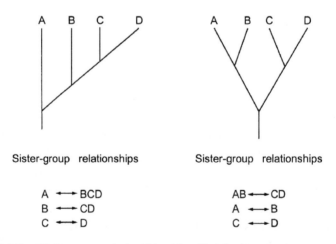

FIGURE 4.9. Sister-group relationships identified for four taxa on two cladograms with different topologies. Sister groups are logically required to be of equal age because they are inferred to have descended from a single most recent common ancestor.

the relative recency of common ancestry implied by the evidence at hand, in most cases it certainly will not represent the complete history of speciation events.

The methodological limitation of cladograms in cladistics—and indeed all phylogenetic methods—is that we have no method for recognizing ancestors, as was discussed in the foregoing paragraphs. Thus, even though one species may be descended from another, as in the case of Recent from fossil, our methods do not allow us to make such determinations. We are not aware of any other scientific technique that is capable of acquiring this knowledge.

These limitations have not deterred some biologists from holding the view that cladistic methods (or other, less rigorous approaches to the study of relationships among taxa) can produce accurate reconstructions of evolutionary history, reveal ancestor-descendant relationships, or depict actual events of speciation. No doubt, any and all scientists would love to obtain such information, and it is certainly possible to contrive models based on molecular clocks and other deterministic assumptions that elaborate scenarios illuminating such patterns and processes in great detail (e.g., Rabosky 2014). However, as discussed in Chapter 2, no matter the strength of such realist desires, definitive answers to such questions remain in the realm of science fiction.

SUGGESTED READINGS

Brower, A.V.Z., and M.C.C. de Pinna. 2012. Homology and errors. Cladistics 28:529–538. [A defense of the equivalence of homology and synapomorphy]

de Pinna, M. C. C. 1991. Concepts and tests of homology in the cladistic paradigm. Cladistics 7:367–394. [A useful review of homology concepts]

Nixon, K. C., and J. M. Carpenter. 1993. On outgroups. Cladistics 9:413–426. [An authoritative and pithy review of outgroup comparison]

Nixon, K. C., and J. M. Carpenter. 2011. On homology. Cladistics 28:160–169. [An attack on the equivalence of homology and synapomorphy]

Patterson, C. 1982. Morphological characters and homology. In: Problems in phylogenetic reconstruction, ed. K.A. Joysey and A E. Friday, 21–74. London: Academic Press. [Within the modern literature, a classic paper on homology concepts]

Rieppel, O.C. 1988. Fundamentals of comparative biology. Basel: Birkhäuser. [An excellent discussion of the philosophical underpinnings of homology concepts]

Weston, P.H. 1994. Methods for rooting cladistic trees. In: Models in phylogeny reconstruction, ed. R.W. Scotland, D.J. Siebert, and D.M. Williams, 125–155. Oxford: Clarendon. [A useful discussion of outgroup comparison]

Williams, D.M., and M.C. Ebach. 2014. Patterson's curse, molecular homology, and the data matrix. In: The evolution of phylogenetic systematics. ed. A. Hamilton, 151–187. Berkeley: University of California Press. [Aa thoughtful consideration of homology inference]

TREE-BUILDING ALGORITHMS AND PHILOSOPHIES

Although systematists have long associated characters with taxa, the relationship between character data and "phylogeny" has not always been obvious (see Sidebar 10). The ideas of Willi Hennig clarified this relationship, and the formalization of these concepts in a quantitative method, via the parsimony criterion, allowed for computer implementation of phylogenetic inference and the feasible solution of previously intractable problems. It is this computational capability that took the study of taxonomic relationships from an almost purely qualitative and speculative enterprise to one dominated by the use of computer software and "objective" methodologies. In this chapter we will examine "quantitative cladistics" in detail, including the issues of fit, parsimony algorithms, and character weighting. We also discuss the use, advantages, and disadvantages of maximum likelihood and Bayesian techniques as alternative approaches to the application of parsimony.

Sidebar 10
"Phylogeny Reconstruction" versus Phylogenetic Inference

Throughout this book we have taken care to refer to representations of the relationships among taxa as "phylogenetic hypotheses," and the process of developing those hypotheses as "phylogenetic inference." All too frequently in the literature we see reference to "phylogenies" being "reconstructed" by this or that method. This terminology is objectionable, for the following reasons. First, phylogeny (or phylogenesis) refers to a process of branching diversification of taxa, just as ontogeny refers to the process of organismal development. Although we can easily watch seeds develop into flowering plants or children grow up, human life spans are not long enough to allow us to observe the process of phylogeny directly. Whatever we know about it has been gleaned from consideration of patterns of similarity and difference among taxa—living or fossil—that we observe at the present time (or have extracted from the literature). Which brings us to the second semantic point: "reconstruction" implies putting some thing back together again according to a plan or self-evident correspondence of pieces, as for example, mending a broken tea cup or replacing the copper in the Statue of Liberty. While we assume that there is a pattern to seek, the particular pattern of relationships among taxa we discover through our systematic efforts is a theoretical abstraction—a conceptual map localizing inferred character-state transformations that systematists conjecture to represent the singular path that the evolutionary history of life might have taken. That pattern is not a diagrammatic likeness of the history of life constructed on an a priori plan, nor a quantitative engineering calculation with confidence intervals, as some of the more algorithmically minded might like to believe, nor is it "the truth." If "phylogeny" (or the execrable "phylogenies," which implies multiplicity and discontinuity of the Tree of Life) could be uncontroversially "reconstructed," an instruction manual would suffice (*Phylogeny for Dummies*, or some such), and there would be no need for this book or any of the theoretical discussion in the other books and journals we have cited.

Background

Phylogenetic analysis, as proposed and practiced by Hennig, involved a restricted number of taxa and a limited amount of homoplasy in any given data set. Analyzed by hand, these matrices needed to express clear patterns in order to be interpretable, and systematists chose those characters—often representing uniquely derived features—that seemed to support natural groups. In other words, the characters selected often straightforwardly identified a unique scheme of relationships. For this reason, Hennig's method was at times misapprehended as implying *clique* analysis, an approach in which groups are formed only on the basis of characters that are all perfectly congruent (see Sidebar 13 in Chapter 6). The practical drawback of the strict Hennigian approach is that simple calculation often does not discover the multiple phylogenetic solutions that exist for a data set containing incongruent characters, and it may not even find the most parsimonious tree (MPT). The power of the computer offers the only possibility for solving problems in phylogenetic analysis other than those represented by small, highly consistent data sets.

Optimization

Optimization is the fitting of characters to cladograms on the basis of some criterion that is minimized or maximized. This concept is central to implementation of a quantitative cladistic approach (parsimony) and also to maximum likelihood and Bayesian methods. In cladistic analysis, the optimality criterion is the minimization of the number of character-state changes—ad hoc hypotheses—on a cladogram. As such, cladistic optimization is a straightforward application of the principle of parsimony to the selection of a particular hypothesis of phylogenetic relationships. Choice or rejection of alternate topologies is determined by the fit of the data to the hypothesis, that is, by the number of character-state transformations (or "steps") required on alternative cladograms (assuming that character-state transformations are weighted equally). "Required" means that for a given hypothesis, the distribution of character states exhibited by the terminals cannot be accounted for with fewer than the number of steps on that topology. Notice that parsimony neither requires nor implies a limit to the number of "extra steps" that could have occurred in the actual course of evolution because cladistic hypotheses are based on the parsimonious accounting of the observed evidence (Farris 1983)—an epistemological rather than an ontological application of the parsimony criterion. Cladistics accounts parsimoniously for actual evidence (lower-level empirical theories of character-state transformation), not hypothetical, unobserved phenomena.

The optimal cladogram is the "shortest tree"—the one among alternatives that requires the smallest number of character-state transformations—or, if the transformations are differentially weighted, the tree that has the smallest cumulative "weight" or "cost" of implied changes (see Differential Character Weighting, below). The minimal number of character-state transformations can be determined by exhaustively enumerating all possible topologies or estimated by extensive sampling. The tree length will depend on (1) the data, (2) the "costs" assigned to different character-state transformations, and (3) cladogram *topology*.

In the cladistic parsimony framework, hypothetical character-state identities are explained by a cladogram when the cladogram allows attribution of those similarities to common ancestry or, stated less metaphysically, to a common inferred transformation event (Brady 1994). Fit of the data determines exclusively which cladogram(s) among all possible cladograms is considered optimal; fit can be inferred only through optimization.

Computer algorithms capable of employing various optimality criteria are widely available, and one of the chief controversies in modern systematic theory is over choice of the most appropriate optimality criterion for inferring natural classifications.

Optimality Criteria

Parsimony was apparently first used as an optimality criterion in phylogenetic analysis by Edwards and Cavalli-Sforza (1964) and Camin and Sokal (1965) (although Edwards and Cavalli-Sforza's application was to intraspecific variation of human blood groups and, therefore, misapplying a phylogenetic method to a tokogenetic problem). Several variant approaches have been proposed as representing more "realistic theories" of character evolution that place asymmetrical restrictions on the possibilities for character-state transformation. The Camin–Sokal theory of character evolution employed the parsimony criterion but treated character-state transformations as irreversible: once a feature is modified, it can never revert to a more ancestral condition. The Dollo theory (after Dollo's Law of evolutionary irreversibility), as described by Farris (1977), treated all apomorphic states as uniquely derived, with homoplasy being accounted for only by reversal (effectively the opposite of the Camin–Sokal scheme, in which homoplasy is accommodated as independent gains). It is not clear how either of these is more "realistic" than equal-weighted parsimony, and neither theory has received widespread acceptance or application in subsequent decades.

More recently, the term parsimony has been used almost exclusively in reference to approaches that minimize the number of state changes on a cladogram for reversible characters with symmetric transformation costs. There are several algorithmic variants that have been referred to by some authors (e.g., Swofford et al. 1996:416) as "a group of related methods," but these merely represent alternative weighting schemes for character-state transformation within the same optimality framework. We describe the most common ones here.

Additive (Farris) optimization (transformation). Kluge and Farris (1969) and Farris (1970) employed the so-called Wagner ground-plan method of analysis, the name deriving from the American botanist W. "Herb" Wagner, who first described the general approach (Wagner, 1961). Under this criterion, state changes in multistate characters are *additive* (ordered, see Chapter 3), but reversal and multiple origination are not restricted. Additivity assigns additional steps (costs) if implied character-state transformations on the cladogram do not adhere to the state-to-state order of multistate characters as defined in the data matrix (Figure 5.1). Thus, in Figure 5.1, the optimization for additive characters can be accomplished with the algorithms published by Farris (1970), Swofford and Maddison (1987), and Goloboff (1993). The general procedure, widely known as "Farris optimization," after its author James S. "Steve" Farris, is most often employed for morphological characters for which there is an evident transformation series established by criteria such as developmental patterns.

Nonadditive (Fitch) optimization (transformation). Fitch (1971) proposed the use of nonadditive (unordered, see Chapter 3) transformations for the analysis of amino acid and DNA sequence data. *Nonadditivity* allows transformation from any state to any other state—such as nucleotide for nucleotide, amino acid for amino acid—without the imposition of additional steps (costs) (Figure 5.2). Optimization under this "theory" of character-state transformation can be performed with algorithms of the type published by Walter Fitch (1971). The method is frequently referred to as "Fitch optimization" and is now widely used for both molecular data and morphological characters for which the systematist does not wish to assign a specific a priori transformation series. Recall that, as discussed in Chapter 3, nonadditive optimizations are always as short as or shorter than additive optimizations, and therefore may be viewed as more parsimonious, unless there are compelling reasons to provide an ordered transformation series for certain characters.

The term "Farris optimization" has also been associated with "accelerated transformation" (*ACCTRAN*) of characters on a cladogram (examples in Agnarsson and Miller 2008). Optimization, in this sense, refers to alternative approaches to localizing the inferred positions of character-state changes on particular branches of a most parsimonious tree, as opposed to finding the most parsimonious tree(s) for a given set of characters (see Chapter 11).

Alternative parsimony optimization algorithms. If one wishes to impose transformation costs different from the "additive" and "nonadditive" approaches, Farris or Fitch optimization cannot be used to find the optimal state assignments. In such cases, it is necessary to use algorithms such as those first developed by David Sankoff and coauthors (Sankoff and Rousseau 1975; Sankoff and Cedergren 1983). These algorithms (dubbed "generalized parsimony" or "step-matrix" by Swofford et al. 1996) consist of enumerating all possible combinations of state assignments for every node in the tree and its surrounding nodes, calculating partial costs in every case, and choosing the best result. It is, in some sense, a "brute force" approach. Using these algorithms, any set of symmetrical or asymmetrical "costs" can be assigned a priori to the different transformations. This approach readily allows for implementation of any "model" (i.e., specified set of state transformation weights) of nucleotide substitution—something not possible under nonadditive (Fitch) optimization. Maximum likelihood and Bayesian techniques

Character	1 2 3 4 5 6 7
Taxon X	0 0 0 0 0 0 0
Taxon A	1 1 0 0 0 1 1
Taxon B	1 1 0 0 0 2 2
Taxon C	0 0 1 1 0 3 3
Taxon D	0 0 1 1 1 4 3
Taxon E	0 0 1 1 1 5 3

2 trees: length 15 steps, CI = 0.86, RI = 0.85

(a)   (b)

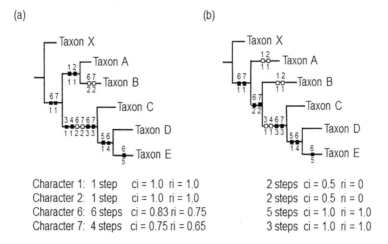

Character 1: 1 step   ci = 1.0 ri = 1.0   2 steps ci = 0.5 ri = 0
Character 2: 1 step   ci = 1.0 ri = 1.0   2 steps ci = 0.5 ri = 0
Character 6: 6 steps   ci = 0.83 ri = 0.75   5 steps ci = 1.0 ri = 1.0
Character 7: 4 steps   ci = 0.75 ri = 0.65   3 steps ci = 1.0 ri = 1.0

FIGURE 5.1. Additive (Farris) optimization (ordered transformation) of multistate characters produces two most parsimonious cladograms with a length of 15 steps for the matrix of data shown in the figure. Cladogram (a) shows the slow (delayed transformation, DELTRAN) optimization of character 6, a 5-step character, with a total of 6 steps, and character 7, a 3-step character, with a total of 4 steps. Notice that because the character coding is not concordant with the most parsimonious tree for the data (the open squares indicating homoplasy), additive coding has added one extra step for each character, but nonetheless each contains substantial synapomorphy content. Cladogram (b) shows the slow (DELTRAN) optimization of both characters 6 and 7 with the minimum number of steps, the homoplasy in the data occurring in the two-state characters 1 and 2. CI = ensemble consistency index, RI = ensemble retention index,_ci = consistency index, ri = retention index.

Character	1	2	3	4	5	6	7
TAXON X	0	0	0	0	0	0	0
TAXON A	1	1	0	0	0	1	1
TAXON B	1	1	0	0	0	2	2
TAXON C	0	0	1	1	0	3	3
TAXON D	0	0	1	1	1	4	3
TAXON E	0	0	1	1	1	5	3

1 tree: length 13 steps, CI = 100 RI = 100

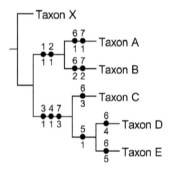

Character 6: 5 steps, ci = 100 ri = 0.00
Character 7: 3 steps, ci = 100 ri = 100

FIGURE 5.2.   Nonadditive (Fitch) optimization (unordered transformation) of multistate characters produces a single most parsimonious cladogram with a length of 13 steps, using the same data as in Figure 5.1, with characters 6 and 7 fitting the cladogram perfectly (the closed circles indicating the absences of homoplasy). All states for character 6 are autapomorphic and uninformative with regard to group formation; therefore, its retention index is 0.00. Under either optimization (accelerated transformation, ACCTRAN, or DELTRAN), state 3 of character 7 is synapomophic for taxa C, D, and E, whereas states 1 and 2 are autapomorphic. Comparing the results of nonadditive optimization with additive optimization (Figure 5.1), it can be seen that the nonadditive tree is shorter but that information on grouping as implied by the multistate characters is lost in favor of treating states as unique to terminal taxa.

usually employ differential transformation cost matrices as well, approaches that are discussed in more detail in sections below.

Optimization of characters, then, is not the application of different methods but rather represents a choice among different rationales for assigning transformation costs among states. Ideally, such a choice should be made on the basis of evidence, as under the arguments presented by Lipscomb (1992) for treating characters as additive when evidence for transformation exists in the form of observed similarity, or on the inferred probability of nucleotide change, as implemented in maximum likelihood algorithms. Once the choice is made, the algorithms developed by Camin and Sokal, Farris, Fitch, and Sankoff are the tools used to find the most parsimonious tree(s) given those transformation costs. Note that the particular optimization approach, or optimality criterion, selected is determinative of the outcome. Different optimality criteria may imply different optimal trees.

The various parsimony-based optimality criteria can be summarized as follows:

Additive (ordered; Farris)	Change costs increase with deviation from specified state-to-state order.
Nonadditive (unordered; Fitch)	Change costs are equal for any state-to-state change.
Camin–Sokal	Reversals are of infinitely high cost.
Dollo	Multiple origins of same derived character state are of infinitely high cost.
Sankoff	Costs for any state-to-state change are assignable on a differential basis.

Sidebar 11
Why Distance-Based Methods Fail

As we have stated many times, the optimal cladogram is the one that requires the smallest number of implied character-state transformations. Phenetic methods, as discussed in Chapter 1, group taxa by degree of overall similarity, as measured by patterns of shared character-state identity.

From a mathematical perspective, the difference between taxa measured on a phenogram is the *Euclidean distance*, while difference between taxa on a cladogram is measured by the *path length*. We will use a simple example to illustrate the distinction between these.

```
      1   2
A     1   1
B     2   2
```

Examining the above matrix, it is clear that taxon A and B differ by a path length of two steps (one step in character 1, one step in character 2). If we think of the data matrix as a two-dimensional space with each character as an axis and the states of the characters as points along the vector of that axis, then taxon A sits at the point (1,1) and taxon B is at (2,2). (In such a scheme, each new character adds an additional dimension, so a matrix of N characters resides in N-dimensional space). The path length can be mapped by counting the distance moved along the axes of the characters that differ between the taxa. This measure is also referred to as the *Manhattan distance* because, like walking on city blocks, the shortest path from point A to point B requires traveling on the ordinal axes (the streets) rather than taking the diagonal line that would cut through the middle of the buildings. Euclidean distance, by contrast, is the straight line between two points in N-dimensional space. Thus, the Euclidean distance between A and B is the hypotenuse of the right triangle (the square root of the sum of the squares of the individual vectors; 1.44 units, in this simple case). At first glance it would seem that Euclidean distances are a shorter—or "more parsimonious"—summary of the character information than the path length (Manhattan) distances.

However, trouble arises because algorithms that build trees from Euclidean distances require that the branch lengths (number of implied character state transformations—not time!) from the root to each terminal taxon be the same, a property called *ultrametricity* (Farris 1981). If this is not the case, then taxa that exhibit relatively few character-state transformations will be grouped together on the basis of a lack of change—by symplesiomorphies; at the same time, those with many autapomorphies may be clustered as distinct from the rest. Although cladograms are not usually drawn to represent differential branch lengths (trees with branch lengths reflecting implied numbers of character state transformations are usually called *phylograms*), the method is perfectly capable of accurately representing patterns of character-state transformation, whether there are many or few transformations on a given branch. Indeed, those character-state transformations are portrayed on many published cladograms, especially those based on the analysis of morphological data.

METRICITY AND THE TRIANGLE INEQUALITY

The use of computer algorithms requires a mathematical model on which tree construction can be based. The models used are *metrics*, which fit the triangle inequality.

The metric used in the cladistic analysis of discrete character data is the Manhattan, or *path-length distance*, where, d equals the distance between two taxa, and i, j, and k are taxa, thus:

$$d(i,j) \leq d(i,k) + d(j,k)$$

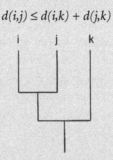

As illustrated with this cladogram, the distances between taxa i and k and between taxa j and k may be different, but their sum must be greater than the distance between taxa i and j.

Phenetic approaches, on the other hand, have employed *ultrametrics*, which are clustering-level distances, and measure Euclidian distances. Under this model,

$$d(i,j) \leq max\ d(i,k), d(j,k)$$

As an ultrametric, the distances between taxa i and k and between taxa j and k must be equal and both must be greater than the distance i and j. The important difference between the path-length distance and the Euclidean distance is that the former allows for variable implied rate of change among branches, whereas the latter requires a uniform rate of change throughout the tree. Indeed, it is claimed in many phenetic writings that classifications must be based on ultrametric trees because this is the only way that information on "levels" in the classification can be correctly stored (Farris 1982; see also discussion in Carpenter, 1990).

If we name the "internal nodes" in the phenogram as a and b,

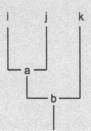

it is easy to see that the requirement that the distance between taxa i and k equal the distance between taxa j and k can be fulfilled only if the implied amount of change from a to i is the same as that from a to j. Adding taxa l, m, n, and so on outside the group of three would require that the implied amount of change from b to k be the same as that from b to i and from b to j. When some of the branches are "longer," the inequality cannot be satisfied for all triplets. This happens, obviously, when there are unequal numbers of character state transformation, implying unequal rates of evolution.

Consider now a tree for which branch lengths (not levels) indicate amount of difference (with numbers by the branches indicating branch lengths):

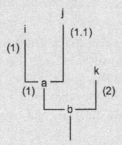

Note that the branch aj is slightly longer than ai. But then, all three instances of the ultrametric inequality are not satisfied:

$$dij \leq max\ (dik, dkj) \rightarrow 2.1 \leq max\ (3, 3.1) \quad \text{TRUE}$$
$$dik \leq max\ (dij, dkj) \rightarrow 4 \leq max\ (2.1, 4) \quad \text{TRUE}$$
$$djk \leq max\ (dik, dij) \rightarrow 4.1 \leq max\ (4, 2.1) \quad \text{NOT TRUE}$$

In such a case, where the observed distances (the actual amount of difference) between taxa are unequal, it will not be possible to produce a tree diagram with the observed distances perfectly retrievable from the distance levels. But path-length distances as used in cladistics will produce a perfect fit. In fact, as shown by Farris (1979a, 1979b), whenever the "levels" model produces perfect fit, so will the "path-length" model. Additional discussion of the application of the triangle inequality can be found in Wheeler (1993).

Storing Trees

While the typical means to visualize patterns of hierarchical relationship among a group of taxa is a branching diagram, these patterns can be equally well represented as nested sets in a Venn diagram (an approach used for representation of relationships among taxa early on by, e.g., Milne-Edwards [1844]), which can be represented in linear text by *parenthetical notation* (sometimes also referred to as the "*Newick format*," after a lobster restaurant in Dover, New Hampshire, where the notation is purported to have been initially jotted down on napkins). This approach, which is how computers store tree topologies, uses nested sets of parentheses to indicate the inclusive nature of groupings. Examples are shown in Figure 5.3. Note that the number of left- and right-hand parentheses is "balanced" in the examples. Computer phylogenetics programs always produce

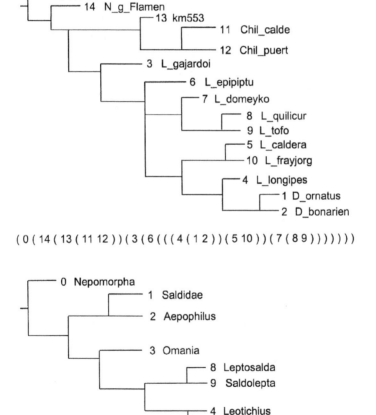

(0 (14 (13 (11 12)) (3 (6 (((4 (1 2)) (5 10)) (7 (8 9)))))))

(0 (1 2) (3 ((8 9) (4 (5 6 7)))))

FIGURE 5.3. Examples of parenthetical notation and corresponding cladograms (note that branch lengths in this figure are arbitrary and convey no information).

notation that is balanced, although they may be able to interpret trees accurately for which not all groupings have balanced parentheses.

Finding Optimal Trees

We focus first on cladistic, or parsimony-based, tree searches. As authors of other textbooks admit, even as they advocate alternative approaches (cf. Wiley and Lieberman 2011; Baum and Smith 2013), one of the many advantages of cladistic methods is that they are comparatively straightforward to explain and understand. After we describe the challenges of discovering most parsimonious trees, we will compare and contrast cladistic parsimony to model-based alternatives.

Exact parsimonious solutions for small data sets can be discovered by enumeration of all possible trees. Because the number of trees increases as a geometric function of the number of taxa, however, it is not possible to calculate exact solutions for larger data sets. There is no known mathematical solution for such problems (see Day et al. 1986), and no matter how much computing power is available, for more than 20 or so taxa, there are more possible trees than can ever be examined (Figure 5.4).

Most currently used phylogenetic algorithms adhere closely in principle to the Wagner method originally proposed by Kluge and Farris (1969) and Farris (1970) for calculating trees. To improve the success rate for finding optimal solutions, many variant implementations of Farris's original approach have been developed, increasing immensely the speed and effectiveness with which optimal trees can be sought.

Farris's Wagner method, like most others, produces "unrooted trees," often referred to as *networks*. Because the parsimony criterion of counting character-state transformations does not require a priori considerations of polarity or synapomorphy (as long as transformation costs are symmetrical), it is not necessary to root the scheme of interrelationships in order to optimize the character states for all internal nodes on the tree. Thus, localizing, and minimizing, the hypothesized positions of character-state transformations on the internal branches of the tree under a given optimality criterion is the core problem of numerical cladistics. The states for the terminal taxa are already known!

A Wagner tree (network) is formed initially by the sequential addition of taxa. Because different addition points produce trees of different lengths, taxa are added in those positions that increase the length of the tree as little as possible. Using this approach, the method attempts to end up with trees of minimum global length. In practice, it is nearly impossible to find the tree(s) of minimum length using the simple Wagner algorithm. Wagner trees, however, are generally

1	1
2	1
3	3
4	15
5	105
6	945
7	10,395
8	135,135
9	2,027,025
10	34,459,425
11	654,729,075
12	13,749,310,575
13	316,234,143,225
14	7,905,853,580,625
15	213,458,046,676,875
16	6,190,283,353,629,375
17	191,898,783,962,510,625
18	6,332,659,870,762,850,625
19	221,643,095,476,699,771,875
20	8,200,794,532,637,891,559,375
21	319,830,986,772,877,770,815,625
22	13,113,070,457,687,988,603,440,625

FIGURE 5.4. The numbers of possible bifurcating rooted trees for 1–22 taxa (from Felsenstein 1978, based on calculations by Cayley 1856 and Schröder 1880. See also Kidd and Sgaramella-Zonta 1971)

much shorter than arbitrary trees (those generated at random, for example) and thus constitute a better starting point for further refinement through the use of branch swapping (see below in Phylogenetic Algorithms).

As the taxa are added to the growing network, the length produced by adding a taxon at each position can be calculated. The length calculation is the time-consuming aspect of phylogenetic computation; current computer programs use fast indirect methods to calculate length.

Note that different addition sequences of the taxa may lead to different trees. Consider the data matrix in Figure 5.5, where taxon X represents the outgroup. Forming a tree with a simple Wagner algorithm, which adds the taxa in the order in which they appear in the matrix, (XAB), C, D produces an incorrect tree (Figure 5.5a). That is because the first three characters determine a unique tree (X ((A B) (C D))), whereas the last two characters are homoplastic. However, if only X, A, and B have been placed in the tree, when C is to be added, the characters that join C and D will appear as uninformative (i.e., autapomorphic, requiring

Character	1	2	3	4	5
TAXONX	0	0	0	0	0
TAXONA	1	0	0	0	0
TAXONB	1	0	0	1	1
TAXONC	0	1	1	1	1
TAXOND	0	1	1	0	0

(a)   Simple Wagner algorithm = 8 steps

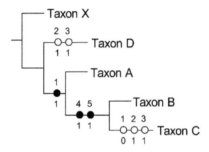

(b)   Randomized addition sequence = 7 steps

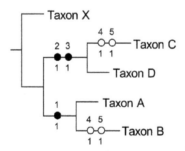

FIGURE 5.5.   Matrix of five characters and five taxa, with taxon X as the outgroup, where (a) using a simple Wagner algorithm may fail to find a correct solution but (b) randomizing the taxon addition sequence achieves the most parsimonious solution.

the same number of steps for any possible placement of C), since D is still not placed in the tree. If the taxa are instead added as (XAC), D, B, the shortest tree is obtained (Figure 5.5b). In other words, the evidence that the Wagner method considers at each step is partial, and characters that are actually homoplastic in the most parsimonious tree may appear initially to be homologous. Thus, a taxon position optimal for all of the evidence might not appear as optimal for only part of it (Goloboff 1998).

Phylogenetic Algorithms

There are now scores of phylogenetic algorithms available for finding trees under the optimality criteria discussed above. The existence of variant approaches to solving essentially the same problem derives not from a plethora of optimality criteria but rather from the fact that finding exact solutions is unfeasible for all but relatively small-sized data sets (i.e., no more than 15–20 taxa). Finding the best trees for large data sets, particularly those with substantial homoplasy, is *very* difficult. Thus, as one might expect, there are many ways to implement solutions, and finding ways to speed up the tree-searching process is an area of active, ongoing research.

The following descriptions of algorithmic approaches to solving phylogenetic problems are general. More elaborate sources are available (e.g., Goloboff 1994, 1996; Swofford et al. 1996; Felsenstein 2004; Chen et al. 2014; Yang 2014; Warnow 2018) for those who are interested. Note that the mechanics of these algorithms are a means to the end of inferring an optimal tree, and improving the efficiency of the algorithms does not equate to the discovery of "better" trees. Finding optimal trees can be, in part, a function of the thoroughness of the search, but, more importantly, is determined by the optimality criterion selected.

Algorithms can be divided into two classes on the basis of the solutions they produce: *exact* versus *heuristic* (approximate).

Exact-solution algorithms are of two types. First, *exhaustive search* algorithms examine every possible cladogram for a given set of taxa. Given the geometrical increase in possible trees as the number of taxa grows, as a consequence the maximum number of allowable taxa in the data set will probably be 12. Second, *branch-and-bound* (implicit enumeration) algorithms perform exhaustive searches within limits (bounds) established during the process of searching for minimum-length cladograms rather than examining all possible cladograms.

A brief explanation may be helpful for understanding how the branch-and-bound approach works. A fully bifurcating tree with n taxa has $2n - 3$ branches. Generating all possible trees would require that the first three taxa be joined, then a fourth taxon be added in the three available positions, then a fifth be added in the five available positions, and so on. The length of the tree would be calculated for all possible topologies as each new taxon is added to the ever-growing tree. Because the length of a tree can never decrease as more taxa are added, if the length of any partial tree exceeds that of the best complete tree(s) found so far (say four taxa produced a tree longer than five), it is unnecessary to continue with the addition of more taxa on the topology represented by that longer partial tree. This approach allows for the rejection of many trees by examining/evaluating only a few partial trees. For example, if the lengths for two positions for the fourth taxon in a partial tree exceed the length for the entire tree when the fourth taxon is attached at the third possible position, this will allow for the rejection of all the cladograms derivable from the former two topologies: two-thirds of all possible topologies can be rejected as unparsimonious just by optimizing the subtrees formed by the three possible addition points for the fourth taxon. In this way, the method can enormously reduce the number of cladograms that need to be examined/evaluated, thereby increasing the number of taxa to at least 18 for which a guaranteed optimal solution can be found (see also discussion in Swofford et al. 1996).

Beyond 18 or so taxa we must rely on *heuristic (approximate solution) algorithms*. The simplest among these is the *Wagner algorithm*, which creates trees de novo by successively adding taxa to a growing network. In its most basic implementations, the Wagner algorithm is very fast but may not produce most parsimonious trees for data that are not highly congruent, as illustrated in the example in Figure 5.5. Nonetheless, trees produced by a simple Wagner algorithm may be used as a starting point for finding more optimal results.

Traditionally, trees shorter than those produced by the Wagner algorithm were found through the use of branch swapping. *Branch swapping* is a process in which branches on a tree are moved in an attempt to find more promising topologies—that is, more parsimonious trees. Two approaches are commonly applied: *subtree pruning-regrafting* and *tree bisection-reconnection*. The latter approach examines more trees and therefore takes longer, but it is also likely to produce better results. The efficiency and effectiveness with which branch swapping finds shorter trees will depend on how close the input tree is to minimum length, the effectiveness of the branch-swapping algorithm itself, the amount of homoplasy in the data, other aspects of data structure, and—of course—the speed of the computer. Detailed discussions of algorithmic approaches can be found in Goloboff (1996), Swofford et al. (1996), and Gladstein (1997).

Because branch swapping does not produce an exact result, one might wish to have a method for determining when a search has found an optimal tree. This can be done as follows: If the swapper performs 20 replications (distinctive taxon addition sequences) and all of them find the same tree(s), then one can be reasonably certain that the shortest tree has been found. If, however, the swapper performs multiple replications, and one or a few replications find the shortest tree(s) while most of the others find trees of different (greater) lengths and topologies, then additional replications may be advisable to further explore the "tree space."

Even branch swapping, as described above, has limitations. Kevin Nixon (1999) and Pablo Goloboff (1999) have described additional algorithmic tools for finding shortest trees. Among these are the parsimony ratchet, tree fusing, sectorial searches, and tree drifting. These approaches, separately or in combination, allow for the critical analysis of very large data sets in a limited time frame.

The *parsimony ratchet* reweights subsets of characters on trees discovered in simple heuristic searches and uses them as starting points for additional searches that may discover trees that are shorter when the original weights are restored (discussed below in Islands of Trees and Solutions for Very Large Data Sets).

Tree fusing employs an approach wherein subgroups of the same composition are exchanged between different trees.

Sectorial searches analyze separately different sectors of a given tree. Goloboff (1999) described it as a special form of rearrangement evaluation. As opposed to branch swapping, the approach is able to analyze these tree sectors more rapidly than would be the case with the movement of individual branches.

Tree drifting is an extension of branch swapping but places limits on the suboptimal solutions that are accepted for further analysis. Under this approach, the speed with which improved solutions can be found is greatly increased.

According to Goloboff (1999) the tree fusing method can produce optimal trees only when very large numbers of suboptimal trees, or a few nearly optimal trees, are fed to it. Those trees can be found most readily through the use of sectorial searches and tree drifting. Thus, these algorithms, when employed in combination through the program TNT (Goloboff et al. 2008) bring a new level of efficiency to the analysis of very large data sets.

No matter what heuristic algorithmic approach is employed, certain strategies applied beforehand may help speed up the analysis. Recommended strategies include the following:

1. Make sure that there are no taxa with identical character-state distributions or taxa that differ from one another only by autapomorphies. These are viewed as identical by the computer and result in multiple equally parsimonious trees that are stored for swapping. Remove redundant taxa prior to analysis to reduce the number of possible topologies and make the search more efficient. These taxa may be replaced later on without affecting the inferred tree length (if identical) or most parsimonious topology (if differing only by autapomorphies).

2. Assess the data matrix for taxa with a lot of missing data. Such taxa may be optimized to join different sister taxa in alternative trees, again

resulting in multiple equally parsimonious trees. A common source of difficulty is complementary missing data, such as taxon A with only morphology and taxon B with only DNA sequence in the data matrix. Removal of one or both of such taxa may improve resolution and speed of the analysis. See Brower (2018b) for a case study.

Most available phylogenetic programs contain algorithms that produce comparable results. They may not produce them in comparable times, however, for the reasons noted above. Sources of phylogenetic software are listed in the Appendix at the end of this volume.

Differential Character Weighting

In Chapter 4 and in the Optimization section above, we discussed differential costs of character-state transformations. It is important to make a distinction between differentiated transformation weights among the states of a single character and differentiated weights applied to different characters in a data matrix. Quantitative cladistic analysis lends itself to weighting, some optimality criteria involve weighting, and most computer phylogenetics programs make provision for weighting. Thus, it will be helpful to gain some perspective on this subject, which has caused much intense discussion and considerable confusion.

First, let us dispel the notion that it is possible to construct an "unweighted" character matrix. By virtue of selecting some characters and excluding others, the researcher is de facto giving many potential but rejected characters the weight of zero. Focus on character discovery among particular structures and their parts can also disproportionately emphasize the importance or weight of those structures in the resultant cladogram. Some examples of alternate character sources that could be differentially sampled include cranial versus postcranial morphology in vertebrates, larval versus adult morphology in holometabolous insects, and even "molecules versus morphology" in most organisms. Likewise, a character coded with n states will require at least $n - 1$ steps on the most parsimonious tree, so characters with more states have more "weight" in the analysis. Such weighting is an unavoidable consequence of the formalization of attributes of a group of taxa into characters and states. What is usually meant by an "equal-weighted matrix" is one in which all possible character-state transformations have a cost of 1.

But some systematists choose to apply differential weights beyond these implicit forms of weighting. Traditionally, such explicit character weighting has been viewed as being of two types: *a priori*, the weights being applied in advance, without consideration of the fit of the characters to a most parsimonious tree, or *a posteriori*, the weights being related to the fit of the characters in a preliminary, or perhaps prior, analysis. These approaches might also be labeled *subjective* and *objective* weighting, respectively.

A Priori (Subjective) Weighting

Weighting criteria such as relative morphological complexity, functional importance, adaptive significance, and probability of state transformation in DNA sequence data are inherently subjective. To understand why, we might consider the result of giving complexity higher weight than simplicity. In such a case, secondary loss might be weighted less because it represents a reduction in complexity. Nonetheless, loss characters can be excellent group-defining attributes, as for example, the absence of limb elements in snakes or the loss of wings in lice or fleas. Experience suggests not only that many apparently "simple" characters define groups perfectly but also that complexity is subject to varied conceptions depending on the observer.

An unvarnished approach to subjective character evaluation was expressed by Guttmann (1977:646): "Only those features and characters whose functions are known, and for which the value of the adaptational changes can be assessed, can be utilized in phylogenetic construction. A phylogenetic theory can neither be constructed by comparison of morphological configurations nor by character analysis." Functional systems provided the basis for Cuvier's system of embranchements (his five major groups of animal taxa) in the early nineteenth century (Appel 1987), but exclusive reliance on such characters was rejected by Geoffroy, Darwin, and others in subsequent decades.

Such a priori views are not restricted to discussions of morphological characters. A clearly stated vision concerning the primacy of molecular data was expressed by Sorensen et al. (1995): "Phylogenetic reconstruction using nucleotide sequencing is thought to be superior to, and definitely more objective than, that based upon morphology . . . This is because, in general, nucleotide substitutions are random, non-selective events, as opposed to trying to determine how to code and weight morphological characters, which are de facto a result of selection." If nucleotide substitutions were truly random, we might expect the results of their analysis to be without hierarchic structure. Yet, hierarchic structure is exactly what Sorensen and all others performing phylogenetic analysis seek. Note, also, that if variation in sequence data were random, then one would probably not want to differentially weight those characters with respect to one another, even if one did choose to throw out the morphological data. However, DNA exhibits complex patterns of nonrandom variation, as one would expect of the molecule that encodes the developmental program for building all those adaptive morphological features. As is emphasized in the section Separate versus Combined Analysis, Chapter 6, there is strong empirical support for the view that congruence exists between morphological and sequence data. When data from these two sources are analyzed simultaneously, the ratio of phylogenetic signal (synapomorphy) to noise (homoplasy)—that is, the ability to recover hierarchy from the data—is improved. Thus, the statements of Sorensen et al. and others concerning the factual superiority of sequence data would seem to represent little more than an opinion, and more particularly an opinion that is not supported by the evidence.

The quotations from Guttmann and Sorensen above offer mutually exclusive perspectives on systematic data, one suggesting that adaptive significance is of paramount importance, the other suggesting that valid character data must be nonadaptive. Such arguments invariably rely on a priori appeals to the value of characters. In our view, the "goodness" of characters as indicators of relationship cannot be judged a priori without theoretical presuppositions that introduce potential bias or subjectivity. Instead, a more objective criterion is to assess the value of individual characters with respect to congruence among the other characters in the data set. Otherwise, ad hoc hypotheses of adaptive significance and selective neutrality could be invoked ad infinitum. In our view, unless features are obviously homoplastic due to parallel evolution, adaptive convergence, or horizontal transfer, as detected by incongruence with other character distributions, their particular adaptive or nonadaptive qualities are irrelevant to their potential value as systematic characters. Thus, weighting strategies that apply "objective" criteria are the only ones to which we will devote further attention here. We will discuss weighting of nucleotide data further in Maximum Likelihood, below.

A Posteriori (Objective) Weighting

Three approaches discussed by Goloboff (1993) in his review of character weighting can be labeled as "objective."

Weighting based on character compatibility. Characters are *compatible* with one another if they do not imply support for conflicting topologies. The most obvious implementation of this weighting approach is "compatibility analysis" itself (see Sidebar 13 in Chapter 6), as first proposed by Le Quesne (1969; see also Sharkey 1989). *Compatibility analysis* requires that all characters used to construct a phylogenetic scheme be perfectly consistent; that is, all must be in absolute agreement concerning the pattern of relationships they support. Such an approach could be described as an all-or-none weighting strategy. As the level of homoplasy increases, the level of compatibility decreases, such that the method explains fewer and fewer of the original observations. Perhaps it is for this reason that as data sets have grown larger and more homoplastic, use of compatibility-based methods has declined. Goloboff has noted that because characters with fewer incompatibilities may nonetheless be more homoplastic, a compatibility-derived weighting scheme is not necessarily based on a global best fit between characters and trees, a violation of all parsimony-based optimality criteria.

Successive approximations weighting. Goloboff discussed two varieties of weighting based on character consistency, in which the more homoplastic characters are those that receive lower weights. Probably best known among this type of approach is *successive approximations weighting* (SAW), first proposed by Farris (1969) as a method of evaluating the strength of characters as they define a given tree. The method functions as follows:

1. Compute the most parsimonious tree(s) for a data set with equally weighted characters.
2. Weight the characters in the data set relative to their agreement with the cladogram(s) in question, for example, by the value of their consistency index (discussed further in Chapter 6).
3. Recompute the tree(s) based on the characters with their newly assigned weights.
4. Continue steps 2 and 3 iteratively until the weights become stable (i.e., do not change from one computational iteration to the next).

This approach assigns equal weights to the characters for purposes of computing the "input" (initial) tree(s). Successive approximations weighting then determines the values (weights) that *should* be assigned to the characters on the tree(s) that the equally weighted characters defined.

If multistate characters are included in a result subjected to SAW, they should be recoded in additive binary format. Under this approach, consistencies for all state changes are evaluated independently, and the weighting function will therefore be applied on a comparable basis for all information on character transformation (Farris 1969; Carpenter 1988a). It is not clear, however, that it would be practical or meaningful to recode DNA sequence data into additive binary format.

Originally, SAW was defended by Farris as being consistent with the logic of parsimony because the weights are assigned based on the consistency of the characters. The procedure does not, however, implement a consistent *optimality criterion* by which the fit of data to a tree can be measured directly: SAW can be applied only to data on the basis of a cladogram produced by prior analysis with those data (see further discussion in Chapter 6).

As stated above, weighting based on character consistency assumes that all characters were given equal weights at the outset, in order to discover an initial set of consistency values that then potentially change as the analysis is iterated. This may be viewed as a parsimonious approach in the absence of evidence to the contrary. The validity of the assumption of prior equal weights can be scrutinized through the application of SAW, but the cladograms produced by that method are often not changed, only "filtered" (selecting one from among a set of equally most parsimonious trees found by the initial equal-weight analysis). The reasoning behind the equal weights assumption has, however, been challenged by Goloboff (1993) on the grounds that all characters do not contribute equally in determining cladogram topology and therefore should not be given equal weights for the purpose of computing cladograms. For this reason, he proposed a second approach based on character consistency.

Implied weighting and fittest trees. Goloboff (1993) extended the logic of weighting on the basis of consistency to an approach that selects a preferred cladogram according to the "implied weights" of the characters as they are distributed on alternate topologies. Goloboff (1993:85) argued that this approach is superior to SAW by noting that the *fittest tree*, the one that is shortest under the weights it implies, is *self-consistent*, resolving conflicts among characters in favor of those that have less homoplasy and therefore provide stronger phylogenetic signal. Conversely, a tree that is not shortest under the weights the characters imply is *self-contradictory*, because conflicts are resolved in favor of those characters the tree is telling us not to trust (i.e., the ones with the greatest amount of homoplasy).

The idea of self-consistency also applies to the results of SAW, where the iterative weighting procedure is repeated until a stable result is achieved. That result is self-consistent, with the weights of characters leading to a preference for the tree they imply. Although SAW has generally been applied in cases in which a parsimony analysis produces multiple trees (see Chapter 6), as pointed out by Goloboff (1993), self-consistency of a single tree might just as reasonably be evaluated through the application of SAW. The starting point for trees subjected to SAW is based on calculations using characters given prior equal weights, whereas when searching for fittest trees under the implied weights procedure, weights are assigned during the tree-searching process. Whereas SAW will almost always reduce the total number of trees to be considered from the number produced by an equal weights parsimony analysis (see also Chapter 6), the implied weighting approach generally produces a limited number of trees as part of the analysis itself.

The weight of a character, then, can be directly related to the fit of that character to the tree, as measured, for example, by the consistency index. Under Goloboff's approach, the value to be maximized is *total fit*, the sum of the fits of all characters to a tree. To relate fit to homoplasy, Goloboff's implied weights approach requires the application of a weighting function to downweight homoplasious characters, as does SAW. If that function is linear, it treats the same step difference as equally important regardless of the amount of homoplasy (Figure 5.6a), which would be the same as if all characters were given prior equal weights. This might be called the traditional approach under parsimony. If all characters were given higher prior weights, the slope of the line would become steeper, but the weighting function would not differ for characters with different amounts of homoplasy.

Goloboff argued that when calculating fittest trees, as Farris (1969) had for SAW, a concave weighting function (Figure 5.6b) appeared to be the most desirable because it gives the highest relative weight to characters with maximum group-defining value. Such a function gives distinctly and conspicuously lower weights to characters with limited group-forming value (i.e., those with the most homoplasy). Under this criterion, the addition of an extra step to an already highly homoplastic character changes the value of the character much less than adding an extra step to a character with no homoplasy at all.

The remaining possibility, a convex weighting function (Figure 5.6c), produces what Goloboff referred to as the "absurd" possibility of weighting characters with more homoplasy more heavily than those that are perfectly consistent with the

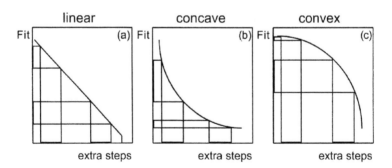

FIGURE 5.6.   Types of weighting functions. (a) Under a linear weighting function, the same step difference is equally important regardless of the amount of homoplasy (i.e., fit and homoplasy have a linear relationship, therefore the slope of the line changes with the weights of the characters but not with respect to homoplasy). (b) Under the concave weighting function, the value of a character decreases relatively rapidly as homoplasy increases from its minimum value and decreases slowly as homoplasy approaches its maximum value, as seen from the slope of the curve. (c) Under the convex weighting function, the value of a character decreases relatively slowly as homoplasy increases from its minimum value and increases rapidly as homoplasy approaches its maximum value, as seen from the slope of the curve (adapted from Goloboff 1993).

tree they define. The *consistency index* (see Chapter 6) is a concave function and therefore provides the initial basis for the weighting function. The concavity of this curve, and therefore strength of the function, may be greater than one wishes, however. Thus, a *constant of concavity* (k) can be added to change the severity of the concave weighting function. A value of 0 for the constant of concavity maintains the concavity of the consistency index, whereas higher values reduce the strength of the weighting function. The "fittest trees" method of Goloboff uses either Farris or Fitch optimization to calculate state assignments and measure the homoplasy for a given (single) character, exactly as for finding shortest trees. The fittest tree(s) may not be the shortest tree(s) because fit maximizes the sum of the average unit consistency index (ci), as shown in the following formula:

$$\sum ci = \sum \left(\frac{m}{s} \right)$$

whereas the shortest tree maximizes the ensemble consistency index (CI):

$$CI = \frac{\sum m}{\sum s}$$

In both formulas, m is the minimum number of steps for a character on the cladogram, and s is the actual number of steps for the character on the cladogram. Both Goloboff's method and SAW often lead to discovery of a single tree, or a reduced set of trees, compared to an equal-weighted parsimony search. We will further discuss the consistency index as a measure of fit in Chapter 6.

Goloboff et al. (2008) executed a large-scale study to examine the effects of weighting on phylogenetic results. These authors used the implied weights approach under a range of concavity values (k). Their results showed that jackknife resampling frequencies (see Jackknifing, Chapter 6) increased when characters were downweighted according to their homoplasy; that is, more highly homoplastic characters received lower weights. The highest jackknife scores were achieved under a weak constant of concavity (10–20), as opposed to the stronger values often used with implied weights or successive weighting (e.g., k = 3). This result implies a more consistent phylogenetic result for a given data set than would be obtained if the data were not weighted against homoplasy.

Finally, we note that despite concerns about subjectivity of a priori weighting discussed above, efforts such as those described here have to a large extent been eclipsed by model-based approaches. The latter employ weights via a priori specified character-state transformation costs, claimed by proponents to be more "realistic" than parsimony-based weights. We discuss these issues further in the Maximum Likelihood and Bayesian Inference sections, below.

Islands of Trees and Solutions
for Very Large Data Sets

The use of branch-swapping algorithms may produce less than optimal results for some data sets (Maddison 1991) because once an algorithm has found a promising topology—a local optimum or "island"—it may neglect other equally promising topologies simply by virtue of its mode of operation. Many tree-searching algorithms are "greedy," which means that once they find a topology of a certain length, they retain and compare only other trees that are the same length as that tree and discard longer trees. This can lead to a sort of localized myopia if tree space is relatively flat with multiple optima. The most effective solutions to this problem were initially conceived to be (1) the use of different starting points (i.e., a tree slightly longer than the shortest tree found without the use of exhaustive branch swapping or a series of "random addition sequence" starting trees) from which to begin branch swapping, and (2) the use of more exhaustive approaches to branch swapping. The former solution has been the more widely employed. Nonetheless, because of the island phenomenon, it is possible that starting to swap on a tree of 100 steps will result in being stuck on a suboptimal island, whereas starting to swap on a randomly generated tree of 1000 steps might ultimately produce shorter trees. The result in the latter case would, however, almost certainly take much longer to produce.

When data sets become large, say more than 200 taxa, the total number of possible solutions is immense, and thus recovery of any solution—let alone effectively surmounting the island phenomenon—can be problematic. One description of the computational state of affairs as of early 1998 was offered by Soltis et al. (1998). These authors showed that combining data sets increased the amount of "signal" in the data and that solutions could therefore be found in days rather than the weeks or months that had previously been required with the software they were using. Regardless of improvements in computational power, the speed with which most phylogenetic algorithms arrive at solutions is directly related to strength of phylogenetic signal; this property had long been known, although it was not mentioned by these authors. What Soltis et al. found, however, was that whereas individual DNA-sequence data sets with little phylogenetic signal often precluded achieving a computational result, combining these data sets increased the phylogenetic signal, decreased computational times dramatically, and found optimal solutions. This counterintuitive result—that a larger data matrix is easier to analyze than smaller ones—may not be a general phenomenon: it likely depends upon the particular attributes of the data set in question.

Some authors have suggested reverting to what we view as theoretically inferior methods of analysis for cases in which available parsimony algorithms were apparently incapable of processing the data at hand. An example is the use of "Neighbor Joining" (Saitou and Nei 1987)—a phenetic technique characterized by Swofford et al. (1996) as purely algorithmic—for no other reason than it produces a single tree in a limited period of time. Neighbor Joining enjoys this favor despite significant flaws: it requires that discrete data first be converted to distance data, the form of the tree is influenced by the order of taxon input, and the single tree produced can be suboptimal, whereas in fact many optimal trees may exist for the data set, as was shown by Farris et al. (1996). Sometimes, a lack of complete resolution may simply reflect lack of evidence, or existence of conflicting evidence, rather than being a limitation of a particular method. For further discussion and examples, see discussion of DNA barcoding in Chapter 7.

A heuristic subsampling approach to addressing the problem of very large data sets was proposed by Farris et al. (1996). Instead of relying solely on the effectiveness of branch swapping to find solutions, their "Jac" program used jackknife resampling procedures (see Chapter 6) and a fast parsimony algorithm to implement large numbers of computations in a relatively short period of time. The results, in the form of a consensus, include only groups that are well supported by the data and exclude cladograms containing groups totally lacking in support. As noted by the authors, poorly supported groups cannot survive resampling and are therefore automatically eliminated from the results. The rationale for the Jac approach was that the use of conventional parsimony algorithms required weeks of computational time for some data sets, and the consensus of the thousands of trees resulting from the parsimony analysis was very poorly resolved. The Jac approach produces results that include only well-supported groupings that would be useful as predictive classifications. It also provides a relatively efficient way to find a better "starting tree" for branch swapping on very large data sets.

Several authors have implemented heuristic searching shortcuts that speed up the process of discovering shortest trees and which essentially obviate the use of the Jac approach described in the foregoing paragraph. Nixon (1999) has shown that the parsimony ratchet, a weighting method that can be easily implemented with existing phylogenetic software, has the ability to produce solutions for very large data sets much more rapidly than traditional searching methods. The parsimony ratchet is implemented through the following steps:

1. Generate a starting tree (e.g., a Wagner tree), followed by some level of branch swapping.
2. Randomly select a subset of characters, each of which is given additional weight (e.g., add 1 to the weight of each selected character).

3. Perform branch swapping (e.g., tree bisection-reconnection) on the current tree using the reweighted matrix, keeping only one or a few trees.

4. Set all weights for the characters to the "original" weights (typically, equal weights).

5. Perform branch swapping again on the current tree (from step 3), holding one (or few) trees.

6. Return to step 2. Steps 2–6 are considered to be one iteration, and, typically, 50–200 or more iterations are performed. The number of characters to be sampled for reweighting in step 2 is determined by the user. Weighting between 5 and 25 percent of the characters provides good results in most cases.

Because the parsimony ratchet samples many tree islands with a few trees from each island, it provides much more thorough evaluation of tree space than collecting many trees from few islands. Using the parsimony ratchet in combination with readily available microcomputer hardware and Goloboff's TNT parsimony program can produce results in a few hours, whereas those same analyses previously required months or years for analysis, as was the case in the study of Soltis et al. (1998). Comparative test runs indicate efficiency increases of 20–80 times over what was previously possible. A concurrent decrease in the length of optimal trees is also often observed. Note that owing to the NP-complete nature of optimality-based searches, constraints of 1998 remain constraints today.

Special Considerations When Dealing with Molecular Data

In Chapter 3 we introduced the concept of optimization alignment (or direct alignment) as implemented in the phylogenetics program POY. Under this approach the process of tree formation and the optimization of nucleotide homologies are performed as part of a single operation. Application of this approach is most obvious in those cases where gene regions are length variable, although it need not be restricted to the analysis of such data. The algorithmic approaches used in POY are not radically different than those described above for other parsimony-based programs.

The rationale for employing POY can be viewed from two perspectives. If the insertion of gaps, the extension of gaps, and the substitution of nucleotides (either as transition or transversion events) are all viewed as having equal costs, then POY can be employed in a fashion similar to what might be done when using other phylogenetic programs. Faivovich et al. (2005) described such an approach in detail in their analysis of hylid frog relationships. As stated by those authors, the theoretical justifications for the application of different costs for opening a gap and extending a gap are not obvious. Faivovich et al. apparently felt likewise about applying different costs to transition and transversion events.

If one wishes to view transversion events as being less likely than transition events, and if one wishes to assign differential costs to opening gaps, extending gaps, and deleting nucleotides—as has been argued by many authors—then one might wish to employ POY using a "sensitivity analysis" (W. C. Wheeler 1995). Under this approach, data are analyzed under a matrix of possible parameters, allowing for the assignment of differential costs to the various types of changes discussed above. The most "desirable" result under such an approach is not self-evident, however, such that some measure of optimality that spans the range of parameter settings must be applied. This has usually been the incongruence length difference (see Chapter 6) or a variant thereof, such as the meta-retention index (Wheeler et al. 2006b).

Rooting: Additional Discussion

When transformation costs are symmetrical for a given data set (as is almost always the case in phylogenetic analyses today), different positions of the root for a given network are exactly the same; that is, alternative rootings are transparent to the parsimony criterion and produce trees of the same length under "standard" parsimony or equivalent fit under Goloboff's method of implied weights. Therefore, the explanatory power of all possible rootings of a given network is the same.

To understand this concept, consider the following unrooted tree, which is depicted graphically in Figure 5.7:

(0 (0 (1 (1 (1 (1 (0 0)))))))

Which state is "apomorphic," 0 or 1? The answer depends entirely on the position of the root, at which time we can read the polarities of each transformation off the tree. Note that no matter where this tree is rooted, the distribution of character states cannot be explained without at least two character-state transformations 0 ↔ 1. Patterns like this are common in sequence data, in which it is not obvious from inspection what the "ancestral" nucleotide at a given site ought to be. As noted in Chapter 4, the position of the root and the concomitant polarity of character-state transformations are usually fixed by including an outgroup established by recourse to a previous, higher-level cladistic analysis. In this instance, if the root is placed in the middle of the tree, with state 1 as

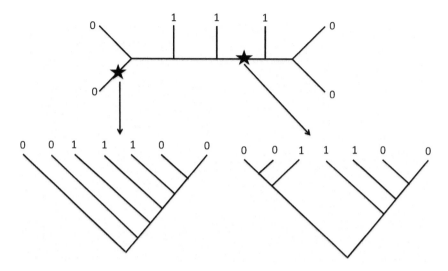

FIGURE 5.7. An unrooted tree (network), depicting graphically the character distributions indicated in the parenthetical notation in the text, and rooted trees for 2 of 11 possible positions for the root. Depending on where the root is placed, the implied ancestral state is either (left) 0 or (right) 1.

plesiomorphic, then there must be two independent gains of state 0. If the root is placed at either end with state 0 as the condition at the root, then the tree implies a reversal.

Alternative Approaches to Phylogenetic Inference

Three-Taxon Statements

A novel, parsimony-based approach to deriving phylogenetic results from discrete character data was proposed by Nelson and Platnick (1991) under the name "three-taxon statements." Under this method a given character matrix is recoded as the number of three-taxon statements the characters represent. The recoded matrix (Figure 5.8) can then be analyzed using conventional cladistic algorithms. Nelson and Platnick suggested that the method might represent a more precise use of parsimony and that it might in some cases reduce the number of most parsimonious trees or provide resolution of nodes that are ambiguously resolved by standard parsimony analysis. Discussion of the method can be found in Kitching et al. (1998), Williams and Siebert (2000), and Williams and Ebach (2008).

Matrix 1   Matrix 2

              Characters   Three-taxon statements

                                                   1   2   3   4   5   6

Taxa   1   2   3   4   5   6   Taxa   ab ab ab ab ab abc

A   1   0   1   0   0   0   A   11 ?0 11 0? 0? 000

B   1   0   0   0   1   1   B   11 0? 0? ?0 11 11?

C   0   1   1   1   0   1   C   0? 11 11 11 ?0 1?1

D   0   1   0   1   1   1   D   ?0 11 ?0 11 11 ?11

Original data:
(1)   (2)

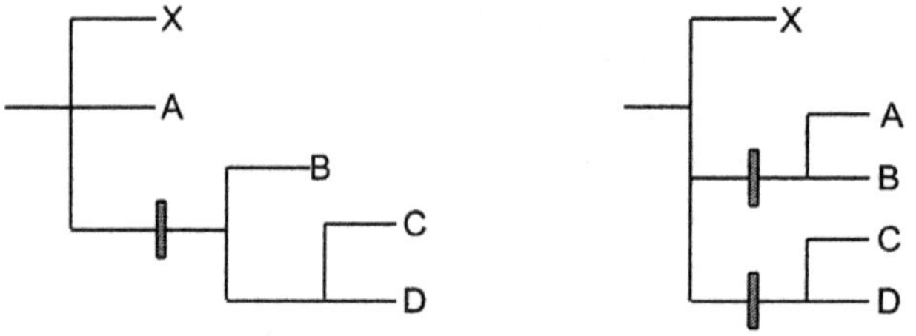

FIGURE 5.8. Coding of data for the three-taxon method. Matrix 1 equals original data; matrix 2 equals the same data coded for analysis by the three-taxon method (from Nelson and Platnick 1991). Each original character requires columns in the new matrix equivalent to the number of three-taxon statements it implies. Thus, character 1 implies C(AB) and D(AB), with the fourth taxon of each alternative coded as missing. Character 6, by contrast, implies 3 three-taxon statements, A(BC), A(BD), and A(CD). The original data produced two equally parsimonious cladograms. The three-taxon method produces only cladogram 1.

As with any alternative approach to widely accepted methods in phylogenetic analysis, the three-taxon method has not gone uncriticized (e.g., Farris et al. 1995a; de Laet and Smets 1998). The primary criticism by those authors was that the three-taxon method does not minimize the number of ad hoc hypotheses of homoplasy but instead treats units of evidence—characters and their states that are interrelated—as independent. Another problem with three-taxon methodology is that it ignores complementary implied groups (?001) in the expanded matrix, the only justification for which is that these are autapomorphies that do not imply groups. Of course, determining which state is derived requires prepolarization of character states, which, as just mentioned, is often not performed prior to tree search in standard parsimony. Although it was quite controversial from a theoretical perspective in the 1990s, the three-taxon method seems to have been seldom employed in analysis of empirical data. In spite of this, controversy continues (Farris 2012; Ebach et al. 2013; Williams and Ebach 2017).

Maximum Likelihood

A practicing systematist ought to understand the methods that she or he uses, and it is also important to understand the methods used by others to facilitate critical evaluation of results. Although we do not advocate maximum likelihood (ML) or other model-based methods in systematics, we describe them here to provide the reader with a basic understanding of their workings and assumptions. Interest in the ML approach to phylogenetic inference (introduced in Chapter 2) derives primarily from the work of Joe Felsenstein (1973, 1978, 2004), who was of the opinion that although few prior attempts had been made to apply statistical inference procedures to the estimation of evolutionary trees, ML (or some other statistical inference method) *must* be used to infer them (see also Cavalli-Sforza and Edwards 1967; Sober 1988; Swofford et al. 1996; Yang 2006; 2014; Warnow 2018). We, and many other proponents of parsimony, are not convinced why this should be the case. Nonetheless, with the advent of accessible software that implements its methodologies, ML has grown in popularity in the past two decades, particularly among those involved in the analysis of DNA sequence data.

The appeal of ML would seem to derive from the aura of quantitative rigor attributed to statistically based techniques for data analysis and the widespread enthusiasm for probabilistic models of sequence evolution among workers who gather and analyze such data. Proponents of likelihood argue that their methods provide a superior, or "more realistic" assessment of the fit between the models and the evolutionary phenomena they are intended to represent. Another superficially appealing attribute is that, owing to their elaborate character-state transformation weighting schemes, ML methods tend to produce one or a few optimal trees, a desirable result in the minds of many authors. In addition, support values attributed to nodes in model-based trees (see Chapter 6) are often higher than those in cladograms based on parsimony. By contrast, parsimony analyses often produce many trees, the consensus of which may be largely unresolved, a result that perforce suggests that internal nodes have lower support values. As noted earlier, lack of resolution in such instances may well be a more appropriate inference if the data do not support unequivocal resolution. It depends, we suppose, on whether the researcher is more concerned about failing to discover marginally supported hierarchical patterns or "discovering" them where they do not, in fact, exist. In this respect, parsimony may be considered the more conservative methodology and philosophy.

We now briefly describe the methodology of ML and then compare and contrast it to the cladistic parsimony approach. Calculating a maximum likelihood tree, as described by Swofford et al. (1996:431), involves the following steps:

1. Start with aligned sequence data.
2. Designate a model of nucleotide substitution based on some "plausible" starting tree (produced using parsimony or a distance-based algorithm such as Neighbor Joining), which may include such parameters as base frequencies, probabilities of different state-to-state transformations, the proportion of invariant sites, and a rate heterogeneity parameter for the sites that are not invariant.
3. Use a heuristic search algorithm to examine alternative topologies.
4. For each topology, compute the likelihood of the observed distribution of character states among taxa for individual nucleotide sites as the sum of the probabilities of every possible reconstruction of ancestral states, given the model.
5. Compute the likelihood of the entire topology as the product of the likelihood of all sites.
6. The preferred topology is the one that has the highest likelihood. This is a very small number, usually expressed as the easier-to-visualize log likelihood, in which case the optimal value recovered in the search is the negative number closest to zero (which, ironically, might be viewed as a "minimum").

As opposed to strictly algorithmic methods such as Neighbor Joining, ML and parsimony both select the preferred topology based on an optimality criterion that requires assessment of multiple alternative trees by iterative, often heuristic, searches, and both have the capacity to incorporate a variety of differential weights, either among characters and/or among states of individual characters. However, that is where the similarities between ML and the parsimony approach end.

Let us briefly examine some of the assumptions of ML analysis that differ from parsimony-based cladistic analysis. Fundamentally, because the ML optimality criterion is maximizing the likelihood of the tree given the model and the data, one must first have an explicit model of evolution for each character in the data in question. Model parameters are often based on average values for some quantity estimated from the data set, such as transition-transversion ratios and differential inferred rates of change among nucleotide sites. Other parameters may not reflect any empirical quantity at all. As Felsenstein (2004:219) said, "there is nothing about the gamma distribution that makes it more biologically realistic than any other distribution, such as the log normal. It is used because of its mathematical tractability." For a given model, then, it is not unreasonable to wonder, "how much is based on observation and measurement of accessible phenomena, how much is based on informed judgment, and how much is convenience?" (Oreskes et al. 1994:644).

Interestingly, several authors (Goldman 1990; Tuffley and Steel 1997; Steel 2002; Goloboff 2003) showed quite a long time ago that when models which treat all characters the same (an "equal weights" assumption) or models which treat every character differently (a "no common mechanisms" assumption) are employed, parsimony and ML give the same results. Thus, the ML models that give different results than parsimony are those that treat groups of characters (such as different codon positions in a protein-coding sequence), or different types of character-state transformations (such as transitions versus transversions), as homogeneous classes that differ from one another by simple numerical parameters. Swofford et al. (1996) and others have discussed the nature of the most commonly employed models, which, as noted, mainly relate to differential character-state transformation costs among nucleotides. Note that while characters and their states may be assigned different parameters, all taxa are treated the same: thus it is not feasible to model, for example, different base compositions or hypothetical rates of change on different branches.

Another important fact to recall is that ML is not "a model" but a framework for analyzing data employing a particular model that must be specified and may fit the data rather poorly (Gatesy 2007). Clearly the number of possible models is virtually infinite. Yet, the criteria by which this multitude of possibilities may be judged are not clear (see Sidebar 12), and in practice, most users of likelihood select 1 of 56 standard alternatives available in popular implementations such as PAUP* (Swofford 2003), PhyML (Guindon and Gascuel 2003), TREEFINDER (Jobb et al. 2004), or RAxML (Stamatakis 2006, 2014).

The more complex and parameter-rich a model is, the better a fit to the data it is likely to have, and the higher the optimal likelihood score for the preferred tree. However, a model with too many parameters is considered to have low statistical power and to be prone to error. A simpler model is therefore preferred if it provides an adequate fit to the data. In general, the choice among model alternatives is based on assessing the amount of increase in the likelihood score obtained by making the model more complicated. If this increase is deemed significant, then the more complex model is preferred. Various tests to compare models have been devised based on likelihood ratios, the Akaike information criterion, or other methods (Posada and Crandall 1998; Alfaro and

Huelsenbeck 2006; Ripplinger and Sullivan 2008). Alternate model-selection criteria often favor different models for the same data, which in turn can lead to different ML trees. The ambiguity stemming from the lack of a single clear optimality criterion is reminiscent of the methodological multiplicity that caused phenetics to founder in the 1970s.

Sidebar 12
What Is "Branch Length"?

In cladograms, branch lengths, counted perhaps as the number of synapomorphies optimized as supporting a given clade, are not considered to convey any particular significance—the salient information is the branching order, which reflects the inferred pattern of sister-group relationships. In model-based trees, by contrast, the estimation of branch length is claimed to be the extra piece of evidence that makes such inferences superior to cladistic hypotheses of relationship. One might suppose, therefore, that branch length is a clearly defined concept. However, such is apparently not the case.

It has been suggested by some (e.g., Swofford et al. 1996) that branch length is equivalent to time. However, except on ultrametric trees (see Chapter 11), this can certainly not be true: the entire point of emphasizing branch length as a quantity worth worrying about is that the length of branches of sister taxa might not be equivalent (which, of course, they would be, if branch length equated to time since divergence). Felsenstein (2004:70) said, "Branch lengths are numbers that are supposed to indicate for a given branch how many changes of state have occurred in the branch." This tepid statement is quite similar to our own glossary definition. It is telling that even one who believes the estimation of branch lengths to be integral to phylogenetic inference did not provide an unambiguous definition of the concept in his magnum opus.

As noted, on a parsimony tree, such a quantity might be measured by the number of unambiguous synapomorphies that occur between two consecutive nodes, or the number of autapomorphies on the branch leading to a terminal. However, even a straightforward count of inferred steps could be complicated by alternative optimizations of homoplastic characters. In the simple case introduced in Figure 4.1, for example, if character 5 is optimized as independent gains of state 1 in taxa B and D, then the branches leading to B, C, and D exhibit 2, 1, and 3 steps, respectively. Alternatively, if character 5 is optimized as a reversal, then the unsupported node (BCD) gains a synapomorphy, and the branch lengths of B, C, and D are 1, 2, and 2. The topologies are identical, but the branch lengths are different in these two scenarios. The more homoplasy in a data set, the more opportunities exist for such optimization ambiguities that lead to alternative branch length scenarios. And that's just in the simplest case, where there is a single most parsimonious tree.

In ML and Bayesian approaches, a model is intended to correct for potentially spurious apparent branch lengths via the Procrustean exercise of shortening empirically long branches by downweighting the characters that make them unusually transformation rich and at the same time lengthening short branches by hypothesizing unobserved character-state transformations. Different models assign different weights to different types of transformations, and so branch length becomes a function of the particular model selected rather than a parameter of the evidence. Such quantitative hairsplitting might improve precision (models often yield a single tree when multiple equally parsimonious trees are inferred from the same data), but precision is not accuracy, and there is no reason to believe that statistical models generate results that are any more defensible as "realistic" or "true."

The deterministic treatment of character-state transformation weights in evolutionary models employed by ML is problematic for several reasons. *First*, aside from base frequencies and proportion of invariant characters, which can be determined directly from the data matrix, most of the model parameters are topology dependent. Thus, the evidence used to assign relative weights of transformation types and to determine branch lengths—to build the model—comes from the pattern of character-state transformation implied by some initial topology, the provenance of which is rarely even mentioned in works employing model-based methods. The model, in turn, determines the transformation probabilities for all sites, for all taxa, for all subsequent analyses. Even if the initial tree is a good one, there is clearly a degree of circularity here, not to mention an inconsistent shift in optimality criterion between the model-construction and tree-inference phases of the analysis. Some have defended this practice as a form of "reciprocal illumination" (e.g. Baum, 2017). It is in actuality more akin to a successive approximations approach, although one that treats characters as belonging to uniform classes with stereotypical properties rather than as individual sources of evidence.

Second, as noted above, ML models tend to contain relatively few parameters and thus are not able to "realistically" represent the intricacies of evolution in any testable manner (Felsenstein 1978). Maximum likelihood does not perform well, for example, when evolution occurs at different rates in different lineages, a not implausible eventuality (Kolaczkowski and Thornton 2004). Wenzel (1997:36) observed, as have others, that "methods [such as ML] are self-fulfilling and do not provide independent evidence of their legitimacy. Trees built according to a given model cannot refute the model, and therefore it is clear that the model . . . is beyond testing." Parsimony-based cladistic methods, on the other hand, evaluate the evidence transparently by comparing implied numbers of character-state transformations and test the plausibility of individual hypotheses of homology via character congruence (see Chapter 2). As pointed out by Siddall and Kluge (1997), even if evolution did not occur, parsimony would still be applicable to the inference of a hierarchy reflecting the Natural System (although not "justified," according to Kluge [2001]).

Third, although models of change for molecular sequence data can be easily constructed, if not easily tested with regard to their realism or reality, the same cannot be said for models of change for morphological characters. For example, on what basis does one determine the probability of origin of a novel derived character, for example, insect wings? Or change from one state of a morphological character to another, for example, tetrapod forelimbs to wings? And how would one compare probabilities across characters, especially when our empirical observations make it abundantly clear that amounts (and presumed underlying rates) of change for different characters *are* different? Lewis (2001) has developed a simple ML model for morphological characters, but this is based on the Tuffley–Steel approach mentioned above, so its results would apparently mirror those achieved by parsimony analysis. The means and rationale for applying more complex ML methods across the spectrum of available character data types remain unspecified.

Fourth, while ML methods usually produce as output a single putatively optimal tree for a data set that may yield multiple equally parsimonious trees, this does not necessarily mean that ML is providing a more precise (or accurate) interpretation of the evidence. Indeed, Steel (1994) and subsequent authors (Chor et al. 2000; Zhou et al. 2006) have shown that under a given model, there may be multiple likelihood maxima that imply different topologies. As Zhou et al. stated (p. 200), "The ML criterion alone may not be sufficiently enough [*sic*] to determine the true phylogeny for certain problems even if we are able to obtain a truly globally optimal tree under the ML criterion!"

Finally, we note that the greater complexity and increased quantification of "modern methods" is not per se a virtue. Model-based analyses rely on their purported higher degree of realism for whatever improvements they are believed to exhibit over parsimony analyses. But since there is no way to corroborate independently the degree of realism of a ML model, this is merely a subjective preference. Cladists use parsimony not because they think it is a realistic model of character-state transformation, nor because they naively believe that parsimonious trees are true representations of phylogenetic history, but because parsimony provides the simplest—and most informative—explanation of the pattern of change when the true process that caused the pattern is unknown (Farris 1983). In the cladistic framework, character-state transformations can be mapped onto a cladogram, the amount of support or homoplasy a character adds to a given topology can be easily assessed, and its evidential meaning is readily apparent. The evidentiary support for ML trees, by contrast, is enshrouded in a fog of statistical parameterization that obfuscates the relationship between the topology and the evidence (and model) from which it was inferred. This indictment may be readily supported by the simple exercise of attempting to reproduce the results of a published empirical ML analysis. All too often, the methods are so inadequately described and ambiguous that this basic tenet of science and desideratum of the early pheneticists—repeatability—is not feasible.

Bayesian Phylogenetic Inference

Because both branch lengths and substitution parameters must be estimated for every alternative topology, ML is much more computationally expensive than parsimony, and large data sets cannot be analyzed easily with parameter-rich, "realistic" models (Swofford et al. 1996; Yang 2014). The Bayesian approach to phylogenetic inference uses ML as its optimality criterion but avoids the extensive calculations by employing an approximation technique called the Markov chain Monte Carlo method, usefully reviewed by Huelsenbeck et al. (2002), Holder and Lewis (2003), and Nascimento et al. (2017). The preferred Bayesian topology is the one supported by the distribution of sampled trees with the highest posterior probabilities, a value that is calculated (or approximated) by evaluating the data in light of subjective prior probability values for the substitution model, branch lengths, and even the topology itself. The same flaws described above relating to ML character-state transformation models thus apply equally to the Bayesian approach. Additional criticisms of Bayesian phylogenetic inference relate to problems with assumptions of flat prior probabilities of topologies (Pickett and Randle 2005) and apparent overestimates of branch probabilities (Suzuki et al. 2002; Simmons et al. 2004).

As Nascimento et al. (2017:1446) noted (in a review *advocating* Bayesian methods),

> Models implemented in Bayesian software are becoming increasingly complicated, and the priors and model assumptions made in those programs are not always clear to the user. Analyses are often conducted using default priors, which may not be appropriate and may lead to biased or incorrect results. Likewise, oversimplified likelihood models may produce biased results, while overcomplicated models may lead to loss of power as well as insufficient computation.

Finally, many researchers view Bayesian inferences of any kind as purely subjective statements of the individual beliefs of the person drawing the inferences rather than as objective claims about the nature of the evidence (Sober 2015). This philosophical incompatibility between likelihood and Bayesian inference is an issue that one might suppose would concern those who comingle the two approaches in their phylogenetic toolkits.

Statistical Inconsistency and Long-Branch Attraction: The Achilles' Heel of Parsimony?

For four decades, advocates of ML (e.g., Felsenstein 1978, 2004; Swofford et al. 1996; Warnow 2018) have rationalized their rejection of parsimony in favor of likelihood on the grounds that the former may be *statistically inconsistent*—a systematic error in which an incorrect result is arrived at with greater and greater statistical support with the addition of more data. The standard argument, introduced in Chapter 2, says that, given a particular distribution of character states, cladograms produced under the optimality criterion of cladistic parsimony may not accurately represent the true pattern of relationships among the taxa and support for a wrong topology will increase as additional characters are added. This is a theoretically valid claim. Swofford et al. (1996:429), in their discussion of the likelihood approach, offered the opinion that parsimony ignores information on branch lengths when evaluating a tree, whereas ML treats changes on long branches as more likely than those on short branches (see Sidebar 12). This "important component" of the method explains, according to Swofford et al. (1996:430), the "consistency" of the ML method in cases in which parsimony is inconsistent.

It should not be forgotten, however, in the evaluation of this argument, that parsimony does provide at least one empirical measure of branch length: the number of character-state transformations supporting a given internal branch (which, as noted, may vary depending upon preference for alternate optimizations of accelerated or delayed transformation). Nonetheless, the lengths of those branches are relevant to cladistics only to the extent that they represent the minimum number of character-state transformations occurring in a most parsimonious solution. Autapomorphies, by contrast, contain no phylogenetic information. As framed by Swofford et al., the "parsimony ignores branch length" argument is therefore irrelevant and misleading as a criticism of the parsimony approach.

The Swofford et al. argument might alternatively be interpreted as meaning that ML uses the information provided by all of the characters in terms of the length of the branch—that is, the "evolutionary rate" along the branch—while parsimony treats every character independently and thus ignores the information on rate change along a given branch. This point is also irrelevant as a criticism of parsimony—or conversely, as a presumed benefit of ML—for it was shown 40 years ago by Farris (1979) that path-length distances as applied under the parsimony criterion always store information on genealogy and similarity as effectively as possible. Thus, no information is lost in a parsimony analysis compared with a likelihood analysis, irrespective of the comments of Swofford et al.

As we have acknowledged several times before, the observation that parsimony *could* be statistically inconsistent is technically correct. No cladist has ever pronounced parsimony to be an infallible approach, and we suspect that most would restrict their claims regarding verisimilitude to epistemological statements about least falsified hypotheses. Nevertheless, to understand the gravity and scope of the inconsistency problem fully, we need to examine two issues that the ML advocates typically do not mention: (1) Under what circumstances is parsimony statistically inconsistent? and (2) Is ML immune to this problem?

As to the first point, the only identified scenario that can lead to inconsistency is "long-branch attraction" (LBA), whereby branches with large numbers of autapomorphic characters evolved in parallel on separate "long branches" group the terminal taxa bearing them together, even though those taxa are not actually each other's closest relatives. Figure 5.9 shows the paradigmatic four-taxon cartoon of this problem. If the "true tree" has a short internal branch and two long nonadjacent branches (tree 1), then independent homoplastic changes on each of these branches may be incorrectly (but more parsimoniously) interpreted as synapomorphies, causing them to form a group (tree 2). When additional characters exhibiting the same pattern are added, the "long branches" become united by increasing numbers of false "synapomorphies"—hence the inconsistency. However, if the two long branches are actually sister taxa (tree 2), then the parsimony method performs better than ML in discovering the "true tree," and ML can even behave inconsistently, preferring topology 1 when the true topology is 2, because of "long-branch repulsion" (Siddall 1998).

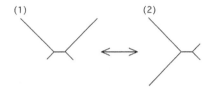

FIGURE 5.9. Cartoons of unrooted four-taxon trees, with branch lengths indicating relative amounts of character-state change. When the internal branch is short relative to the branch lengths of two nonadjacent taxa (1), parsimony may favor topology (2). On the other hand, maximum likelihood may favor (1) when the correct topology is (2).

The source of the inconsistency idea, like ML itself as a phylogenetic technique, sprang from the work of Felsenstein (1978), as noted earlier in this chapter. Although Felsenstein's inclination was to believe that the parsimony approach—in the sense that it is generally applied in this book—could produce erroneous results, he also stated as an alternative that "the conditions which must hold in order to have lack of consistency may be regarded as so extreme that . . . [this] result may be taken to be a validation of parsimony or compatibility approaches" (Felsenstein, 1978:402). Thus, even the prime mover of ML as a phylogenetic tool felt that in almost all cases, ML was unnecessary and redundant relative to parsimony (a theoretical prediction borne out by evidence—see below). Despite ongoing assertions of the pervasive and pernicious effects of LBA (Anderson and Swofford 2004; Bergsten 2005; Lartillot 2015; Kjer et al. 2016; Qu et al. 2017; reviewed in Brower 2018c), there remains some question as to whether LBA actually occurs or whether it has been assumed to occur so it may be invoked in the rather circular argument attempting to justify the likelihood approach.

The lack of substantiated cases (not to mention the lack of a clear understanding of what "branch length" even means—see Sidebar 12) suggests that LBA may be quite rare in empirical studies with appropriate taxonomic sampling. When apparent sister taxa with long branches are grouped together in phylogenetic analyses, all things being equal, the origin of the "accelerated evolutionary rates" of these lineages is more parsimoniously interpreted as a synapomorphy that unites them than as homoplastic parallel rate acceleration (Whiting et al. 1997). So, is parsimony forcing such taxa together, or is ML forcing them apart? There is no empirical means to determine whether parsimony or ML is behaving inconsistently when they favor different alternatives under such circumstances.

Although Anderson and Swofford (2004) stated that "few examples of LBA involving real sequence data are known," Bergsten (2005) reviewed LBA and argued that "presently suggested examples of LBA are no longer merely a few" based on its frequent invocation as a keyword in the Web of Science database. Of course, wealth of citations does not connote wealth of either argument or evidence: just because something is invoked frequently in the literature does not mean it is an actual, empirically supported phenomenon. Indeed, as suggested earlier, it may be that as a keyword, LBA was invoked more as an expectation than as an observation, as a ready-made justification for an a priori methodological preference.

Bergsten's concern stemmed from the perceived incongruence of a particular gene or molecular data partition with other evidence or a preconceived notion of what the "correct" topology ought to be (approaches to partitioning versus combining data are discussed in Chapter 6; the distinction between *species trees* and *gene trees* is discussed in Chapter 7). He explored case studies from mammalian orders and genera of cynipid gall wasps and provided an involved protocol for testing LBA as an explanation of the perceived incongruence. The goal of these machinations seems to be to rationalize discarding data—either taxa or characters—from the analysis in order to recover a preferred hypothesis. While we think that reexamination of the data can provide insights into improved character coding that may decrease homoplasy, we are not convinced that simply labeling perceived anomalous results as instances of LBA is anything more than an ad hoc excuse for retaining a favored unparsimonious alternative, in spite of the evidence. We reiterate that, in many cases, thorough taxonomic sampling and selection of characters appropriate to the analysis may diminish the unruly behavior of the data.

Even if we concede that LBA could be a problem for the parsimony approach (which the previous examples and discussion suggests is at best a controversial proposition), there is a second, more general problem implicit in the consistency issue. When we encounter the standard claim that parsimony may be inconsistent, while ML is not, we need to remember that "maximum likelihood" in such statements represents a class of methods, not any method in particular. The deliberately obfuscated assumption underlying this claim is that in order to be statistically consistent, the ML model selected must be one that accurately matches the true pattern of character-state transformation in the data. Pronouncing that ML is consistent is thus equivalent to saying that parsimony with the correct character-weighting scheme is guaranteed to yield the true tree—a clearly untenable truism. Many of the simulation studies purported to support the superiority of ML over parsimony (e.g., Huelsenbeck 1995; Yang 1997) have compared ML methods against parsimony using data simulated under the same ML model that was subsequently used to build the tree, effectively setting parsimony up to fail. However, when actual relationships among taxa are concerned, there is no way to know whether a particular model is correct for a given data set or not, leaving ML in the same predicament that its advocates claim afflicts parsimony. As Alfaro and Huelsenbeck (2006:95) stated, "Models that poorly fit the data can lead to inconsistent behavior of likelihood methods and produce poor or misleading estimates of all the parameters of the phylogenetic model including topology, branch lengths, and substitution rates, as well as influence bootstrap values and posterior probabilities" (embedded citations omitted). So, a given ML model may be just as theoretically prone to inconsistency as parsimony, and there is no way to tell whether either method is yielding "correct" topologies that correspond with reality.

In the absence of independent measures of metaphysical reliability, how can we assess the performance of one method versus another? Rindal and Brower (2011) conducted a meta-analysis of 1000 empirical articles in the journal *Molecular Phylogenetics and Evolution* to test whether model-based methods outperform parsimony, as claimed by many advocates of the former. They found that when parsimony and a model-based approach were used on the same data set, the results did not differ appreciably more than 99 percent of the time. While this result is no guarantee that either method has converged upon the truth, it clearly shows that if parsimony is routinely inconsistent, then so are the results from ML (or Bayesian) analyses.

To sum up, the oft-repeated but manifestly misleading claim of parsimony's potential for statistical consistency was originally conceived and continues to be employed to prejudice ignorant persons against the cladistic approach. Parsimony is only theoretically likely to behave inconsistently under particular, probably rare, biological circumstances, and ML models may also be prone to inconsistency (see further discussion in Brower 2016a, 2018c). Although it is possible to identify incongruence among characters or data partitions, there is no way to tell whether or not a particular method is behaving consistently with a given empirical data set—a claim relying on a correspondence theory of truth. As discussed earlier, ML-based methods possess several further inherent and conspicuous limitations, such as empirical opacity and the lack of a clear optimality criterion (or even a means of selecting one), and no advantages that parsimony does not also enjoy. In analyses of DNA data from actual organisms, model-dependent methods almost always produce results very similar to those of parsimony. And, paradoxically, unlike approaches used in other fields of science, where data inform theory, with ML-based approaches, theory informs data (see discussion under Statistics, Probability, and Models in the Historical Sciences, Chapter 2). Even after decades of opportunity to offer a sound philosophical reason for rejecting parsimony and adopting less parsimonious model-based methods, other than metaphysical arm-waving about "realism," the statistical inconsistency subterfuge remains the only criticism offered by parsimony's detractors to justify their less parsimonious, more assumption-laden alternatives. Given the demonstrable baselessness of the inconsistency issue and the lack of any other substantive empirical argument, we conclude that parsimony remains the theoretically and practically superior means to infer patterns of phylogenetic relationship.

Finally, one might imagine that if, as Rindal and Brower (2011) found, parsimony and model-based approaches almost always give similar results, that such findings might imply that agreement among competing methods (or alternative models) offers some sort of corroboration or greater chance that the inferred tree is "correct." Kim (1993) made such a claim, and syncretic beliefs of this sort are apparently widespread in the systematic community, to judge from the frequency with which multiple analytical methods are employed in published empirical studies. However, as Brower (2000a) and Giribet et al. (2002) argued, there is no more reason to put greater credence in a phylogenetic result discovered under different optimality criteria than there is to think that passing multiple statistical tests increases the probability that a hypothesis is true. Such confirmatory results are pseudoreplicated (Hurlburt 1984) and do not represent independent tests. Thus, one well-considered, carefully conducted and clearly articulated phylogenetic method is sufficient. Naturally, in our view, that method should be parsimony.

SUGGESTED READINGS

Farris, J.S. 1970. Methods for computing Wagner trees. Systematic Zoology 19:83–92. [The classic paper on phylogenetic algorithms]

Farris, J.S. 1983. The logical basis for phylogenetic analysis. In: Advances in cladistics. Vol. 2, Proceedings of the Second Meeting of the Willi Hennig Society, ed. N.I. Platnick and V.A. Funk, 1–36. New York: Columbia University Press. [A critical discussion of the logic of phylogenetic inference]

Goloboff, P.A. 1993. Estimating character weights during tree search. Cladistics 9:83–91. [A useful discussion of weighting in general and implied weights in particular]

Goloboff, P.A. 1999. Analyzing large data sets in reasonable times: solutions for composite optima. Cladistics 15:415–428.

Swofford, D.L., G.J. Olsen, P.J. Waddell, and D.M. Hillis. 1996. Phylogenetic inference. In: Molecular systematics, 2nd ed., ed. D.M. Hillis, C. Moritz, and B.K. Mable, 407–514. Sunderland, MA: Sinauer. [An extended and eclectic discussion of approaches to inferring phylogenetic trees]

Wheeler, W.C. 2012. Systematics: A course of lectures. Chichester, UK: Wiley–Blackwell.

EVALUATING RESULTS

Many systematists would like to know "how good" (well supported, robust, believable) their trees are. In this chapter we will examine the tools used to evaluate the quality and plausibility of results of phylogenetic analyses. Some are broadly applied and generally understood as valid. Others have been less widely applied, although compatible with the philosophy of science advocated in Chapter 2. Still others appear to have serious limitations or do not conform to the hypothetico-deductive philosophy we have advocated.

Fit Statistics

The term "fit" has been widely used in the phylogenetic literature to indicate the degree to which data conform to (or are explained by) a cladogram. A character with n states that has a "perfect fit" to a given cladogram has $n - 1$ steps. The most commonly used measure of fit applied to discrete character data is the *consistency index*, or ci, originally proposed by Kluge and Farris (1969).

This is a rather simple measure, but one of considerable value. It is computed using the following formula:

$$ci = \frac{m}{s}$$

where s is the observed number of character-state changes (steps) and m is the minimum possible number of changes given the number of states observed. Values for ci range from 1.00 for perfect fit to near 0.00 for the worst possible fit. An example calculation of the ci is shown in Figure 6.1.

The number of homoplastic steps in a character is then

$$s - m = h \text{ (homoplasy)}$$

The consistency index (ci) measures the degree of homoplasy of a single character. The "ensemble consistency index" (CI; Farris 1989) measures the cumulative homoplasy of the entire data set. The latter measure is simply the sum of the minimum number of possible changes for all characters divided by the sum of the observed number of changes for all characters in the most parsimonious tree. This value is frequently reported in the legends of figures illustrating empirical results and is often viewed as an indication of the quality of the data. That said, it has been shown that characters exhibiting many extra steps, and data matrices with low CI values, can still add resolution and support to a phylogenetic hypothesis (Källersjö et al. 1999).

Because the ci does not have a minimum value of zero, Farris (1988, 1989) introduced the "rescaled consistency index" (rci), which has a value of zero when a character has as much homoplasy as possible, a desirable scaling feature when used in conjunction with some character-weighting functions. Computing the rci requires determining the maximum number of observable steps, g, for a character on a cladogram. Note that although it is theoretically possible for a character state to change back and forth multiple times on a single branch over time, there would be no observable evidence of such an occurrence. The maximum number of observable changes is therefore an epistemologically parsimonious value. Calculating this value can be accomplished by the method shown in the bottom of Figure 6.1, whereby the maximum number of observable changes for a two-state character is equivalent to the number of terminal taxa in which the character can change. The rci is then computed as follows:

$$\text{rci} = \frac{g - s}{g - m} \cdot \frac{m}{s}$$

Mickevich and Lipscomb (1991) argued that *concordance* (congruence) between the transformation series and the hierarchy of the tree is a better measure of how well multistate characters fit a cladogram than the ci. However, there is no widely accepted formula or measure that can be applied for the concept of concordance (see Sidebar 13).

Characters (nonadditive) | Characters (additive)

Taxa	1	2	3	4	5	6		1	2	3	4	5	6
X	0	0	0	0	0	0		0	0	0	0	0	0
A	1	0	0	0	1	1		1	0	0	0	1	1
B	1	0	0	0	0	1		1	0	0	0	0	1
C	0	1	0	1	2	1		0	1	0	1	2	1
D	0	1	1	1	2	2		0	1	1	1	2	2
E	0	1	1	0	2	2		0	1	1	0	2	2
m	1	1	1	1	2	2		1	1	1	1	2	2
s	1	1	1	2	2	2		2,1	1	1	2	2,3	2
ci	1.0	1.0	1.0	0.5	1.0	1.0		0.5,1.0	1.0	1.0	0.5	1.0, 0.67	1.0

Tree length = 9 — Tree length = 10
CI (ensemble consistency index) = 0.89 — CI (ensemble consistency index) = 0.80

Computing character consistency index

Character 1 $ci = \dfrac{m}{s} = \dfrac{1}{1} = 1.00$ $ci = \dfrac{m}{s} = \dfrac{1}{1} = 1.00$ (tree 2)

Character 4 $ci = \dfrac{m}{s} = \dfrac{1}{2} = 0.50$ $ci = \dfrac{m}{s} = \dfrac{1}{2} = 0.50$

Character 5 $ci = \dfrac{m}{s} = \dfrac{2}{2} = 1.00$ $ci = \dfrac{m}{s} = \dfrac{2}{3} = 0.67$ (tree 2)

Computing ensemble consistency index, non-additive character coding

$$CI = \frac{\Sigma M}{\Sigma S} = \frac{1+1+1+1+2+2}{1+1+1+2+2+2} = \frac{8}{9} = 0.89$$

FIGURE 6.1. Computing consistency indices. The data matrices at the top of the figure provide data for six characters relating to an outgroup and five ingroup taxa; the most parsimonious trees for these data are shown below the matrices. On the left side, the character data are treated as nonadditive; on the right, they are treated as additive. The nonadditive data produce a single tree; the additive data two trees. The rows below the character indicate the values for m (minimum number of steps), s (actual numbers of steps), and ci (consistency index for each of the characters); also indicated are the length of the trees and the CI (ensemble consistency index). The states and nature of change in the characters are indicated on the trees; open circles represent reversals; numbers above the line represent character numbers; numbers below the line represent character states. The ci is then computed for characters 1, 4, and 5. Finally, the CI is computed for the tree in which the characters are treated as nonadditive.

Sidebar 13
Similar Terms, Similar Definitions: The Meaning of Compatibility

The degree to which character distributions are in agreement with one another is a central property in the logic of cladistics. This property is frequently described in the literature by the terms *congruence, concordance,* and *consilience.*

The etymologically similar term *compatibility* will also be found, but it has a distinct connotation. Compatibility is associated with the work of Walter Le Quesne (1969), who outlined the concept of "true cladistic characters" (see also Estabrook et al. 1977). Le Quesne's approach, which is usually referred to as "compatibility analysis" or "clique analysis," is based on the assumption that the best-supported groups are those that are defined by the maximum number of perfectly compatible characters, or *maximal cliques.* A clique of compatible characters perfectly defines the same hypothesis of grouping (i.e., none of them show any homoplasy in the preferred topology). At first glance, this approach may appear indistinguishable from "cladistics" as we have described it. However, the following not-so-subtle differences, as described by Farris and Kluge (1979), make it clear that the two approaches are indeed quite different.

First, for any given set of taxa, there may be many maximal cliques. This might not be a problem, but there are no agreed-on criteria by which to choose among these cliques if they contradict one another. *Second,* once the maximal clique is formed on the basis of certain character information, it is not clear what the fate of the remaining (incompatible) character information should be. Therefore, a large percentage of the available characters might be effectively thrown out and have no contribution to any hypotheses of grouping. This problem is particularly true for molecular data sets, in which most characters often exhibit at least some homoplasy. *Third,* the greater the homoplasy in the data set, the more proportionately limited the support for any given clique will be because the "true" characters must uniquely support a single topology.

In spite of the favorable comparisons of compatibility analysis with cladistics on the part of its proponents, clique techniques were largely abandoned with the rise of computerized algorithms that can accommodate homoplastic data matrices. It would seem that most systematists do not want to discard the preponderance of their data because it is not perfectly compatible with a single topology. Indeed, Mari Källersjö and colleagues

(1999) showed that homoplastic characters contain most of the evidence that allows resolution of relationships in some large molecular data sets. Nonetheless, the concept of compatibility still occasionally surfaces in the literature, and we therefore want the reader to be aware of its attributes and limitations.

Measures of Synapomorphy

Measures of synapomorphy are less frequently reported than the consistency index. Farris (1988, 1989; see also Archie 1989b) used the term *retention index* for a measure that determines the *fraction of potential synapomorphy retained as synapomorphy on a cladogram.*

The retention index (ri) for a single character is computed as follows (Figure 6.2):

$$\text{ri} = \frac{g - s}{g - m}$$

where g is the maximum number of steps possible for a character, s is the observed number of steps, and m is the minimum number of steps. An example calculation of the ri is shown in Figure 6.2. The RI (ensemble retention index) can be computed for all characters on a cladogram as well. A value of 1 indicates a character that is completely consistent with a cladogram, whereas smaller values approaching 0 indicate that a higher proportion of the maximum homoplasy possible for a character is present. Characters that are autapomorphic have ci values of 1 and ri values of 0, indicating no synapomorphy content. A character may have a relatively low ci but still have a relatively high ri. Both of these indices may be used as a basis for establishing a posteriori character-weighting schemes (see Chapter 5 and below).

Resolution of Branches

The desideratum of systematics is a fully resolved hypothesis of relationships among all taxa, which is usually envisioned as a strictly bifurcating hierarchy (Hennig 1966; Brady 1985; Brower 2000b). Complete resolution of this type does not by itself entail the evolutionary mechanism that every branching event occurred by the bifurcation of a parental lineage into two daughter lineages, as implied, for example, by the oft-reproduced figure 4 of Hennig (1966) (but see Sidebar 3 in Chapter 2). Instead, full resolution of a cladogram indicates that nonadditive character codings

Taxa	1	2	3	4	5	6
m	1	1	1	1	2	2
s	1	1	1	2	2	2
g	2	3	2	2	3	3
ri	1.0	1.0	1.0	0.0	1.0	1.0

Tree length = 9
RI (ensemble retention index) = 0.86

additive character codings

Taxa	1	2	3	4	5	6
m	1	1	1	1	2	2
s	2,1	1	1	2	2,3	2
g	2	3	2	2	5	3
ri	0.0,1.0	1.0	1.0	0.0	1.0, 0.67	1.0

Tree length = 10
RI (ensemble retention index) = 0.78

Computing g, maximum possible number of steps

Character 1 — nonadditive

A B C D E
1 1 0 0 0
$g=2$
0
X(0)

Character 5 — nonadditive

A B C D E
1 0 2 2 2
$g=3$
2
X(0)

Character 1 — additive

A B C D E
1 1 0 0 0
$g=2$
0
X(0)

Character 5 — additive

A B C D E
1 0 2 2 2
$g=5$
2
X(0)

Computing character retention index

Tree 2

Character 1 $ri = \dfrac{g-s}{g-m} = \dfrac{2-1}{2-1} = \dfrac{1}{1} = 1.00$ $ri = \dfrac{g-s}{g-m} = \dfrac{2-1}{2-1} = \dfrac{1}{1} = 1.00$

Character 5 $ri = \dfrac{g-s}{g-m} = \dfrac{3-2}{3-2} = \dfrac{1}{1} = 1.00$ $ri = \dfrac{g-s}{g-m} = \dfrac{5-3}{5-2} = \dfrac{2}{3} = 0.67$

Computing ensemble retention index, additive character coding

$$\text{Tree 2} = \frac{\Sigma g - \Sigma s}{\Sigma g - \Sigma m} = \frac{(2+3+2+2+5+3)-(1+1+1+2+3+2)}{(2+3+2+2+5+3)-(1+1+1+1+2+2)} = \frac{17-10}{17-8} = \frac{7}{9} = 0.78$$

FIGURE 6.2. Computing retention indices. Using the same character data and cladograms as in Figure 6.1, we can compute the retention index, ri, for the characters and the cladograms. The necessary values for computing ri, m (minimum number of steps), s (actual numbers of steps), and g (maximum possible number of steps), are shown at the top of the figure, followed by the tree lengths and the ensemble retention indices (RI). The computation of g for characters 1 and 5 is shown in the center of the figure. The nonadditive character data produce smaller values of g because characters can change from one state to any other state in a single step (e.g., 0 → 2 = 1 step), whereas for additive characters the same change represents two steps. Retention indices, ri, are computed for characters 1 and 5; the RI is computed for the matrix of characters treated as additive.

the data contain patterns of synapomorphy that support groups at every level of the hierarchy, maximizing the tree's information content regarding relationships among its taxa. Of course, a given tree is only as good as the data (and analysis) that underlie it, and it is for this reason that cladograms are typically referred to as "hypotheses" of relationships (see Sidebar 4 in Chapter 2), always subject to revision with the acquisition of additional evidence or reevaluation of existing evidence via *reciprocal illumination*.

There are two reasons why a cladogram may not exhibit a strictly bifurcating pattern of relationships (DeSalle and Brower 1997). First, there may be no character that implies resolution for a given node. The following simple data matrix contains no information that resolves relationships among taxa A, B, and C: all three share state 1 for character 2, which suggests that they form a clade with respect to O, but there is no further information regarding whether the pattern of grouping should be (A(B,C)), (B(A,C)), or (C(A,B)).

O   00
A   01
B   01
C   01

The second reason why a cladogram may not exhibit a strictly bifurcating hierarchy is conflicting character information (homoplasy). In the following simple data matrix, it can be seen that character 3 supports the grouping (A(B,C)), character 4 supports (B(A,C)), and character 5 supports (C(A,B)).

O   00000
A   01011
B   01101
C   01110

In this case, there are three equally parsimonious optimizations of the character-state transformations (six steps each), yielding three alternative, fully resolved, equally parsimonious, and contradictory trees; in each of them, one character represents a synapomorphy and the other two are inferred to be homoplasies. The parsimony criterion does not provide a means to select among these alternatives without application of a supplemental criterion such as differential character weighting.

Note that in neither of these instances does the lack of resolution among A, B, and C imply that A, B, and C underwent simultaneous radiation from a common ancestor. Such a hypothesis is an ontological claim about the process of evolution based on the absence of evidence. Polytomies in cladistics are viewed merely as indications of a lack of resolution due to missing or conflicting data (Nelson and Platnick 1980; see also Sidebar 3 in Chapter 2).

Multiple Equally Parsimonious Cladograms

As shown above, a given data matrix may imply more than one optimal tree owing to ambiguities in implied patterns of character-state transformation. This is not a limitation of the parsimony method but an accurate representation of the information content of the data. Some alternative analytical methods always produce a single tree as an artifact of their algorithms, without regard to the number of equally plausible trees the data might actually imply. We mentioned the "Neighbor Joining" method in this regard in Chapter 5 (Islands of Trees and Solutions for Very Large Data Sets). Many phenetic techniques have the same property. Because of the complexity of their weighting schemes, maximum-likelihood-based methods also usually produce a single most likely topology. In neither case do the advantages of having an unambiguous answer outweigh the fact that the answer tells us little about the attributes of the data themselves. The application of parsimony in cladistic analysis yields hypotheses that reflect patterns in the data as closely as possible. Although cladistic analysis of a given data matrix may produce results in the form of one or a few trees, the numbers of trees for another data set may be much higher. The following techniques have been proposed for use in further evaluation of results containing multiple trees.

Consensus and Compromise Techniques

The consensus approach was proposed by Edward Adams (1972) as a way of combining information, from rival classifications or rival trees, for the same set of organisms. The combination of trees from alternative sources, such as morphology versus molecules, as originally conceived by Adams, will be addressed in the Separate versus Combined Analysis section below. In this section, we will discuss consensus methods as a means to summarize information from multiple equally parsimonious cladograms (*fundamental trees*) resulting from analysis of a single data set. Note that the set of fundamental trees from a given analysis may include arbitrary resolutions that are not supported by any characters or are supported by mutually exclusive optimizations of conflicting characters, as in the two examples above.

Since the original introduction of Adams's (1972) approach, a number of additional techniques have been described, traditionally and somewhat confusingly lumped under the term *consensus tree*. All members of this class of techniques search for groups of taxa common among a set of equally parsimonious (or otherwise optimal) fundamental trees, and they differ only in the strictness of their requirement for the degree of correspondence in the topology between the summary and the original trees. The muddled history of terminology concerning consensus and compromise approaches was reviewed by Kevin Nixon and James Carpenter (1996). They suggested that application of the term "consensus tree" should be restricted to the "strict consensus" technique, which produces an output tree containing only clades found in *all* input trees. They recommended calling the output from other techniques "compromise trees," because even though those approaches often produce more completely resolved output trees than does the strict consensus, those trees do not summarize exactly the congruent groupings from all of the input trees.

The following definitions and Figure 6.3 summarize the attributes of available techniques. In the definitions, *component* refers to a node in the cladogram and all of its less inclusive branches (= clade).

Consensus technique (strict): Only the components occurring in all input trees are found in the resultant tree; or, in set theory terminology, the consensus tree represents the intersection of the clades implied by the fundamental trees. This method was used by Schuh and Polhemus (1980), who attributed the concept to Nelson (1979). The results were later termed "Nelson trees" by Schuh and Farris (1981) and "strict consensus trees" by Sokal and Rohlf (1981). Because one of the compromise tree methods below is also referred to as a "Nelson consensus," the "strict" appellation is preferable.

According to the logic of Nixon and Carpenter (1996), all of the following "consensus" methods are actually compromise techniques.

Combinable components (Bremer): The resultant compromise tree contains only components found in at least one of the input trees but is compatible with all of the trees (Bremer 1990). When completely resolved trees are compared, the combinable components and strict consensus trees are the same. The method differs from strict consensus only when one or more of the fundamental trees contains unresolved nodes that are resolved in others. Combinable components trees present these as resolved; strict consensus trees present them as polytomies. Secondary comparison of consensus trees by the combinable components method (e.g., Lanyon 1993) can lead to erroneous overresolution (DeSalle and Brower 1997). Combinable component consensus is also the technique by which "supertrees" are constructed (see Supertrees, below).

General (Nelson, after Page [1989]): The resultant tree contains all replicated components and all nonreplicated components combinable with the replicated components and with each other (Nelson 1979).

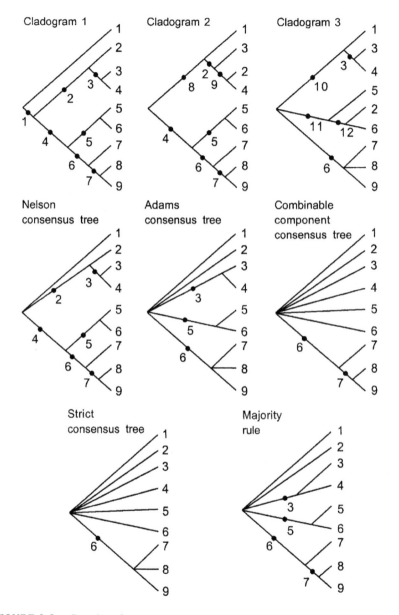

FIGURE 6.3. Results of consensus and compromise techniques. Top row: input cladograms; the numbers on the internal branches indicate components shared in common among the fundamental trees. Bottom two rows: consensus and compromise results under five methods, with the components recovered by the consensus technique from the fundamental trees being indicated by the numbers on the internal branches (modified from Bremer 1990).

Adams: Unstable components and taxa (occurring at different nodes on the various fundamental trees) are pulled down to the first node on the cladogram that summarizes the agreement among the different topologies (Adams 1972). This method may imply groups that appear in none of the input trees.

Majority rule: The resultant tree contains the components found in over 50 percent of the input trees (Margush and McMorris 1981). This is the weakest of all possible techniques because components incongruent with the majority rule compromise tree may exist in the 49.99 percent of the trees not used to form the compromise. Nevertheless, it is this sort of "consensus" that serves as the basis for many of the resampling procedures used to assess tree quality (discussed below in Measures of Support and Stability for Results), and despite its manifest shortcomings, majority rule is probably the most commonly encountered method in the phylogenetic literature for representing a summary of multiple fundamental trees or resampling measures of "support."

Strict consensus and compromise results have been criticized as not providing the most efficient explanation of the data on which the input cladograms were based (e.g., Miyamoto 1985). Bremer (1990) argued that they are nonetheless useful for establishing classifications because they summarize information on groups (by identifying taxonomic congruence) for which there is support in the data. They are also useful in those not-infrequent situations where multiple cladograms exist and where other parsimony-consistent techniques do not resolve the choice because the structure of the data themselves do not permit it.

As the most conservative summary of agreement among fundamental trees, the (strict) consensus may often not be well resolved, but common information on grouping will nonetheless be transparently represented. Compromise techniques all fail—in different ways and to a greater or lesser degree—to preserve information on components found in all of the input trees. As noted above, the results of Adams's technique may even include groups not found in any of the input trees, and the majority rule approach may ignore information in direct conflict with its result.

Thus, as pointed out by Nixon and Carpenter (1996), the consensus of a set of equally parsimonious fundamental cladograms contains only those branches that unambiguously possess support under all possible most parsimonious optimizations of the character data. This will not be true of the results of compromise techniques. Therefore, the consensus is the obvious preference among all techniques proposed because its results are the only ones that convey unequivocal evidential support. Assessing consensus is thus another aspect in which the parsimony approach is conservative.

Sidebar 14
Topology: The Shapes of Cladograms

Should we expect most parsimonious cladograms to have a particular
pattern of branching, for example, *symmetrical* versus *pectinate*? On the
grounds of efficiency of information retrieval, practitioners of phenetics at one time argued for the desirability of techniques producing trees
with symmetrical branching, but it has been long argued (Strickland
1841; see Sidebar 17 in Chapter 8) and more recently formally demonstrated (Farris 1976a) that there is no reason to expect trees to exhibit
any particular pattern of symmetry or asymmetry. Seeking a topology
that results in efficient parsing of taxa is essentially a desideratum for a
special-purpose classification that has no relationship to the empirical
evidence. There is thus no rational basis to include tree shape as a criterion for choice of method. Whether branching patterns actually take
one form more commonly than the other is irrelevant for phylogenetic
methods, but it is clear that asymmetry in trees is quite common. Asymmetrical cladograms imply a greater degree of "nesting" of synapomorphic character states and therefore a greater degree of implied cumulative
character support for the more distal branch points than would be the
case for a symmetrically branching tree with the same number of taxa
(see, e.g., Mickevich and Platnick [1989] concerning the information
content of classifications).

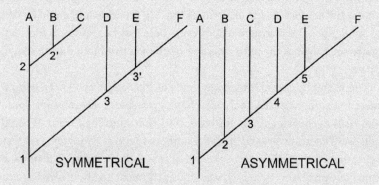

FIGURE 6.4. Comparison of symmetrically versus asymmetrically
branching cladograms, showing the greater degree of nesting of
synapomorphies in the asymmetrical topology on the right.

Treelike structures depicting phylogenetic relationships have been drawn in many different forms. The myriad possible cladograms for even a limited number of taxa make it desirable to draw cladograms in a standard format to facilitate ease of comparison of branching patterns. One convention—although not universal—is for cladograms to always branch to the right as shown in Figure 6.4, so that the observer can predict the position of taxa forming successive sister-group relationships. Note that both of these alternatives are supported by four character-state transformations.

Successive Approximations Weighting

Carpenter (1988a) recommended successive approximations weighting (SAW) as an auxiliary criterion to select from among—and thereby reduce—the numbers of equally most parsimonious trees often found when using parsimony algorithms in phylogenetic analysis. He argued that the results would be superior to those of consensus approaches because they would yield one or a few trees that described the data optimally, rather than a single, poorly resolved tree of suboptimal explanatory power. The approach for implementing successive weighting has already been described in Chapter 5.

Many authors have applied the SAW approach advocated by Carpenter and shown that it often greatly reduces the total numbers of most parsimonious trees. However, an apparently confounding result may be a SAW tree different from any of the most parsimonious input trees to which successive weighting was applied. Such a result implies an inconsistent optimality criterion and could be considered ambiguous and disregarded; however, there may be reasons to consider one or more such trees as the most desirable of all trees produced during analysis of the data.

As an example, consider the arguments of Platnick et al. (1991), who, while conducting a phylogenetic analysis of haplogyne spider relationships, found 10 cladograms of equal length on the basis of 67 characters for 43 taxa. When these results were successively weighted, the authors found 6 cladograms with stable weights, none of which were in the initial set of 10. Of those 6, Platnick et al. deemed one preferable because it was among the most highly resolved and therefore the most informative of the 6 and was—when weights were restored to their original equal statures—only one step longer than the original 10, whereas others among the 6 were as much as three steps longer. The choice of that tree might find further justification on the basis that it was also the one produced under the "fittest tree" approach subsequently published by Goloboff and discussed in

Chapter 5 (Platnick et al. 1996). Whereas nearly all previously published analyses had accepted the most parsimonious tree(s) as the answer with the greatest
explanatory power, and therefore to be preferred, Platnick et al. (1991) argued
that treating all characters as possessing equal weight biased the result in favor
of some characters of lesser explanatory power. This example shows, once again,
how differing assumptions about character coding, even as interpreted a posteriori, can change our understanding of relationships. As Hennig said (1966:103),
"where convergences occur or are suspected, only the most subtle distinction of
the individual characters and the most discriminating evaluation will prevent us
from false conclusions."

Data Decisiveness

As we saw in Chapter 2, the argument that homoplasy in and of itself limits our
ability to conduct phylogenetic analysis under parsimony is erroneous. Goloboff (1991) developed a measure of *data decisiveness* as a way of distinguishing amount of homoplasy from confidence in results. Data decisiveness shows
that there is greater reason to prefer a result in which one tree with relatively
high homoplasy is generated, as opposed to a result comprising thousands of
equally parsimonious and nearly most parsimonious trees, even though those
trees have a relatively high consistency index. Substantially homoplastic data may
be relatively decisive; that is, they may allow for a clear choice among cladograms.
This conclusion agrees with Farris's (1983) observation that although the aim of
parsimony is to minimize hypotheses of character-state transformation, parsimony does not require the number of actual character-state transformations to
be minimal. We may therefore conclude that a large amount of homoplasy does
not, ipso facto, imply lower plausibility of results (see also Källersjö et al. 1999).

Separate versus Combined Analysis:
Total Evidence versus Consensus

Some authors have argued that systematists should treat data from different
sources, or sets of characters evolving under putatively different evolutionary
models, as separate categories of evidence for a given set of taxa (Bull et al. 1993;
Miyamoto and Fitch 1995; Edwards et al. 2016). Contemporary model-based
phylogenetic analyses often begin by assessing data from different sources to
identify such data partitions and craft alternative model parameters to fit them
(e.g., Lanfear et al. 2017). Some examples of contrasting "kinds" of data might

include larval versus adult morphology, morphological versus behavioral features, morphology versus molecules, or one gene versus another. Assessment of
the degree to which data from different sources converge upon the same hypothesis of relationships for a given group of taxa has been a central problem for
systematics for a long time (Dyar and Knab 1906). Examination of arguments
for and against different methods for addressing this issue allows for the identification of essentially two opposing approaches—indeed philosophies.

Total Evidence

The concept of "total evidence" stems from the philosophical work of Rudolf
Carnap (1950). Generally speaking, the principle is that the more evidence
brought to bear on a given question, the sounder the resulting inference is likely
to be. Proponents of this point of view in systematics argue that all data available
for a set of taxa should be pooled and analyzed simultaneously under a single
optimality criterion. The rationale is that only through such an approach can
maximal agreement among data for the set of taxa be fairly assessed and a globally parsimonious solution be inferred. The term *total evidence* was introduced
to systematics by Arnold Kluge (1989; Kluge and Wolf 1993), who appropriated it from the inductivist philosophical literature (Rieppel 2005b). *Combined
analysis* and *simultaneous analysis* also have been recommended as possibly more
appropriate labels (Nixon and Carpenter, 1996) and *concatenation* has become a
popular term for the combination of DNA sequences from separate gene regions
into a single matrix row for such analyses.

Consensus

An alternative viewpoint—sometimes called *taxonomic congruence*—is that different data sets for the same set of taxa should be analyzed separately, and the
common result should be determined via the application of consensus methods
to the topologies generated by those separate analyses. The primary arguments
used to support this approach are that (1) "independent" data should be subject
to independent analyses (e.g., Bull et al. 1993); (2) discrete character data and distance data cannot be analyzed simultaneously in their native form and therefore
demand the use of consensus (de Queiroz et al. 1995); and (3) DNA sequence
data may "swamp" morphological data if the two are analyzed together because
of the much greater number of data points involved (Kluge 1983). All three of
these arguments are fallacious.

To clarify the history of usage, it should be noted that the term *taxonomic
congruence* was used by Mickevich (1978) in reference to the degree to which

classifications based on different data sets imply the same patterns of relationship, an idea originally proposed in the literature on life-stage concordance in holometabolous insects (Dyar and Knab 1906) and later addressed by pheneticists (e.g., Rohlf 1963; Michener 1977). The work of Mickevich and others (e.g., Schuh and Polhemus 1980; see also Schuh and Farris 1981) used consensus techniques in attempts to refute the claims of pheneticists that their classificatory techniques produced more informative and stable results than cladistic techniques. In the work of Mickevich, the data were of ostensibly different types—for example, morphological and allozyme; in the case of Schuh and Polhemus, subsets of the total data set were selected at random. Thus, the usage of "taxonomic congruence" as a synonym of "consensus" in such cases confounds approaches with very different aims and methods.

The differences between the three approaches are clarified in Figure 6.5. It should be obvious from what has been said so far, that "total evidence" and "consensus" are approaches to judging data and inferring reliability of hypotheses of relationship. 'Taxonomic congruence,' in the sense of Mickevich, refers to an approach for judging the efficacy of methods. Indeed, for Mickevich (1978), the question was not whether different data sets would tell the same story—she assumed it—but rather how to judge the effectiveness of different approaches to classification for recovering that story.

Having clarified the terminology and outlined some basic tenets, how then might we assess the arguments for and against the total evidence and consensus approaches?

Evaluating the Total Evidence Approach

The theoretical rationale for total evidence is the same argument as that for parsimony as an integral part of method in science (Kluge 1989). As discussed in Chapter 5, performing analyses that allow every character-state transformation to speak equally to the end result is the least assumption-laden mode of phylogenetic inference. There also are sound empirical reasons for combining data, such as the observation that combined analysis of data sets with substantial homoplasy may produce results showing greater congruence than separate analyses of the same data sets. A particularly striking example of this phenomenon was demonstrated by Gatesy and Baker (2005) in their reanalysis of genomic data from various yeast species originally published by Rokas et al. (2003). They found that combining data from 47 genes that each individually yielded the "wrong" tree produced the "correct" one—bearing out Farris's (1983) prediction that homoplasy is random and phylogenetic signal will prevail even if homoplastic character-state transformations greatly outnumber synapomorphies.

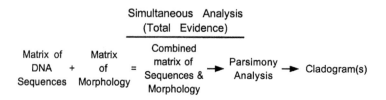

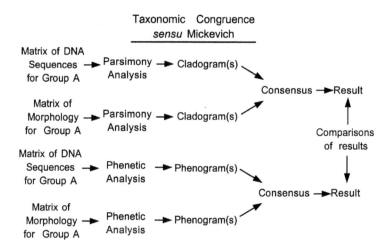

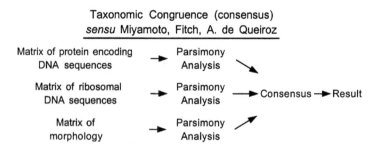

FIGURE 6.5. Comparison of three analytic approaches to cladistic analysis: *total evidence* (simultaneous analysis) as originally proposed by Kluge and Wolf; *taxonomic congruence* as used by Mickevich to judge the ability of methods to produce congruent classifications from different data sets for the same taxa; and *consensus* as used by Miyamoto, Fitch, and A. de Queiroz to evaluate the conformity of multiple data sets to a preconceived "true" answer.

Arguments against total evidence are of four types. *First*, cladograms derived from certain data sets may not reflect the true phylogeny for the group, a "fact" that may be obscured by the total evidence approach (Bull et al. 1993). Our discussion of the philosophy of science in Chapter 2 leads us to reject this argument as vacuous because we can never know the true tree (and if we did we would have no need for the results of phylogenetic analysis). We will revisit this argument in the contentious realm of gene trees and species trees in Chapter 7.

Second, it has been argued that because not all data types can be combined, such as, for example, distances, which are not attributes of taxa and therefore are not combinable with discrete character data, some data may have to be excluded from a total evidence analysis (de Queiroz et al. 1995; Miyamoto and Fitch 1995). Including such data would, perforce, require the use of consensus techniques. The exceptional cases necessitating the use of consensus certainly offer no necessary rationale for analyzing all data subsets separately and then seeking the final answer via consensus. In those individual cases where analysis via consensus is necessary, then the method should be applied if it is believed that the non-character-based data are worth considering. Some authors (e.g., Brower et al. 1996) have suggested that such data should be ignored, and as we have pointed out elsewhere in this volume, distance data derived from DNA-annealing and allozyme electrophoresis have largely disappeared from the systematics literature.

Third is another incarnation of the now thoroughly debunked statistical inconsistency issue: Kubatko and Degnan (2007) and Roch and Steel (2015) have argued that concatenated gene sequence data may produce incorrect and inconsistent hypotheses of relationship with respect to their preferred coalescence techniques, and Quinn (2019) has suggested that this provides a sort of domino-effect rationale for rejecting both concatenation and cladistic parsimony. As we have discussed at length in Chapters 2 and 5, the accusation that a particular empirical method could be inconsistent is a truism, as any method *could be* inconsistent. Determining that the method actually *is* inconsistent presumes a priori knowledge that some alternative hypothesis of relationships that the method does not recover is true. Concern about these matters when we do not already know the answer amounts to metaphysical hand-wringing. "Whatever *is* may *not be*."

Fourth is the debate over what, from a philosophical perspective, actually constitutes "total evidence." Rieppel (2005b, 2007a) has criticized what he calls "cladistic *instrumentalism*," the assumption that data can be uncritically amassed in the matrix and sorted out by character congruence in the phylogenetic analysis, a position espoused by Kluge (2003) and Wheeler et al. (2006a). This objection revolves around rather esoteric arguments concerning character recognition, homology, and ontological grounding. Our evaluation of phenetic methods in

Chapter 1 (and elsewhere) showed that its practitioners applied an uncritical approach to operationalism based almost entirely on raw observation, which was criticized from the outset as naive, but it seems clear enough from the discussion in Chapter 3 that cladists view characters as something more than raw observations and that not every potential attribute of an organism will prove to have value as a taxonomic character. Thus, the "total evidence" for a given phylogenetic question would seem to be all the characters that the systematist has hypothesized to have potential evidentiary value, just as the *New York Times* publishes, "all the news that's fit to print." We might observe that although Rieppel's preoccupation with causal grounding of characters is certainly a worthy philosophical problem (cf. Vergara-Silva 2009), it is misplaced in this context, since the total evidence perspective is less concerned with the absolute amount of data actually or potentially available and more concerned with how one treats the data one has in hand. To avoid such semantic confusion, perhaps *simultaneous analysis* is a better label for the approach.

Evaluating the Consensus Approach

The *primary* argument for consensus seems to have been that the aim of phylogenetic studies is to find the "true phylogeny," and that not all data sets are equally capable of revealing that result (Swofford 1991; Bull et al. 1993; de Queiroz et al. 1995; Miyamoto and Fitch 1995). By extension, if all data are combined in a single matrix, then the correspondence—or lack thereof—of subsets of the data to the true phylogeny will be obscured. Knowledge of the true phylogeny, according to these authors, may be derived from simulations, from studies of well-supported phylogenies of natural groups, and from known phylogenies of laboratory bred or domesticated organisms (Miyamoto and Fitch 1995). These naive realist authors failed to recognize that a methodological recipe for inferring the true phylogeny of the millions of extinct and living species can never be obtained by extrapolation from experimental phylogenetics (Sober 1993). Genealogical relationships will be most clearly hypothesized through the application of systematic inference free from a priori assumptions about the uniformity of models. It should nevertheless be pointed out that advocates of the simultaneous analysis approach often employ data partitioning as a means to explore patterns of variation in the data (e.g., Miller et al. 1997; Judd 1998; Brower and Garzon-Orduña 2018).

A *second* argument offered in favor of consensus is that *independence* is more likely for characters from different data sets than for characters from the same data set and that the necessity of independence among data sets justifies using the consensus approach. Yet, the independence of partitioned characters will be no greater using consensus than would be the case in a combined data set, and therefore the argument does not in and of itself offer a justification for the consensus approach. Indeed, DeSalle and Brower (1997) pointed out that if data partitions are viewed as "independent," then the characters contained within those partitions must de facto be nonindependent, a violation of the assumptions of most systematic methods (Farris 1983; Felsenstein 1983) and a direct contradiction of the logical premise of the independent partitions idea. This perspective also leads to the counterintuitive conclusion that the larger a data partition is, the less weight its individual characters should bear in determination of the hypothesis of relationships via consensus. Once again, genealogical relationships will be most unambiguously hypothesized through the application of systematic inference free from a priori assumptions about nonindependent, and therefore misleading, information content of various character subsets (as suggested by Hennig's auxiliary hypothesis).

A *third* argument, stemming from ideas promoted by Felsenstein and Sober (see discussion of statistical inconsistency in Chapters 2 and 5), suggests that parsimony may provide misleading results if there is a high degree of homoplasy. In this view, if some data sets (subsets) are highly homoplastic, they should be analyzed separately because when combined in a simultaneous analysis, the result could lead to an erroneous conclusion. We have argued against this logic at every turn, beginning in Chapter 2. As pointed out above, the inconsistency concern has no methodological or empirical basis. Indeed, the work of numerous authors (e.g., Wheeler et al. 1993; Brower and Egan 1997; Soltis et al. 1998; Gatesy and Baker 2005; Schuh et al. 2009, among many others) has indicated that although individual molecular data sets for a given group of taxa may produce what might be viewed as meaningless or "wrong" results because of homoplasy, missing data, or the absence of group-forming variation, when combined these same data sets produce a much stronger phylogenetic signal.

Fourth, studies combining disparate data types have failed to support the belief that sequence data will overwhelm or "swamp" morphological data and therefore influence the result disproportionately (e.g., Chippindale and Wiens 1994; Miller et al. 1997). If we assumed that all sites in a DNA sequence varied in an informative way across the set of taxa being analyzed, then large amounts of sequence data might overwhelm morphology in any given analysis. All empirical evidence shows, however, that the number of informative nucleotide sites is probably on the order of 20 percent or less for most studies and that most molecular data are substantially homoplastic. Many combined morphology–DNA sequence data sets contain a proportion of morphological characters roughly equal to 20 percent of the total number of nucleotides being analyzed. Even in the genomics era, morphology remains fundamental to phylogenetic inference as a source of characters, as the means to identify the taxa sampled, and, of course, as a means to sustain the relevance of fossils to the study of the history of life. In sum, the total evidence (simultaneous analysis) approach takes data partition size to be irrelevant and considers character congruence to be the final arbiter in judging the phylogenetic information content of data.

Arguing against consensus, some authors have emphasized that the approach does not produce results that optimally describe the data because the consensus result is usually less well resolved than any one or all of the input phylogenies. Thus, information on characters is lost (Miyamoto 1985), even though information on shared components of the topology is not. This sort of taxonomic congruence discards or obscures character support for the topology in a manner akin to phenetic distances, a clearly undesirable obfuscation of the empirical evidence.

In summary, the only strong argument for taxonomic congruence would seem to be in the applications such as that used by Mickevich (1978), as a test of the efficacy of methods. All other arguments for consensus as an analytic technique appear to be based on false or untestable preconceptions about the nature of the data and of the expected results. Indeed, a substantial body of empirical data is now accumulating with the strong suggestion that simultaneous analysis of molecular + molecular data sets or molecular + morphological data sets produces results with higher consistency and retention indices, the strongest indicators of hierarchic structure within the data. None of these conclusions, however, negate the value of consensus among multiple fundamental cladograms for revealing groupings for which unequivocal character support exists in a data set and for the recognition of those groups in formal classifications.

Supertrees

With the proliferation of phylogenetic hypotheses based on DNA sequences has come an attempt to stitch together trees containing different samples of taxa into a single "supertree" via topological "consensus" meta-analyses (Baum 1992; Sanderson et al. 1998; Bininda-Emonds et al. 1999; Wilkinson et al. 2005; Warnow 2018). The methods use a taxonomic congruence approach with a combinable component consensus criterion so that, for example, a tree reflecting relationships (A(B,C)) and a tree reflecting relationships (B(C,D)) can be combined as (A(B(C,D))). A single grand hypothesis of "the Tree of Life" is a profound desideratum, and supertrees offer an intuitively appealing means to synthesize phylogenetic information, yet the method relies on compromise approaches, such as combinable component "consensus" technique, discussed above. Accepting a topology at face value without consideration of support for its various components ignores underlying patterns of homoplasy that might emerge if the evidence were analyzed in a globally parsimonious framework (Barrett et al. 1991). Gatesy et al. (2002) have demonstrated that the supertrees approach can lead to unparsimonious hypotheses that are incongruent with the cladogram implied by the data when they are analyzed as a "supermatrix," and Goloboff (2005) has pointed out some other disconcerting flaws with recent efforts to formalize supertree meta-analytical procedures. Nevertheless, the method continues to garner support (Cotton and Wilkinson 2009; Li and Lecointre 2009; Warnow, 2018).

Measures of Support and Stability for Results

Approaches for evaluating support or stability of phylogenetic results have been devised on the assumption that we might wish insight beyond the application of parsimony alone. Some measures purporting to provide such insights are statistically based, others are not. Brower (2006b) drew the distinction between support and stability as follows: *Support* is the degree to which the data in the original matrix favor the most parsimonious topology over alternatives and is a parameter of the data. *Stability*, by contrast, is the degree to which *pseudoreplicated* subsamples of the original data discover the same branching patterns that are present in the most parsimonious topology. We will discuss one support measure and several measures of stability. Note that none of these measures provide a significance test for the "truth" of a phylogenetic hypothesis—they merely measure or estimate the evidentiary weight of the evidence that underlies a preferred topology.

Branch Support

As a way of determining support for branches implied by the original data, Bremer (1988, 1994; see also application in Davis 1995) proposed a measure based on the number of extra steps required to lose a branch in the consensus of nearly most parsimonious trees. Although long referred to as "Bremer support," Grant and Kluge (2008) observed that Goodman et al. (1985) employed the same concept three years earlier. *Branch support* (BrS, so abbreviated to distinguish it from the abbreviation for bootstrap) is a parameter of a given data set rather than an estimate based on *pseudoreplicated* subsamples, such as a bootstrap or jackknife value. Analogs of the parsimony-based BrS method have been developed for maximum likelihood by Meireles et al. (1999), and Lee and Hugall (2003).

The BrS concept can be stated alternatively as the difference in length between the most parsimonious tree and the shortest tree in which the branch of interest is not resolved. Branch support values, also referred to as Goodman–Bremer support values, and *decay indices*, are usually reported as an integer for each branch in the consensus of trees one step longer than most parsimonious, two steps longer, and so on (Figure 6.6). Values of BrS range from zero to infinity. A branch with a BrS of zero collapses and forms a polytomy in a consensus of most parsimonious trees. The higher the BrS, the more contradictory evidence would need to be added to the matrix to erode support for a given clade. The sum of BrS values for all branches on the tree is referred to as "total support."

Although BrS is evidently a nonasymptotic, linearly increasing measure of support, the meaning of the number is not entirely clear because of the complexity

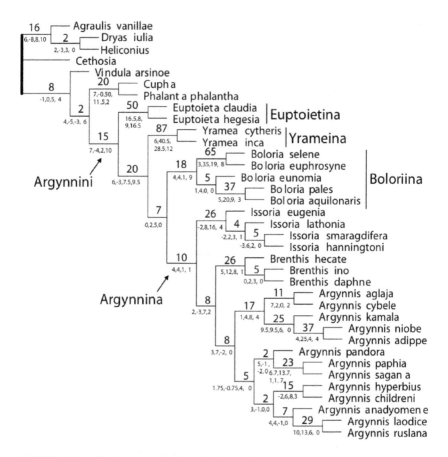

FIGURE 6.6. Cladogram of inferred relationships among members of the Argynnini (Lepidoptera: Nymphalidae) from morphology and three gene regions (Simonsen et al. 2006). Branch support (BrS) values are shown above branches, smaller numbers below branches are partitioned branch support (PBrS) for morphology, mtDNA COI, Ef-1 alpha, and wingless, respectively. Note that the PBrS values sum to the BrS value.

of interactions among character data. Branch support can be calculated either by searching for anticonstraint trees (the shortest tree that does not contain clade x) or by retaining sets of trees incrementally longer than the shortest tree and observing which nodes collapse at each increment. Both of these methods can fail to discover the shortest alternative trees and so may overestimate the BrS value.

An interesting property of BrS values is that they can be calculated for individual characters as well as for an entire data set, and the values are additive (the sum of BrS values of all characters is equal to the overall BrS value). Baker and DeSalle (1997) described *partitioned branch support* (PBrS) as a measure for assessing the contribution of different data partitions to support for the topology generated by simultaneous analysis. Partitioned branch support allows a means to assess the local degree of congruence or incongruence for data partitions at individual branches, rather than simply for the tree as a whole. Values of PBrS can be either positive or negative, depending on whether a particular partition supports or contradicts a given branch. A negative PBrS value suggests that a particular clade is not supported by that data partition (or that the partition supports an alternative hypothesis of grouping).

The additivity of PBrS values also allows the comparison among data partitions of "total partition support" for the entire tree. Rather than comparing topologies of individual partitions (e.g., "gene trees") via taxonomic congruence, total partition support allows assessment of the contributions of each partition, or combinations of partitions, to the overall support for the tree in a total evidence context. This is a rather straightforward means to infer the evidentiary value of various partitions relative to one another. See Brower and Garzon-Orduña (2018) for an example of how such evidence may be presented.

Gatesy et al. (1999b) discussed the emergent evidentiary synergy of "hidden support" among partitions as assessed by this approach. Brower (2006b) provided a review of the methods that included instructions and heuristic examples on how to calculate BrS and PBrS values.

Jackknifing

The jackknife produces pseudoreplicated data matrices in which the character data are removed from the original data matrix for one (or more) taxon; the new, subsampled matrix is analyzed; and then that taxon is replaced and another withdrawn for the next sample, until all such combinations of samples have been analyzed. Jackknife results are reported as a proportion of the cladograms in which a given clade occurs (Figure 6.7). The use of a taxon jackknife was proposed by Lanyon (1985; see additional discussion in Siddall 1995) as a method for assessing inconsistencies in distance data. It also has been used as

```
tread
( 0  ( 14  (( 13  (11  12 ))  (3 5 6 10  ( 7 ( 8 9 )) ( 4  ( 1 2 ))))))) ;
proc/ ;
```

CONSENSUS TREE: 8 NODES (CUTOFF VALUE: 50)
Frequency index: 0.528

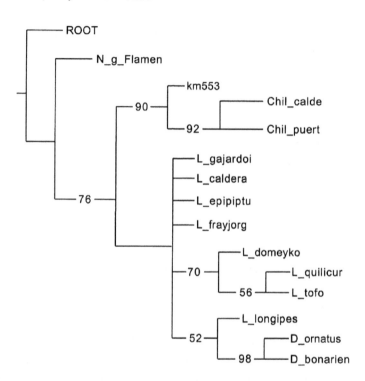

FIGURE 6.7.   Jackknife results for a small data set as produced through the use of Goloboff's program NONA and a jackknife utility program, based on 50 replications. The consensus contains those clades for which there was jackknife support in more than 50 percent of the trees (the cutoff value). The numbers on the nodes represent the percentage of trees in which the nodes occurred.

a fast heuristic method for finding shortest trees in large data matrices (Farris et al. 1996; see Chapter 5), but here we discuss it as a measure of tree stability. Of all data subsampling and randomization methods so far proposed, the taxon jackknife may offer the greatest appeal because it does not attribute support to data where none actually exists, and it is not affected by the presence of autapomorphies. A character jackknife that analyzes repeated subsamples of characters for all taxa can also be employed, but most enthusiasts of character resampling seem to prefer bootstrapping.

Bootstrapping

The *bootstrap*, as applied to phylogenetic analysis by Felsenstein (1985), functions by "resampling characters with replacement" (another pseudoreplicated subsample approach). It is probably the most widely used method for assessing "support" (actually stability) in phylogenetic trees, and has been employed as a tool for testing the "significance" of branches in methods like Neighbor Joining and maximum likelihood, approaches that tend to find a single shortest tree (see discussion in Chapter 5).

A bootstrap analysis begins with the construction of a pseudoreplicate data matrix by randomly sampling columns of characters from the actual data matrix until the same number of characters as in the original matrix has been drawn. In such a pseudoreplicated sample, some characters will not be represented, and others will be drawn more than once, effectively increasing their relative weights. This new matrix is analyzed, and results are stored. Multiple resamplings are conducted, often hundreds or thousands of them. The bootstrap value for a given branch is reported as the percentage of the resulting trees in which that group occurs. This measure is sometimes misconstrued to imply a confidence interval for the results.

Although the bootstrap has been widely applied in the phylogenetic literature, its drawbacks have just as frequently been overlooked. One limitation is that the results are negatively affected by the presence of unique characters (autapomorphies) that support no groups (Carpenter 1996). The method assumes that all characters are drawn from a statistically uniform distribution, but it is apparent that morphological data matrices are not constructed to be statistically uniform, and it is doubtful (and not testable) that molecular data exhibit that property either.

If a clade is unambiguously supported by one character, its bootstrap value will often be less than 60 percent; if supported by three or more, its bootstrap will typically be 100 percent. Davis (1995) demonstrated that there is a strong linear correlation between bootstrap values and BrS values when support is relatively modest. The linear relationship no longer holds when bootstrap values hit their maximum of 100 percent, of course—BrS values continue to increase indefinitely, as additional characters supporting a given clade are discovered, but the bootstrap's asymptote at 100 percent means that it is impossible to tell from that measure whether a given branch is supported by 10, 100, or 1000 synapomorphies (see Table 6.1). That fact should convince even the most ardent advocates of bootstrapping that BrS is a superior means to reflect the relative plausibility of a given branch in large genomic data sets.

Finally, it is easy for consumers of phylogenetic trees to interpret bootstrap percentages as statistical measures of confidence ("probabilities") for the

TABLE 6.1 Bootstrap and branch support values for "backbone" nodes of an anchored phylogenomic tree for skipper butterflies

METHOD	NODE													
	1	2	3	4	5	6	7	8	9	10	11	12	13	14
Bootstrap	100	100	100	100	100	100	100	100	98	100	100	100	100	100
Branch support	4599	5000	2751	2119	117	4062	128	1230	2	2119	2235	1349	3094	1994

Source: Data from Toussaint et al. 2018: figure 3.

metaphysical reality of the branches in question, which they plainly are not, in any standard or statistically defensible sense of the term. As noted previously, bootstrap values also represent the stability of results to perturbation of the data by pseudoreplication, not "support" (Brower, 2006b), as might be garnered from the sampling of additional data. The bootstrap therefore tends to mislead in several different ways.

Randomization Tests

Yet another proposed approach to assessing confidence in phylogenetic results involves randomizing the distribution of character states among taxa. The underlying premise is that data producing a tree that appears to be phylogenetically informative might have structure that is due to chance alone. This approach, independently proposed by Archie (1989a) and Faith and Cranston (1991), reshuffles (permutes) states of a character randomly among taxa, with each character being treated independently during the course of the permutation process. Multiple permutations of the data are performed, cladograms computed, and comparisons made between the lengths of cladograms derived from randomized and original data in order to establish confidence limits, usually at the 95 percent level.

Two types of comparisons have been made: those just of lengths of the resultant trees (PTP, permutation tail probability) and those dealing with monophyletic groups (T-PTP, topology dependent permutation tail probability). Faith and Cranston (1991) claimed that permutation tests provided an "absolute criterion" by which phylogenetic hypotheses can be judged. As pointed out by Carpenter (1992), the methods provide no test of the hypotheses themselves and make the apparently unwarranted assumption that the number and frequency of character states remain constant. Faith (1992:271) subsequently offered a justification of his method in a hypothetico-deductive context, suggesting that "what is passed off as 'corroboration' [in cladistics] does not extend beyond identification of the mpt [most parsimonious tree], and cannot be Popperian corroboration at all." This argument was ridiculed by Farris (1995) as being a misinterpretation of Popper's arguments and a conflation of content of hypotheses and corroboration of hypotheses. Carpenter et al. (1998) concluded that neither PTP nor T-PTP is useful in phylogenetic analysis because they can attribute significance to data that support no resolved grouping. For further debate see Faith and Trueman (2001) and Farris et al. (2001). Permutation tail probability seems to have faded from the recent literature, perhaps because of these arguments, or perhaps simply because many researchers have defaulted to rote use of bootstrap values as a lingua franca measure of "support" despite the misunderstandings concerning its undesirable attributes described above.

Another randomization test for assessing incongruence among data partitions was developed by Farris et al. (1994, 1995b) based on the *incongruence length difference* (ILD; Mickevich and Farris 1981). The ILD is the amount of extra homoplasy required in the most parsimonious tree from the combination of two (or more) data partitions in a single analysis relative to the sum of the most parsimonious tree lengths when the individual data partitions are analyzed separately:

$$ILD = L(\text{combined analysis}) - (L(\text{sum of analyses of partition } 1 \ldots n))$$

Most data partitions have at least some homoplasy due to intrinsic character conflict within the partition. The Farris et al. test samples the combined set of characters into random data partitions of the same size as the original partitions and assesses the ILD value for each of them. If the ILD of the original data is significantly larger than the distribution of randomized partitions, then the data partitions compared are deemed to be *incongruent*. Such a conclusion, of course, leads the researcher back to the philosophical issues about combining versus partitioning discussed above. In many empirical cases, intrinsic homoplasy within partitions plays a far more significant role than incongruence homoplasy among partitions, suggesting that there is less disagreement among "gene trees" than is often feared. See an example in Table 6.2.

Sensitivity Analysis

In analyses of DNA or RNA sequence data that exhibit complex length variation and require alignment (or direct optimization) by a computer algorithm, differences in the relative weights assigned to base pair mismatches versus gaps can result in different hypotheses of relationships among the taxa. There is no agreed-upon standard set of weights for the different kinds of implied changes in the sequences, and so W. C. Wheeler (1995) has employed *sensitivity* analysis as a means to test the robustness of various clades to perturbations in the relative weighting parameters of base-pair mismatches versus gaps (see also Wheeler et al. 2006b). In general, if a clade appears under a broad range of different relative weights, then it is viewed as more robust than clades that come and go under different weighting schemes. Applications can be seen in Schulmeister et al. (2002), Ogden and Whiting (2003), and Laamanen et al. (2005). Not unexpectedly, the method has been a subject of philosophical debate (Grant and Kluge 2003, 2005; Giribet and Wheeler 2007). From our perspective, arbitrarily changing weights is analogous to arbitrarily changing optimality criteria, and therefore the rationale for sensitivity analysis is very much akin to the argument by Kim (1993) that when multiple methods agree, the result is corroborated. We have therefore criticized this approach (Chapter 5; see also Brower 2000a).

TABLE 6.2 Enumeration of partition support and homoplasy for a molecular data set examining relationships among heliconiine butterflies (Brower and Garzón-Orduña 2018)

DATA PARTITION	NUMBER OF CHARACTERS	NUMBER INFORMATIVE	TAXA DELETED	NUMBER OF TREES	TREE LENGTH	INTRINSIC HOMOPLASY	INCONGRUENCE HOMOPLASY	TOTAL SUPPORT
COI-COII	2311	739	3	6	5378	3991 (74.2%)	110 (2.0%)	318.9
16S rRNA	534	109	26	20,592	539	322 (59.7%)	22.5 (4.0%)	4.1
Ef-1 alpha	1189	247	4	>900,000*	931	503 (54.0%)	46.8 (4.8%)	117.0
wg	374	126	19	41,952	487	251 (51.5%)	29.5 (5.7%)	18.9
dpp	315	80	36	101,086	258	114 (44.2%)	20.0 (7.2%)	-1.6
argK	608	140	26	1439	502	261 (52.0%)	26.0 (4.9%)	14.8
cad	810	227	29	532	830	408 (49.2%)	91.5 (9.9%)	11.9
cmdh	790	204	24	56,697	771	374 (48.5%)	28.0 (3.5%)	7.5
ddc	339	97	34	753,393	353	175 (49.6%)	26.5 (7.0%)	5.8
gapdh	722	153	30	75,553	575	281 (48.9%)	36.0 (5.9%)	3.7
idh	678	209	21	6114	775	425 (54.8%)	37.0 (4.6%)	13.8
rps2	418	98	25	67,487	354	173 (48.9%)	21.0 (5.6%)	8.4
rps5	615	130	32	48	451	199 (44.1%)	21.0 (4.3%)	2.0
aact	743	137	36	140	533	216 (40.5%)	33.0 (5.8%)	4.5
cat	739	189	36	1870	657	292 (44.4%)	13.0 (2.0%)	7.7
gts	936	176	36	768	581	200 (34.4%)	19.5 (3.2%)	-0.4
hsp40	539	133	39	156	439	180 (41.0%)	21.0 (4.6%)	1.0
lam	880	170	47	1344	533	170 (31.9%)	30.0 (5.3%)	-0.2
tada3	757	177	26	8182	604	236 (39.1%)	8.0 (1.3%)	3.3
trh	677	208	26	366	923	533 (57.8%)	53.0 (5.4%)	-6.2
vas	722	121	27	196,208	469	213 (45.4%)	21.0 (4.4%)	4.4
hcl	761	124	46	4397	384	140 (36.5%)	11.0 (2.8%)	2.0
All data	16,456	3,998	0	120	18,049	9,657 (53.5%)	725 (4.0%)	541.1

Note: Taxa with no data for a given partition and all but one member of groups of taxa with identical sequences were deleted from calculations of numbers of most parsimonious (MP) trees and tree length for that partition (these deletions do not affect tree length). Total support for individual partitions is the sum of partitioned branch support over all branches. Incongruence homoplasy (D homoplasy) is the number of extra steps between the partition's average length on the 120 MP trees from the combined analysis and the length of the MP trees for the partition analyzed by itself. Fractional values occur because the partitions may require different numbers of steps on alternative trees from the combined analysis. D for all data is the sum of D for individual partitions (also equal to the difference between the sum of the tree lengths from all MP trees of all partitions and the tree length of the combined data set). Intrinsic homoplasy for the combined data is the sum of the intrinsic homoplasies of the individual partitions.

* Search truncated before completion.

A lot of the effort to extract useful phylogenetic information from difficult-to-align sequences seems to represent a sort of Concorde fallacy (continuing to invest in an unprofitable project to try to recoup sunken costs). As we mentioned in Chapter 3, if hypotheses of homology are so ambiguous as to imply major differences in relationships among the taxa under study when alternative alignment weighting schemes are employed, perhaps the best course of action is to seek gene regions that render synapomorphies that are less controversial to interpret as such (Brower and DeSalle 1994). Simply choosing costs that result in an alignment and topology that match some preconceived preference (cf. Wheeler et al. 2017) represents confirmation bias, to say the least. As Platnick (1979:538, quoted at greater length in Chapter 2) said, "if we already knew the correct solutions, we would have no need for the methods."

SUGGESTED READINGS

Brower, A.V.Z. 2006. The how and why of branch support and partitioned branch support, with a new index to assess partition incongruence. Cladistics 22:378–386. [A step-by-step guide with simple heuristic examples]

Carpenter, J.M. 1988. Choosing among multiple equally most parsimonious cladograms. Cladistics 4:291–296. [Discussion of the application of successive approximations weighting]

DeSalle, R., and A.V.Z. Brower. 1997. Process partitions, congruence and the independence of characters: inferring relationships among closely-related Hawaiian *Drosophila* from multiple gene regions. Systematic Biology 46: 751–764. [Discussion of the logic, or lack thereof, of partitioning molecular data]

Goloboff, P.A. 1991. Homoplasy and the choice among cladograms. Cladistics 7:215–232. [Discussion of data decisiveness]

Nixon, K.C., and J.M. Carpenter. 1996. On simultaneous analysis. Cladistics 12:221–241. [Discussion of the total evidence approach]

Section III
APPLICATION OF CLADISTIC RESULTS

SPECIES

Concepts, Recognition, and Analytical Problems

In the middle of the twentieth century, "the species problem" was the predominant philosophical and methodological question in systematics, and the biological literature continues to be replete with discussions of speciation mechanisms and species concepts. In the earlier editions of this book, we chose not to discuss these issues in detail, for several reasons. *First*, there are many other pertinent sources available (e.g., treatises by Mayr 1982; Hey 2001; Coyne and Orr 2004; Wilkins 2009; Richards 2010; and Kunz 2012; edited volumes by Ereshefsky 1992; Claridge et al. 1997; Howard and Berlocher 1998; Wilson 1999; Wheeler and Meier 2000; and evolutionary biology textbooks such as Futuyma and Kirkpatrick 2017). *Second*, the actual mechanisms of species formation are irrelevant to most systematic questions—following Hennig's distinction between *tokogenetic* and phylogenetic relationships, systematists are mainly concerned with the emergent aspects of relationships among species as evidenced by synapomorphies rather than with the polymorphic traits and allele frequencies of populations in the process of speciating. And *third*, although the quest for a universally applicable definition of species is a pervasive desideratum in the biological literature, the formal codification of a species concept applicable across all of biology has proven elusive and may be unnecessary from a pragmatic (or pluralistic) perspective. Systematists deal with recognizable or diagnosable taxa, either at the minimum level (usually called species) or at some more inclusive level. While more inclusive taxa are recognized and diagnosed by synapomorphies that imply their monophyly, as we shall see, species need only be recognizably distinct to serve as valid terminal units for systematic analyses.

We formerly assumed that because nearly any college course in population genetics or evolution deals extensively with issues of species formation, that readers of this volume ought to be familiar already with these general issues of evolutionary biology. However, it is no longer apparent that what is taught in process-oriented courses is revelatory of ideas pertinent to systematics, and indeed, it seems entirely possible that students could emerge from those experiences with misconceptions detrimental to the understanding of concepts discussed here. Some authors have presented systematics (or as they style it, "phylogenetics") as a nested subfield within "genetics" (e.g., Edwards et al. 2016)—a reductionist view analogous to the suggestion that all biology is really chemistry or physics. As we argue in this book, systematics is a coherent discipline with epistemologically distinct aims and methods that is neither subsumed within, nor whose results are entailed by, microevolutionary models and metaphysics. Systematics employs different data, asks different questions, and performs in a distinct, emergent evidentiary framework from the p's and q's of population genetics. It has become increasingly apparent that a clear explication of the systematists' perspective on species concepts and the role of species in systematics is needed. Here we provide a short review of pertinent controversies.

A Metaphysical Quagmire

Darwin (1859:52) said, "I look at the term species as one arbitrarily given for the sake of convenience to a set of individuals closely resembling each other, and that it does not essentially differ from the term variety, which is given to less distinct and more fluctuating forms." The preponderance of the subsequent species-concept literature revolves around two questions: "Do species exist?" and, if so, "How do we define species in a way that 'carves nature at its joints'?" Both of these are questions about the reality of collective entities as they occur through time and space in the world, largely independent of human cognition. Despite Darwin's nominalist reservations, most people (including the authors) interested in studying species believe/assume/hypothesize that the collective entities we call species are real entities worthy of reification as relatively discrete and evolutionarily independent lineages of organisms in nature. Species are the raw materials for field guides, taxonomic checklists, and the referents of names on the tips of many, if not most, cladograms, especially now that sequence data are part and parcel of phylogenetic studies writ large. Simply by virtue of assigning a label to some assemblage of organisms, the systematist makes a theoretical claim about the metaphysical existence of that assemblage as a discrete and integrated thing hypothesized to exist (or to have existed) at a given place and time. When, for example, we talk about the presence of some admixture of Neanderthal DNA in modern human genomes, we require an a priori concept of what both "modern human" and "Neanderthal" mean as distinct entities with separate, identifiable genetic signatures that can be distinguished from one another according to some set of articulated, repeatable criteria. However, as this example intimates, both the clarity and the applicability of species definitions begin to dissolve when greater spans of time and space are brought into consideration.

Historically, enthusiastic taxonomists described nearly every novel specimen, especially in charismatic taxa such as birds or butterflies, as a new species, sub-species, form, or morph. For example, the Neotropical butterfly genus (or sub-genus) *Agrias*, now considered to contain just four species, has over 450 names applied to its phenotypic variants (Lamas 2004). Likewise, any fossil excavated from a new bed was likely to be described as a new species, and if from a differ-ent stratum than similar, previously named forms, perhaps as a new genus. In the 1940s, New Synthesis evolutionist authors Ernst Mayr and George Gaylord Simpson set out to banish what they called typological thinking from the study of contemporaneous geographical variation and the fossil record, respectively. Instead, they advocated population thinking, a view emphasizing individual vari-ation and genetic, ecological, spatial, and temporal continuity rather than the characters favored by systematists. Mayr (1942:120) popularized the "biological species concept," (BSC) which states, "A species consists of a group of popula-tions which replace each other geographically or ecologically and of which the neighboring ones intergrade or interbreed wherever they are in contact or which are potentially capable of doing so (with one or more populations) in those cases where contact is prevented by geographical or ecological barriers." Or shorter: "Species are groups of actually or potentially interbreeding natural populations, which are reproductively isolated from other such groups."

With its central criterion being the easy to comprehend notion of interbreed-ing, the BSC remains a widely taught approach to delimiting species of contem-porary, sexually reproducing organisms. It is not applicable, of course, to asex-ual organisms, fossils, *allopatric* populations, or any taxa about which we have insufficient information about their mating habits, and it suffers from numerous conceptual difficulties, not least of which is the observation that the ability to interbreed is a symplesiomorphy. Despite his focus on phylogenetic relation-ships, Hennig's (1966:18) species concept is very similar to the BSC. According to Hennig, species are "relatively stable, reproductively isolated . . . complexes of individuals interconnected by genealogical relationships."

For fossils (or more generally, for lineages evolving through time), about which evidence of interbreeding is not available, Simpson (1951:289) proposed the "evolutionary species concept" (ESC): an evolutionary species is "a phyletic lineage (ancestral-descendant sequence of interbreeding populations) evolving independently of others, with its own separate and unitary evolutionary role and tendencies." It is immediately evident that this concept is purely theoretical and provides no criteria for recognizing independently evolving units.

There are dozens of other species concepts. For a relatively comprehensive enumeration, classification, and thoughtful discussion of their relative merits, see the review by Richard Mayden (1997), who, incidentally, concluded that the only "primary" concept is Simpson's ESC (as slightly modified by Ed Wiley and himself—see Wiley and Mayden 2000). Kevin de Queiroz has widely promulgated (e.g., 1999:60) "the general lineage concept of species," under which species are "segments of population level evolutionary lineages." De Queiroz considered this to encompass and supersede all other species concepts, but like Wiley and Mayden's concept, it is rather obviously a restatement of Simpson's ESC, and can fairly be seen as a synonym of Ghiselin's (1969) "chunks of the genealogical nexus," as well (see Sidebar 15).

Sidebar 15
Species as Individuals and as Homeostatic Property Cluster Kinds

Since the 1960s, philosophical discussions of concepts and definitions of species have been burdened with efforts to impose metaphysical correctness. Hull (1965) bemoaned "2000 years of stasis" in systematics due to Platonic essentialism, and Ghiselin (1975) proposed "a radical solution to the species problem": that species are individuals (see also Hull 1976). As mentioned in Chapter 2, *individuals* in a philosophical context are spatially and temporally restricted entities that do not have defining properties. In contradistinction, *classes*, or *kinds*, are universal and eternal entities that are defined by their properties. Based on the premise that species are things that originate, evolve, and go extinct through time, with the corollary that they therefore may not have permanent, defining features, it has become a commonplace in the literature to assert that species and monophyletic taxa are individuals (Wiley 1989; Frost and Kluge 1994; Coleman and Wiley 2001; Ereshefsky 2007).

An alternative metaphysical stance is that species are instead homeostatic property cluster (HPC) natural kinds (Boyd 1999; Keller et al. 2003; Rieppel 2010). According to these authors, HPC species are causally integrated processual systems, linked by genetic, epigenetic, and ontogenetic mechanisms (i.e., Hennig's tokogeny). Advocates of HPC species argue

that species-as-individuals concepts relying on phylogenetic origins cannot account for phenomena such as hybrid speciation, while advocates of
species-as-individuals argue that HPC taxa linked by "family resemblance"
could be paraphyletic or polyphyletic.

In our opinion, both of these points of view fall into the essentialism
trap. The essential quality of species-as-individuals is descent from a common ancestor; the essential quality of HPC species is the cluster of properties that defines them—a little bit fuzzier than Platonic ideals but essentialistic nonetheless. Perhaps the very endeavor of defining a concept of
species invites ontological overreach.

As entomologists the taxa we study may be known only from a few dried
specimens on pins; they represent taxonomic concepts that were handed
down from predecessors, often a century or more before. We hold a view
of species that reflects our empiricist philosophical stance articulated in
Chapter 2: species and other putative taxa are hypotheses to be tested with
evidence from their characters.

The trouble with both of the preceding concepts or families of concepts is that
while they may well describe what is really going on or has gone on in nature
in a general/theoretical sense that is compatible with the theory of evolution,
neither provides criteria (other than actual interbreeding) for deciding, in a particular instance, whether a given organism belongs to species A versus species B
or for determining how many species we might recognize in a given assemblage
of organisms. Untestable platitudes about lineages and evolutionary roles do not
help systematists or casual observers of biological diversity determine how to
parse that diversity into natural units. The ESC is general, to be sure: generally
useless as well. Mayden (1997:416–417) saw this as a virtue, not a limitation, and
ardently advocated anti-operationalism:

> Convenience is not a criterion that should be optimized when attempt-
> ing to discover and understand pattern and process in the natural world.
> Operationalism is a fundamental fault of any species concept adopting it.
> Whatever is operational is determined strictly by the perceived reality of the
> viewer. If the viewer's senses perceive only a portion of reality and these are
> expressed in an operational definition of what reality consists of, then we
> will never know otherwise. If, however, the viewer is capable of perceiving
> or conceptualizing all of reality, then all of diversity can be discovered with-
> out placing limits on what can be recognized with an operational concept.

The futility of this metaphysical aspiration speaks for itself—what other "reality" accessible to scientific inquiry is there than "perceived reality"? No matter how sophisticated the tools and methods enhancing our conceptualization of reality may become in the future, systematists will still be constrained by their perceptions. In our more modest, empirical view, following the philosophical principles laid out in Chapter 2, we embrace our perceived reality and prefer species concepts that incorporate tools for identifying and delimiting species as empirical hypotheses, thereby providing us with efficacious working terminal elements for phylogenetic analysis and classification of more inclusive taxa. It is to some of those that we now turn.

Phylogenetic Species Concepts

It is fortunate that cladists employed the notion of a "phylogenetic" species concept based on diagnosability before more metaphysically inclined authors appropriated the term for concepts founded on monophyly or common ancestors. As noted, Hennig's species concept was a version of the BSC, and it fell to his followers to develop a species concept that is well suited to cladistic principles. Among the earliest of the post-Hennigian empiricists was American Museum ichthyologist Donn Rosen, who, in his studies of Guatemalan *Heterandria* fishes (1979:277), came to the realization that "reproductive compatibility is a primitive attribute for the members of a lineage, and has, therefore, no power to specify relationship within a genealogical framework." Instead, he reasoned, "if a 'species' is merely a population or group of populations defined by one or more apomorphous features, it is also the smallest natural aggregation of individuals with a specifiable geographic integrity that can be defined by any current set of analytical techniques." Rosen's concept, sometimes called the apomorphic concept because of its requirement that every recognized species must have its own derived character state, accomplished two key advances for systematics: it proposed a cladistic criterion for recognizing species, and it defined species as the minimal units of analysis, as far as taxonomy is concerned, thus setting a lower bound for systematic inquiry.

This lower bound is not just an arbitrary cutoff but represents an epistemological barrier between population-genetic microevolutionary approaches that measure allele frequencies and the like and the methods of systematics. In the systematic paradigm, taxa bear *characters*, and it is by the particular states of those characters a taxon exhibits that it is recognized, diagnosed, and defined (in the taxonomic sense). *Character states* are usually viewed as the fundamental data of systematics, what Ross (1974) referred to as the "material basis of systematics."

The discovery of fixed character-state differences between two species provides unambiguous empirical grounds for distinguishing them from one another and also implies the cessation of gene flow between them, at least for those characters. Thus, character differences represent empirical evidence of evolutionary divergence (Harrison 1998). The "isolating mechanisms" of the BSC, such as chromosomal inversions, "lock and key" genitalia, or differing courtship repertoires, can be viewed as genetic, morphological, or behavioral characters in this regard.

One might legitimately ask, however, if an approach of accepting some minimal-level taxon based on character differences does not allow for the treatment of males and females as different taxa, or the treatment of different morphs of a polymorphic taxon as distinct taxa. The answer must be that such an approach would be naive, ignoring information that biologists have already acquired and assimilated into background knowledge—the reproductive continuity of parents and their offspring, whether the parents are bisexual, asexual, or hermaphroditic. A second "phylogenetic" species concept, published by Eldredge and Cracraft (1980:92), took this issue directly into account: "a species is a diagnosable cluster of individuals within which there is a parental pattern of ancestry and descent, beyond which there is not, and which exhibits a pattern of phylogenetic ancestry and descent among units of its kind." Here, we see both the criterion of diagnosability and the recognition that actual interbreeding, when observed, is a reliable indicator of conspecificity. Note that Eldredge and Cracraft did not require an apomorphy for each species but rather some "diagnosable" combination of character states. Nelson and Platnick (1981:12) stated these criteria even more clearly: "species are simply the smallest detected samples of self-perpetuating organisms that have unique sets of characters."

Why did Eldredge and Cracraft, and Nelson and Platnick, drop Rosen's requirement that every species have its own apomorphy? For two reasons—one theoretical and one practical. The theoretical reason was articulated by Nixon and Wheeler (1990) in a paper proposing what might be described as the "mature" version of the phylogenetic species concept, which they defined as "*the smallest aggregation of populations (sexual) or lineages (asexual) diagnosable by a unique combination of character states in comparable individuals (semaphoronts).*" Nixon and Wheeler (1990) argued that because species are terminal taxa, the organisms that are members (or parts) of a species, as a single taxon, do not possess the property of monophyly. The members of a species (assemblages of individual organisms) may be diagnosed either by the shared presence of a feature or the shared absence of that feature with respect to their sister species, and it is therefore inappropriate to refer to that feature as a synapomorphy or a symplesiomorphy.

The practical reason for dropping the requirement that each species have an apomorphy is that there are many instances in which a pair of sister taxa differ from one another by the presence or absence of some character state, one condition of which is plesiomorphic (Figure 7.1).

Another important clarification embodied in the Nixon and Wheeler species concept is the concept of *semaphoront*. This is another Hennigian neologism, which refers to the fact that individual specimens representing different life stages or sexes of a particular species may have different character states (Hennig was a dipterist, interested in larval as well as adult morphology). The sum of all observable character states for all such life stages, which may need to be observed from separate specimens, makes up the *holomorph* of the species. Note that characters from DNA sequences, behavior, and other nonmorphological sources are all components of the holomorph. It is clear when we consider the problem of semaphoronts, particularly in taxa with very different life stages such as holometabolous insects, that not every "individual" exhibits all the character states that might diagnose the holomorph. For example, the presence of some feature of the male genitalia is not a "synapomorphy" that unites all the members of a species, since it is absent in the females, as well as eggs and larvae.

Unless specifically stated otherwise, the Nixon and Wheeler (1990) phylogenetic species concept (PSC) is the one we have in mind when we talk about species in this book. We reemphasize that this version of the PSC has no criterion of species monophyly, and indeed, that for those who adhere to this concept, the notion of species monophyly is an oxymoron.

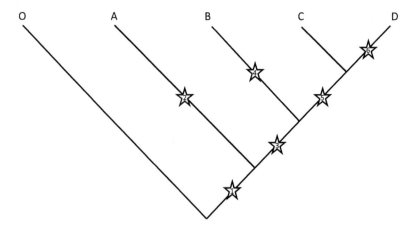

FIGURE 7.1. A cladogram of four species (and an outgroup). The stars represent synapomorphies or autapomorphies for various clades or terminal taxa, respectively. Note that species C is diagnosed by the presence of synapomorphy 5 and the absence of autapomorphy 6—"a unique combination of character states."

The practical aspects of studying the natural world at times confound the seemingly simple issue of connecting different life stages and reproductive communities together. For example, in certain groups of wasps, virtually independent classifications exist for the males and the females because current knowledge does not allow for unequivocal association of the two sexes of the same species. Species discrimination among clonal organisms or suspected hybrids can also be problematic. In fungal systematics, there are formally separate nomenclatures for anamorph (asexual) and teleomorph (sexual reproductive) life stages, reflecting the long-standing historical incapacity to associate asexual vegetative hyphae with their sexual counterparts, "mushrooms" (ascocarps and basidiocarps). Indeed, the solutions to species delimitation adopted by specialists in one group of organisms may be quite different from those applied in another. In some cases the problem will be evident and therefore subject to resolution before detailed systematic studies commence; in other cases solutions to such problems will be clarified only through careful systematic study.

According to the Nixon and Wheeler, PSC species can be recognized by some unique *combination* of character states, although many may also possess derived attributes unique to themselves. At the level of more inclusive taxa, such as genera or families, groups cannot be diagnosed by this method because any and all possible combinations would thereby become diagnosable. This would produce arbitrary classifications, potentially including paraphyletic taxa. Instead, *monophyletic groups of two or more species are recognized and diagnosed on the basis of shared possession of derived character states—synapomorphies—that do not also occur in other groups* (but see discussion of homoplasy in Chapter 4).

Across the community of systematists and other kinds of biologists who may adhere to a diversity of species concepts and criteria, the *circumscription* of species— deciding what breadth of morphological, geographical, or temporal variability falls within a species boundary (sometimes referred to in the taxonomic literature as "splitting" versus "lumping")—is not always clear-cut and may take on a subjective quality (see Sidebar 16). For example, early in the twentieth century, ornithologists recognized some 20,000 species of birds, whereas toward its end, they recognized about 9000 species (Haffer 1997). Presumably, birds have neither speciated nor despeciated during this short period of evolutionary time. What changed, instead, was the "concept" of the minimal diagnosable unit, the criterion by which species were delimited—and the consequent proliferation of subspecies. Particularly owing to the influence of Mayr and his preference for the BSC, with its criterion of interbreeding, many geographically differentiated populations of birds formerly considered specifically distinct, such as the Baltimore and Bullock's Orioles (*Icterus galbula* and *I. bullockii*), were consolidated into single species based on the existence of geographical *hybrid zones* between them, where some interbreeding takes place and individuals with characters of both parental forms may be encountered. The number of recognized bird species is beginning to rise again. This is not so much because truly new taxa are evolving or being discovered but because new evidence and more critical analyses have clarified that what were once thought to be variants within a "species" can be diagnosed as independent entities in nature—taxa inferred to have distinct histories that can be studied with the tools of systematics (e.g., Cracraft 1992). If the aim of systematics is to provide order to the diversity of life, then clearly more order is revealed by recognizing as great a number of terminal taxa as is warranted by the evidence. Ultimately, we may characterize these entities as "separate lineages," hearkening back to terms from the ESC, but that characterization is an empirical result of systematic study based on our "perceptions of reality," not a metaphysical presupposition. We cannot observe lineages in evolutionary time; we must infer them.

Sidebar 16
How Many Species of Ungulates Are There?

If, according to the phylogenetic species concept (PSC), species are "the smallest aggregation of populations ... diagnosable by a unique combination of character states," a logical corollary is that there should be no infraspecific taxa, such as subspecies, because if such a group is diagnosable, it is a species, and if it is not, then whatever differences exist simply represent geographical variation among organisms across the species' range and do not warrant a separate taxonomic status. Joel Cracraft (1983) was an early advocate of the PSC and employed it to narrowly circumscribe many more species of birds-of-paradise than traditional classifications had recognized (Cracraft 1992).

Mammalogists Groves and Grubb (2011) published a PSC-based classification of Bovidae (cattle and antelopes) that increased the number of recognized species in the group from 143 to 279, largely based on the elevation of former subspecies to specific rank. Controversy immediately erupted among competing schools of mammal taxonomists, each camp accusing the other of employing phenetic methods for its species delimitations (Zachos et al. 2013a; Cotterill et al. 2014) and warning of the dire consequences for conservation of both more inclusive (Gippoliti and Groves, 2013) and less inclusive (Heller et al. 2013; Zachos et al. 2013b) circumscriptions. While an empirically minded person might argue (as Darwin did) that the difference between a subspecies and a species is mainly semantic, critics of Groves and Grubb's "taxonomic inflation" found this "splitting frenzy" to be "a bizarre and unacceptable contortion of biological reality" (Zachos and Lovari 2014). The splitters countered that, "a vociferous lobby continues to invoke subjective morphological resemblance to delimit, and especially to lump, individuated lineages of bovids into fictitious taxa, at the cost of ignoring real species" (Cotterill et al. 2014:828). Heller et al. (2014:835) fired back: "A taxonomic method that partitions diversity into arbitrary and meaningless categories with a high accuracy is not useful in biology." The controversy rages on. . . .

As both groups of authors claim to adhere to the PSC, at least as an operational basis for identifying ESC-type lineages, the crux of this dispute seems to pivot less upon differences between species concepts than upon preferences for different types of evidence (morphometrics versus gene genealogies). We can draw three general conclusions: *First*, while subspecies provide an additional hierarchical level in a Linnaean classification and so greater taxonomic resolution than a long list of species within a genus, if relationships among those taxa can be represented phylogenetically, based on fixed character-state differences, then the (debatable) extra information content of the infraspecific rank vanishes, and a strict application of the PSC demands that they be viewed as species. *Second*, nomenclatural instability is confusing for conservationists and other users of classifications, and approaches are needed to reconcile alternative views in biodiversity databases (see Franz et al. [2016] for efforts to align differing views of mammalian classification). However, *third*, controversy is healthy for systematics: in addition to generating entertaining threads of dialectical argument in the literature, iconoclastic hypotheses provoke additional tests with more or different sources of evidence, thereby sparking new phylogenetic research programs for the next generation.

Gene Trees, Species Trees, Phylogeography, and Other Sources of Confusion

Gene Trees versus Species Trees

Although the challenges of huge genomic data sets in the twenty-first century have resulted in calls for "paradigm shifts" (e.g. Edwards 2009) in how we should think about molecular systematics, these existential concerns are far from novel to the present day, and it is instructive to consider their origins, which stem from molecular population genetics. The distinction between "gene trees" and "species trees" was first introduced by Morris Goodman (viz., Goodman et al. 1979), but was implicit in Fitch's (1970) writings on *paralogous* genes, represented as duplications of portions of the genome within a single lineage. A *species tree*, in Goodman's view, was one indicating a "known" scheme of phylogenetic relationships among a group of taxa, implicitly assuming that such a pattern exists in an unambiguous and fully resolved form. A *gene tree*, by comparison, described a scheme of relationships—based on an empirical comparison of a segment of amino acid or DNA sequence among those taxa—that might or might not be congruent with the species tree. In Goodman et al. (1979), the "true" phylogeny—species tree—was assumed a priori on the basis of nonsequence data, although what data Goodman et al. actually used to derive that scheme of relationships is totally unclear in their paper. The discovery of multiple copies of globin genes in tetrapod and particularly mammalian genomes meant, for example, that if one naively compared an alpha globin gene from a human and a cow with a beta globin from a gorilla, the cow and human genes would appear more similar (or "closely related") to each other than either of them is to the gorilla. They might even share synapomorphies with respect to the gorilla, when considered in light of globin gene sequences from an outgroup, such as a chicken or a shark. This is because the gene duplication event of the alpha and beta copies occurred before the divergence of the three taxa, and the pattern being inferred is the gene history rather than the history of the taxa. Of course, all three taxa (and the outgroups) have both gene copies, and the solution to this problem is simply to sample the same ("*orthologous*") gene copy from each of them.

Goodman's distinction between gene trees and species trees was focused primarily on the orthology/paralogy problem, but the concept was expanded by Fumio Tajima (1983) to address the possibility of incongruent phylogenetic patterns between that implied by a given gene region and the "true species tree" for other reasons. Tajima noted that under certain circumstances, such as in large populations with no selection, within-group variation at single-copy loci (i.e., genes that are not part of a family of duplicated regions such as the globin genes) could be greater than between-group variation, even in mitochondrial genes, which are effectively haploid and among which recombination is therefore negligible. Tajima concluded that it is therefore possible to reach incorrect phylogenetic conclusions when such nucleotide sequence data are analyzed by themselves; these results he referred to as *gene trees*. Note that in the long-standing tradition of population genetics (cf. Edwards and Cavalli-Sforza 1964), Tajima conceptually conflated *tokogenetic* and phylogenetic variation in this work.

According to Jeffrey Doyle (1992), further processes that can produce gene trees not congruent with the species tree are (1) *horizontal gene transfer*, (2) *introgression*, (3) *ancestral polymorphism/lineage sorting*, and (4) paralogous gene families, as originally envisioned by Fitch and Goodman et al. This laundry list of mainly tokogenetic problems for molecular phylogenetic inference remains a staple of the literature today.

David Maddison (1997) provided another discussion of the gene tree–species tree congruence issue, again assuming a "true phylogeny" without indicating how it might be known or recognized as distinct from actual inferred hypotheses of phylogenetic relationships based on empirical data. He concluded that phylogenies are like clouds (Figure 7.2). It should be noted that the concerns of many authors expounding on the gene tree–species tree problem are largely metaphysical, based on the premise that the tree implied by some subset of data might be incongruent with respect to "the truth." Of course, the closest approximation to the truth available to systematists making such comparisons is just another hypothesis of relationships, ideally based on some sort of evidence, accepted a priori as a benchmark for comparison. Because of the existence of homoplasy, it will always be possible to partition data in ways that will reveal apparent patterns of incongruence that befog phylogenetic results with ambiguity. This is why many cladists adhere to the total evidence perspective, discussed in Chapter 6. Succinctly articulating the Kantian phenomenological challenge of this issue, Joni Mitchell (1967) lyrically concluded:

> I've looked at clouds from both sides now
> From up and down, and still somehow
> It's clouds' illusions I recall
> I really don't know clouds at all

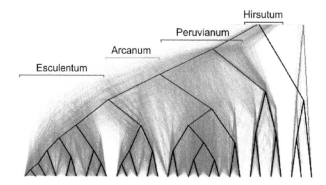

FIGURE 7.2. A "cloudogram" for 15 species of *Solanum* sect. *Lycpersicon* (Solanaceae) (modified from Pease et al. 2016, figure 2B) based on nonoverlapping 100 kilobase genomic sequence windows, produced by the DensiTree package (Bouckaert and Heled, unpublished). The "consensus phylogeny" is shown in black. What insights do you suppose figures such as this are intended to convey?

Preoccupation with the gene tree–species tree dichotomy is thus strongly associated with the conviction that phylogenetic analysis aims to discover the "true tree" for the group of taxa under study and that the evidence in hand might be misleading. While the possibility of error, due to statistical inconsistency or other quirks, is a given in empirical science, the endeavor to diminish such eventualities through "more realistic" models would seem to be on philosophically and scientifically shaky ground, as discussed in Chapters 2 and 5. We might reiterate that the question of whether population-level variation is amenable to phylogenetic analysis had already been questioned by Hennig, who observed that variation at this level is often nonhierarchical, or tokogenetic (see also Davis and Nixon 1992). Many examples of putative gene tree incongruence explained by introgression and ancestral polymorphism would appear to fit in this category. Horizontal transfer has been documented to occur, especially among bacteria, but the fact that we can recognize it as an incongruent pattern at all is dependent on there being an a priori hypothesis of relationships with which the phylogenetic signal contained in some part of the genome disagrees. Thus, the explanation of detected patterns of homoplasy by hypotheses of introgression and horizontal transfer presupposes a predominant underlying phylogenetic pattern.

Of course, phylogenetic hypotheses inferred independently from two or more segments of DNA may be incongruent with one another, in which case at most one of them can be correct, or both may be false, with respect to the "true tree." In such instances, one may erect ad hoc hypotheses to explain why the gene trees differ, or one may simply combine the data to infer a single, general hypothesis of relationships (see "total evidence" in Chapter 6), acknowledging but not obsessing over the existence of homoplasy. The epistemological significance of the gene tree–species tree argument was questioned by Brower et al. (1996:434) when they noted that the "attention paid to them [gene trees] in the literature may reflect more the unease they inspire in our phylogenetic paradigm than their prevalence in nature." Articulating the background knowledge assumptions for the simultaneous analysis approach, they reasoned (p. 442), "If we believe there is a hierarchical pattern, then we should expect the data to reflect that pattern alone ... The best estimate of hierarchical relationships is still derived from parsimonious interpretation of all the data." The discovery of that pattern, whether on the basis of morphological data, behavioral data, or molecular data, ultimately rests on theories of character homology and transformation. If we assume that duplicated portions of the genome—paralogs—cannot on their face be distinguished from one another and will therefore be associated erroneously, we will have compared nonhomologs—a mistake of character analysis. Obviously, if we knew the comparisons were erroneous, we would not make them in the first place (see discussions in de Pinna 1991; Vrana and Wheeler 1992). As Hennig

(1966:121) stated, "'phylogenetic systematics would lose all ground on which it stands' if the presence of apomorphous characters in different species were considered first of all as convergences (or parallelisms), with proof to the contrary required in each case."

The strongest empirical corroboration of a hypothesis of homology is in the discovery of congruence between it and other hypotheses of homology. Therefore, discovery of rampant incongruence might suggest comparison of nonhomologous sequences. In the second edition of this book, we noted that the hand-wringing over incongruence between gene trees and species trees seemed to have abated somewhat in the twenty-first century. At the time, the main concern was whether DNA data were congruent with data from morphology, which to a great extent they were and continue to be. More recently, however, with the growth of comparative genomics, there has been a resurgent clamor of discussion of these issues, now in regards to phylogenetic discordance among different genes, and lately we find that the gene tree–species tree argument has been used with increasing frequency to explain perceived incongruence, with the a priori expectation that it is a widespread and potentially confounding issue for phylogenetic inference. Concern has again been raised about statistical consistency, now in regard to the phylogenetic signal emerging from the combined (or *concatenated*) data set relative to the supposed differing signals of separate gene segments or subsets that are presumed both to be homogeneous within but heterogeneous among partitions. See Springer and Gatesy (2016) for a critique of methods intended to solve this "problem" and Bravo et al. (2019) for a more sanguine view.

A case study pertinent to these issues is the report by Kozak et al. (2015) of widespread reticulation among genomes of species in the nymphalid butterfly genus *Heliconius* based on phylogenetic analyses of 22 gene regions. Brower and Garzón-Orduña (2018) reevaluated these data, discovering two chimeric mitochondrial sequences (resulting from Kozak et al.'s data management errors) that were responsible for nonmonophyly of *Heliconius* for that partition. The corrected mitochondrial DNA (mtDNA) sequences strongly supported the monophyly of *Heliconius*, and a parsimony analysis of the corrected, combined data also corroborated prior hypotheses of relationships among members of the group. More significantly, Brower and Garzón-Orduña showed that the purported "incongruence" among the nuclear gene regions was in large part due to missing data and within-gene homoplasy rather than to actual incongruent signal among the genes: using incongruence length difference comparisons (Chapter 6), they found that overall 53 percent of the combined tree length was due to homoplasy within individual gene regions and only 4 percent due to incongruence among them.

At the very least, we may conclude from this result that if chopping up genomes into units of phylogenetic homogeneity were a desideratum that one wanted to pursue, there is more homoplasy—more than ten times as much—within "genes" than among them, and, therefore, that "genes" are not the correct partitions to carve nature at the joints regarding patterns of homoplasy.

The observation that "gene trees" represent nothing more than phylogenetic hypotheses inferred from contiguous stretches of DNA that may or may not have any coherence from the perspective of "phylogenetic signal" is not novel. It has long been understood that the characters embodied by the nucleotide sequence of a given "gene" do not exhibit a unitary pattern of evidence that is free from homoplasy (DeSalle and Brower 1997; see also Springer and Gatesy 2016). In the effort to find homogeneous data partitions, one can subdivide and subdivide, separating introns from exons, first, second, and third codon positions, structural from functional regions of resultant amino acid chains, and so forth. Such endeavors usually reside in the realm of model parameterization, but clearly the effort to identify partition homogeneity leads to a reductio ad absurdum, in which partitions are identified by the tautologous criterion that the included characters offer congruent signal (compatibility—see Sidebar 13, Chapter 6) or until each nucleotide is once again treated as a separate character (which is the way cladistic parsimony views such evidence).

Aside from these theoretical conundrums, there are also practical challenges to consider: as data sets become larger and analytical methods more complex, opportunities for confounding technical errors to remain undetected and the temptation to concoct ad hoc methodological workarounds to hypothetical evolutionary idiosyncrasies burgeon. The results of such efforts could be nonsense, but nonsense couched in such convoluted analytical gymnastics becomes impenetrable to readers and reviewers and therefore immune to critical evaluation, let alone falsification. Preservation of empirical transparency—a basic desideratum of the scientific method—is itself a strong practical reason for a parsimonious approach to assembling and analyzing large data sets.

As we have argued in Chapter 6, from the perspective of systematics, which is concerned with inference of relationships among taxa, special concern about the "gene tree–species tree problem" is unwarranted beyond the properly skeptical view that any piece of evidence could be misleading. The terminology is, nonetheless, still used. We argue that the "gene tree–species tree" distinction should be replaced with terms such as "molecular partition result" and "total evidence result," which shifts the epistemological emphasis from the metaphysical correspondence of evidence to an ineffable "truth," to the congruence of different partitions of available empirical evidence with one another, with the aim of inferring a coherent hypothesis of relationships under the phylogenetic paradigm of a single irregularly bifurcating hierarchy. Note that partitioned Bremer support (discussed in Chapter 6) allows exploration of character and/or partition conflict explicitly in the context of simultaneous analysis of all data.

Phylogeography

The study of intraspecific gene genealogies with the presumption that they convey information about phylogenetic structure of populations is referred to as *phylogeography*. The approach emerged from evolutionary genetic investigations of allozymes and mtDNA in the 1970s and 1980s (Avise et al. 1987). The great heyday of gene tree-based phylogeography occurred from the late 1980s until the mid-2000s, during which time unambiguous trees based on single mtDNA data sets proliferated owing to the development of the polymerase chain reaction (PCR) for amplifying specific segments of DNA in vitro. This deluge of data led to the foundation of new journals such as *Molecular Ecology* and *Molecular Phylogenetics and Evolution*.

While couched ontologically in genealogical terms and claimed by its proponents to be a research paradigm competing with the cladistic systematic approach advocated in this book (cf. Avise 2014), phylogeographical investigations render as largely meaningless fundamental systematic concepts such as synapomorphy and the relationship between *semaphoronts* and the *holomorph*. This is because these studies use individual organisms, rather than species or more inclusive taxa, as their terminal entities, thereby once again conflating tokogenetic and phylogenetic patterns. The epistemological confusion engendered by this category error (or perhaps ignorance of such subtleties altogether) seems often to lead to a sort of instrumentalist pragmatism regarding methods of phylogenetic inference, and many practitioners of phylogeography have employed phenetic clustering to infer hypotheses of relationship (e.g., Hebert et al. 2004; Templeton 2004). Of course, as Hennig (1966) pointed out, there is no reason to assume that tokogenetic patterns are appropriately represented by a strictly bifurcating hierarchy, and therefore the entire phylogeographical endeavor is undertaken on epistemological thin ice.

In animals, mtDNA is a circular plasmid genome, usually about 15 kilobases long, effectively haploid, and almost exclusively maternally transmitted via the cytoplasm of the egg cell in sexual reproduction. Therefore, it does not exhibit biparental tokogenetic patterns and on account of this can be employed to generate "phylogenetic" patterns of the maternal lineage, even among individuals in an interbreeding population. Mitochondrial DNA (along with various ribosomal RNAs) became a preferred study molecule in the late 1970s because it could be separated from nuclear DNA by centrifugation and, with restriction enzymes, cut into pieces that always summed to the total length of the circle, making it amenable to study via restriction fragment length polymorphisms and restriction site mapping. With the advent of PCR, the first generation of "molecular systematists" maintained their focus on the molecules they were already accustomed to studying, and the elision of maternal genealogies with phylogenetic patterns became a paradigmatic, if ill-conceived, commonplace.

John Avise referred to mtDNA phylogeography as the "bridge between population genetics and systematics" (Avise et al. 1987). The concern that patterns of mtDNA variation might over-resolve or be divergent from the "true species tree" (relationship implied by other criteria, such as morphology or patterns of interbreeding) is a source of the theoretical preoccupations with gene tree–species tree congruence discussed above. Avise thus promulgated the tradition of conflating population-level and phylogenetic analyses initiated by Cavalli-Sforza and Edwards two decades before.

Use of PCR to amplify essentially any segment of DNA with ease also allowed tree-based methods to be applied to population-level analyses of nuclear genes. Such loci may exhibit not only ancestral polymorphism in populations but also heterozygosity within diploid individuals, resulting in patterns at intraspecific levels that are often manifestly not treelike. Armed with these new data, and gene genealogical coalescent process models that assume selective neutrality, constant population sizes, and other restrictive and not necessarily realistic aspects of the data (e.g. Kingman 1982; Hudson 1990), an invading army of population geneticists crossed Avise's bridge into the realm of inference of relationships among taxa, establishing what to systematists seems like a parallel dimension of microevolutionary assumption-laden *phylogenetics* (or, to emphasize its roots and perspective, we might say "phylo-genetics"). Systematists understand that intraspecific problems are not compatible with the epistemological assumption of a strictly bifurcating hierarchy (Hennig's tokogeny–phylogeny distinction) nor with fundamental systematic concepts such as synapomorphy, while phylo-geneticists ignore these philosophical nuances and endeavor to contrive ever more realistic models to accommodate the process of speciation, which generally result in narrative scenarios of historical microevolution that are eminently unfalsifiable. As we have discussed in Chapter 2, we consider the latter approach to be a fool's errand.

DNA Barcodes—Uses and Limitations

DNA barcoding, which (at least for animals) usually employs a 640-base pair segment of the mitochondrial cytochrome oxidase subunit I gene to infer clusters of similar individuals, has been touted as a species identification, discovery, and delimitation tool (Hebert et al. 2004; see also Chapter 12).

The least controversial use of DNA barcodes is to identify unknown specimens by comparing their sequences to a library of authoritatively identified "known" sequences. This is especially useful in forensic situations in which a sample is a fragment or residue, or in cases in which morphology does not allow identification (ID) below a relatively inclusive taxonomic level, as with early instar larvae of many holometabolous insects (Sperling et al. 1994; Wu et al. 2017; Yeo et al. 2018). As the amount of available data in public databases such as GenBank continues to grow, the chances that an unknown sample may match a sequence deposited in the database and thereby be identified increases. This may be especially true if the unknown specimen comes from a region where the fauna has been relatively comprehensively sampled, such as North America and Europe. Of course, the accuracy of an ID performed this way is entirely dependent on the accuracy of the IDs of specimens from which sequences in the reference database were derived. It is therefore critical that voucher materials from barcoded specimens are maintained so the ID may be verified or revised based on morphology at a later date, if necessary.

What constitutes a "positive ID" based on a DNA barcode? Identity (a perfect match) or a high degree of similarity between the unknown sequence and one or more reference sequences (hopefully identified as the same thing!). Typically, sequences are compared by pairwise distances in a BLAST search against a large reference database such as GenBank, and results with less than 2 percent difference are considered to be conspecific. Note that the criterion for IDs conducted in this manner is phenetic similarity—often unweighted pairgroup method of analysis (UPGMA) or Neighbor-Joining distances (see Chapter 5). Initially, Paul Hebert and colleagues (2003a, 2003b) proclaimed that a "barcode gap" existed between conspecific and heterospecific individuals—conspecifics differed by less than 2 percent, while heterospecifics generally differed by greater than 4 percent. This has been shown to be wishful thinking (Meyer and Paulay 2005; Cognato 2006; Wiemers and Fiedler 2007; Goldstein and DeSalle 2010), and as more comprehensive samples have accumulated, especially over the geographical ranges of broadly distributed species (Bergsten et al. 2012), the barcode gap is often obscured, both by intraspecific variability and in some instances by close similarity between species considered to be distinct based on alternative criteria.

Other times, an unknown sample may not match anything in the barcode reference database. Such samples could represent new taxa or, more likely, just taxa that have not yet been sampled for their barcodes. It is also possible that sequences from multiple individuals initially considered to belong to the same species based on morphology fall into distinct barcode clusters ("BINs," barcode index numbers; cf. Hausmann et al. 2013) that imply that they are not all the same. In a widely cited early barcode study of Costa Rican skipper butterflies,

Hebert et al. (2004) announced that, based on the discovery of separate phenetic clusters of barcode sequences among 471 individual butterflies sampled, they had discovered "ten species in one" of what was formerly considered to be *Astraptes fulgerator*. Subsequent cladistic analysis (Brower 2006a) showed that several of those phenetic clusters were based on symplesiomorphy and not phylogenetically supported. It remains unclear how many cryptic species occur in this group.

Others investigators have proposed more complex, model-based algorithms to divine species delimitations from barcode data. Even so, equating a barcode identity with a species identity is an exercise in pragmatic operationalism that may or may not bear upon actual species boundaries and relationships—or those empirically supported by the rest of the holomorph. Metaphysical preoccupations about gene trees potentially not reflecting the "true" species tree are of even greater concern when the segment of DNA employed is short and the number of characters correspondingly limited. The best practical approach is to use barcode evidence to suggest hypotheses that may (or may not) be corroborated by other sources of evidence, such as additional genes, morphology, behavior, and so on. We note that Dayrat (2005) coined the term "integrative taxonomy" to refer to this method, as though he was heralding a conceptual breakthrough. In actuality, the consideration of characters from all aspects of the holomorph is what circumspect systematists have been doing all along.

SUGGESTED READINGS

There are thousands of published resources on "the species problem." Here are a few relatively recent ones that will provide an *entrez* to the literature.

Knowles, L.L., and L.S. Kubatko, eds. 2010. Estimating species trees: Practical and theoretical aspects. Hoboken, NJ: Wiley–Blackwell. [A model-oriented approach to phylogenetics]

Mayden, R.L. 1997. A hierarchy of species concepts: The denouement in the saga of the species problem. In: Species: The units of biodiversity, ed. M.F. Claridge, H.A. Dawah, and M.R. Wilson, 383–424. London: Chapman and Hall. [As noted, a thorough comparative introduction to the diversity of species concepts]

Richards, R.A. 2010. The species problem: A philosophical analysis. Cambridge: Cambridge University Press. [A relatively recent review of the quagmire]

Swigart, J.D. 2018. What species mean: A user's guide to the units of biodiversity. Boca Raton FL: CRC Press. [Another recent review]

Wheeler, Q.D., and R. Meier, eds. 2000. Species concepts and phylogenetic theory. New York: Columbia University Press. [A point-counterpoint discussion among proponents of different concepts by some of their authors, such as Mayr and Wiley]

Wilkins, J.S. 2009. Species: A history of the idea. Berkeley: University of California Press. [A historical perspective]

NOMENCLATURE, CLASSIFICATIONS, AND SYSTEMATIC DATABASES

In this chapter we will review the principles and rules governing biological nomenclature and explore two areas that represent the most practical results of descriptive taxonomy and phylogenetic analysis—formal classifications and systematic databases.

Biological Nomenclature and Classification

Nomenclature

Antiopa, Camberwell Beauty, Kiberi-tateha, Le Morio, Mourning Cloak, Rusałka żałobnik, Sorgmantel, Suruvaippa, Trauermantel: these are some of the common names for the Holarctic butterfly species whose scientific name is *Nymphalis antiopa*. Since systematic biology is a global endeavor, it has long been evident that employing a common language to refer to its entities is a basic prerequisite to effective communication among members of the scientific community. Biological nomenclature serves as the language by which we name organisms and groups of organisms and communicate our knowledge about them; indeed, a taxon's name is the critical semiotic link between specimens, taxonomic concepts, literature, and all other information pertaining to that taxon. To preserve its clarity and *stability* of meaning, biological nomenclature is governed by rather elaborate and detailed rules, rooted in the system of binominal nomenclature, that have been evolving since they were first proposed in the mid-nineteenth century.

The works of Linnaeus, *Species Plantarum* (1753) and *Systema Naturae, 10th Edition* (1758), are the chosen starting points for modern biological nomenclature in most groups of plants and animals. Concomitant with the introduction of a consistent *binominal* system of names, Linnaeus organized knowledge of the living world in the form of an irregular hierarchic classification (that is to say, groups at the same level of the hierarchy are not necessarily composed of equal numbers of subordinate taxa). It is the fundamentals of the system of binominal nomenclature and the critical aspects of the Linnaean hierarchy that we will explore in this chapter.

Nomenclature is not an end unto itself but rather a necessary adjunct for the organization and conveyance of information on biological diversity. Scientific names themselves do not embody information, but they provide a common anchor point for concepts that can be found in the literature. Some scientific names and their associated concepts, like *Homo sapiens*, may be well-known to the scientific and general public. Others, like *Capsus ater* or *Malus pumila*, may be meaningful only to specialists. The binominal system that was first used in a uniform way in the aforementioned works of Linnaeus has persisted because of its functionality, because it is the only system that has been universally accepted, and because the entire 250+ year history of biological nomenclature is based on it.

Latin was the language of systematics, as of most scholarly work, in the time of Linnaeus. Although scientific names are not strictly required to be Latin, the codes of nomenclature (see below, Codes of Nomenclature) have traditionally emphasized the formation of scientific names as if they were Latin or latinized, usually from Greek. Latin is still an integral element in both botanical and zoological nomenclature, and given that several million published names for plants and animals are composed of Latin words or roots, it is unlikely to become obsolete in the foreseeable future.

Species in Linnaeus' system are referred to by just two distinctive conjoined words—a *binomen*, rather than by a descriptive phrase, as had been the practice of many of his predecessors. Every species is placed in a genus. The names of genera are Latin nouns; the names of species are Latin adjectives in gender agreement with the nouns (Latin has three genders—masculine, feminine, and neuter) or, alternatively, are Latin nouns or adjectives in apposition (not requiring gender agreement). Generic names begin with capital letters. Specific names are in lowercase, although in botany those based on proper names frequently begin with a capital. Both words are italicized. Because every species must belong to some genus, the "name" of the species is therefore a binomen, as for example:

Homo sapiens (a masculine noun and adjective)
Lycopersicon esculentum (a neuter noun and adjective)

Acacia drummondii (a feminine noun with a masculine adjective in
    apposition—the species was named after a man named Drummond)

The individual words forming a scientific name are often referred to as the
generic epithet and the specific epithet. Scientific names do not include diacritics
(e.g., ~, `, ´, ¨), although they may be hyphenated in certain instances.

Every generic name (epithet) and specific name (epithet) has one or more
authors, the name(s) of the person(s) who first published them. The author of the
generic name may not be the same as the author of the specific name. This aspect
of nomenclature probably causes more day-to-day confusion than any other. Inclusion of author names allows for a more accurate tracing of the history of application
of names in the taxonomic literature. When the author's name is included, names of
similar spelling are usually readily distinguishable from one another as referring to
separate taxa, rather than as misspellings of a name for the same taxon, for example:

Rhinacloa pallipes Reuter
Rhinacloa pallidipes Maldonado

Particularly in the older literature, authors' names that are familiar to experts in
a particular group are often abbreviated, sometimes in what may appear to be an
idiosyncratic manner to the uninitiated, for example:

Homo sapiens L. (for Linnaeus)
Catocala ulalume Stkr. (for Herman Strecker)

Dates of authorship may also be included, to help distinguish among names and
to assist in locating relevant literature:

Macrocoleus femoralis Reuter, 1879
Cyrtocapsus femoralis Reuter, 1892
Psallopsis femoralis Reuter, 1901

The authorship indicated after the binomen refers to the species, rather than the
genus, and this information can be particularly important if a species name has
been shifted from one genus to another as a result of revisionary taxonomic study.
If the name of the author is in parentheses, this indicates that a species is currently
placed in a genus other than the one in which it was originally described by that
author. In the botanical literature, the name of the original author of the species
may be placed in parentheses followed by a second name, the latter representing
a subsequent author who moved the species to its genus of current placement:

Werneckia minuta (Werneck)
Ceratozamia boliviana Brongn.
Zamia boliviana (Brongn.) A. DC.

Typically in zoological catalogs and other listings, names as used by persons other than the original author are written with the name of that person separated from the scientific name by punctuation, usually a colon, as, for example:

Phytocoris marmoratus Blanchard
Phytocoris marmoratus: Stonedahl

The latter listing indicates use of the combination by an author other than the original author of the species name. The word *sensu* (in the sense of) may also fill this role. This does not necessarily imply that the taxonomic concept is different, however.

Codes of Nomenclature

Most early authors subsequent to Linnaeus followed the example of his now famous works and adopted a strictly binominal system of naming. Some did not, and their works are now largely rejected and forgotten with respect to the names they applied to plants and animals.

The discussions of nomenclature in this chapter deal with issues treated in the botanical and zoological codes, which are the "official rule books" of nomenclature for botany (including phycology and mycology) and zoology, respectively. There are also an *International Code of Nomenclature for Cultivated Plants* (2016), an *International Code of Nomenclature of Prokaryotes* (2019), and an *International Code of Virus Classification and Nomenclature* (2018). Students of these areas should consult the appropriate volumes (or websites).

The first formal attempt to add order to the increasing proliferation of scientific names in zoology came in 1842 with the British Association Code, or Stricklandian Code, named after its author, Hugh Edwin Strickland (see Sidebar 17). The first truly international efforts to regulate nomenclature in zoology involved the establishment of the International Commission on Zoological Nomenclature (ICZN), which published the *Règles Internationales de la Nomenclature Zoologique* in 1905. Since that time, the rules of zoological nomenclature have been administered and periodically modified by the ICZN. The most recent *International Code of Zoological Nomenclature* (*Code*), published in 1999 (and freely available online), is divided into a series of articles and recommendations. "Articles" are intended to be followed strictly by those involved in the creation or modification of names, whereas "recommendations" are intended as guidelines that should be followed. The *Code* also contains a glossary providing definitions of terms used in zoological nomenclature, a useful aid in interpreting the *Code* and other writings pertaining to nomenclature. Recommendations for changes to the zoological *Code* and decisions by the ICZN concerning the status of particular names are published in the *Bulletin of Zoological Nomenclature*.

Sidebar 17
Hugh Edwin Strickland (1811–1853): A Prescient
Victorian Systematist

Hugh E. Strickland was a well-rounded naturalist, making significant empirical contributions in geology, paleontology, and systematic ornithology. He also wrote important articles addressing theoretical systematics in two distinct areas: codification of scientific nomenclature and the rejection of a priorism in phylogenetic hypotheses.

As mentioned elsewhere in this text, Strickland was the chair of the British Association committee (which also included a youthful Charles Darwin) responsible for publishing the first set of recommendations for rules to govern scientific nomenclature in zoology, including the principle of priority (Strickland 1842). This contribution arose from his vociferous opposition (Strickland 1837) to the disorderly proliferation of contradictory nomenclatural systems that was beginning to clutter the literature:

> No one is more desirous of the improvement of science than myself; but *reform* implies something more than *change*; and it is precisely because I do not consider that the proposed changes are *for the better*, that I enter my protest against them. On a superficial view of the case, it may certainly appear, that to change a less appropriate scientific name for one that is more so, *is* a change for the better: but what is the result? If, to take the most favorable view of the case, the scientific world should agree to adopt an "improved" nomenclature, yet, even then, all our standard works on natural history would become, in great measure, a dead letter; every museum in the world would require to be relabeled; and the disentanglement of synonyms (already a sufficiently laborious though necessary duty) would become almost hopeless. But if, as would almost certainly be the case, these "improved nomenclatures" should be only partially adopted, the disentanglement of synonyms would then become *quite* hopeless, and the curse of Babel would be entailed on the scientific world.

Strickland's quote is remarkably apt in regard to the modern controversy regarding Linnaean nomenclature and the Phylocode (see Phylogenetic Nomenclature, below).

Strickland's second important contribution was as an early advocate of viewing the pattern of relationships among organisms as an irregularly bifurcating hierarchy. In England in the 1820s and 1830s, there was a tremendous blossoming of speculation about the Natural System—the pattern of groups subordinated within groups. A particularly vocal faction led by William Sharp MacLeay and including Nicholas Aylward Vigors, William Swainson, and Thomas Horsfield advocated a numerologically oriented classification scheme known as the quinarian system, based on symmetrically nested rings of five taxa. Strickland (1841) found the quinarian system to be artificial, and bluntly said so:

> The natural system is an accumulation of facts which are to be arrived at only by a slow inductive process, similar to that by which a country is geographically surveyed. If this be true, it is evident how erroneous must be those methods which commence by assuming an a priori system, and then attempt to classify all created organisms in conformity with that system. The greater part of these arrangements are based on the assumption that organic beings have been created on a regular and symmetrical plan, to which all true classifications must conform. Some naturalists have attempted to place all animal species in a straight line, descending from man to a monad. This theory assumes that each species (excepting the two extremes) has two and only two direct affinities; one, namely, with the species it precedes, and the other with that which follows it. Others, perceiving the existence in many cases of more than two direct affinities, have assigned the most mathematical symmetry to the different parts of the system by maintaining the prevalence throughout of a constant number, such as 2, 3, 4, 5, or 7. In applying these views to facts, they have of course found numerous exceptions to the regularity of their assumed formulae, but by adducing the extermination of some species, and our ignorance of the existence of others, and by applying a Procrustean process to those groups which were either larger or smaller than the regulation standard, they have removed the most glaring objections to their theory, and have with wonderful ingenuity given their systems an appearance of truth. But when the unprejudiced naturalist attempts to apply any one of these systems to Nature, he soon perceives their inefficiency in expressing the real order of affinities. The fact is, that they all labour under the vital error of assuming that to be symmetrical,

which is in an eminent degree irregular and devoid of symmetry. The Natural System may, perhaps, be most truly compared to an irregularly branching tree, or rather to an assemblage of detached trees or shrubs of various sizes and modes of growth.

Here, the shape of the Natural System as an irregularly bifurcating hierarchy is plainly described as an empirical result, 18 years prior to the publication of the *Origin of Species* and a year before Strickland's young friend, Charles Darwin, mentioned the tree metaphor in his notebooks (Darwin 1909). This statement has obvious implications for the claim of some twenty-first century systematists (e.g., Kluge 2001) that inference of systematic relationships is not "justified" without the prior assumption of "descent, with modification" (see Sidebar 5, Chapter 2).

In an ironic and gruesome tragedy of the industrial age, a 42-year-old Strickland was struck and killed by a train while hunting for fossils in a railroad embankment. What further contributions to systematics and the still-embryonic field of evolutionary biology were forestalled by his untimely death?

The regulation of names in botany has a history similar to that in zoology, with the adoption of a set of laws at the Paris International Botanical Congress in 1867. The *International Code of Botanical Nomenclature* (*Code*) was published in 1952, supplanting all prior codes and implementing a more orderly approach to the formation and regulation of names. Like the zoological *Code*, the botanical *Code* is divided into articles and recommendations. The most recent *Code of Botanical Nomenclature,* now referred to as the *International Code of Nomenclature for Algae, Fungi and Plants (Shenzhen Code),* was published in 2017. Revised editions of the botanical *Code* appear at six-year intervals corresponding to the International Botanical Congresses. All recommendations for changes in the botanical *Code* are published in the journal *Taxon.* The astute reader may have noticed that the taxa treated by this work are a polyphyletic assemblage of multicellular eukaryotes that are not animals.

According to its preamble, the *International Code of Zoological Nomenclature* was designed with the intent of promoting "stability and universality in the scientific names of animals and to ensure that the name of each taxon is unique and distinct." The botanical *Code* "aims at the provision of a stable method of naming taxonomic groups." As we have discussed elsewhere, however, stability for its own sake is not necessarily a desirable property. The names covered by the *Codes* can be divided into the following four general groups. A name proposed in a given general group has equivalent standing and can be employed at any taxonomic level contained within that group.

Suprafamilial names: Names above the family-group level (e.g., orders, classes) are unregulated in zoology, need not be based on generic names, and do not usually have standardized endings (e.g., Aves, Ditrysia). In botany, it is recommended that such names be based on a nomenclatural type (generic name) and have standardized endings (see below).

Family-group names: This group includes all names above the level of genus group, up to and including superfamily; the most commonly included ranks are tribe, subfamily, family, and superfamily (except for the last in botany). Names in this group have standardized endings in both botany and zoology (with eight allowed exceptions in botany for names used by Linnaeus, e.g., Compositae [= Asteraceae] and Gramineae [= Poaceae] and are based on generic names [see below under The Linnaean Hierarchy]).

Genus-group names: This group includes generic and subgeneric names, to which the same rules apply.

Species-group names: This group includes species and subspecies names in zoology and also additional infrasubspecific names in botany, with the same rules applying at all levels.

Although the *Codes* for zoology and botany differ in their organization and in particular details, the basic tenets of the *Codes*, stated in the form that they are found in the zoological *Code*, can be outlined as follows:

1. Priority
2. Publication and availability
3. Typification
   a. Species-group types
   b. Genus-group and family-group types
4. Homonymy
5. Synonymy

Priority

The principle of priority is no less than a basic tenet of science—the first person to publish a concept or idea is its author—and according to some its application may be traced to Henry Oldenberg, the first secretary of the Royal Society. An early, clearly articulated statement of the principle was articulated by Linnaeus, in his *Critica Botanica* (1737): "priority in time confers precedence." *Priority* in nomenclature dictates that the first name published in reference to a taxon is the name that will be used thereafter. One might ask, "What is the problem? Doesn't every taxon have just one name"? The answer is "not necessarily." Imagine, for example, different authors working in different places, not aware of one another's activities. Under such circumstances the same species may have been described from two to several times. Or, possibly more commonly, a given species shows great variability, which was not understood at the time of its initial discovery. The same or different authors might apply different names to the taxon, unaware that they were dealing with morphologically very different males and females of the same species, for example. Furthermore, during the early nineteenth century, there was prolonged debate over whether historical names deemed inappropriate should be replaced with better ones. The potential chaos stemming from such actions was a major impetus for Strickland's efforts to codify nomenclature (Witteveen 2016; see Sidebar 17).

Almost all the time, the earliest published name applied to a taxon takes priority over others and is viewed as the correct name. The concept of priority sometimes runs afoul of nomenclatural (or nomenclatorial) stability when the oldest name is not the one that has been most widely used. Such circumstances are dealt with by the *Codes*. The modern *Codes* all have provisions for setting aside long-unused names, even though those names would be favored by strict application of the rule of priority. In zoology, such names may be placed on a list of "rejected" names by appeal to the ICZN. Names are "conserved" in botany in deference to their long-standing usage.

As an example of applying the rule of priority:

> *Lygaeus saltitans* Fallén was published in 1807. The genus *Chlamydatus* Curtis was described in 1833, with a new species *marginatus* Curtis. In 1858 Fieber described a new genus *Agalliastes*, including *saltitans* (Fallén), among other species. The type of *Agalliastes* was fixed as *saltitans* by Kirkaldy in 1906. Considering the two to be conspecific, Flor synonymized *marginatus* Curtis with *saltitans* Fallén (a simple case of priority of the older specific epithet over the younger). At the level of the genus, *Chlamydatus saltitans* (Fallén) is the combination that must be used on the basis of priority of the generic name *Chlamydatus*, and the generic name *Agalliastes* becomes an objective junior synonym of *Chlamydatus* (see Synonymy, below). Note that it is the date of publication of the genus, rather than the species, that is decisive in the latter instance.

For ranks above the species level, priority is important when the circumscription of groups changes—when the number of included taxa is increased or decreased based on a taxonomic revision. When two genera are lumped together into a single genus, the older generic name is used and the more recent name is viewed as a subjective junior synonym. For example, a recent cladistic analysis (Fri̇ç et al. 2007) showed that species of blue butterflies in the familiar Eurasian genus *Maculinea* Van Eecke 1913 form several subordinate clades within another genus with an older name, *Phengaris* Doherty 1891.

Availability

Whereas priority is a comparatively objective criterion, availability is much more nebulous (Figure 8.1). In the context of the *Codes*, most names would be considered "available" if they meet the following four criteria:

1. appear in a work published after 1753 for plants and 1758 for most animals;
2. meet the criteria for "publication" designated in the *Codes*;
3. are written in the Latin alphabet (in this day and age, the English alphabet); and
4. are binominal (if referring to species).

The *Codes* also require that newly proposed names be accompanied by a description of the taxon (the text of which for botany must be written in Latin or, as of 2012, English) and have a designated type (see below, Typification). A name that does not meet the criteria of availability is referred to as a *nomen nudum*. Such names formally do not exist as far as the rules of nomenclature are concerned, and their originators get no acknowledgement as authors. A nomen nudum becomes available if it is subsequently published according to the above rules by the original "author" or anyone else. That person becomes the author of the name. Novice authors who may have described new taxa as part of their dissertation or thesis work should be aware that those documents are not considered "published," and names in them are nomina nuda.

Of these four criteria of availability, publication is the most difficult to characterize adequately. Although many works from the eighteenth and nineteenth

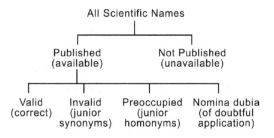

FIGURE 8.1.   The categories of scientific names (adapted from Blackwelder 1967).

centuries have presented problems with regard to satisfying the criteria of publication, no traditional printed works present the difficulties being encountered in this age of online publications. The 2017 botanical *Code* requires a publication to be disseminated to botanical institutions with publicly accessible libraries as printed-paper copies or as a PDF that has an ISSN or ISBN serial number. The requirements of the 1999 zoological *Code* have been revised twice since the code was published (International Commission on Zoological Nomenclature 2008, 2012). Most recently, an Official Register of Zoological Nomenclature has been initiated, which requires the registration of new names in its database, known as ZooBank. Names must be published in a fixed form, either in widely available print or electronic format (online nomenclatural databases that can be updated are not "fixed"—permanent and unchangeable—and therefore do not meet the requirements of publication). As noted, taxonomic descriptions in graduate theses and dissertations do not meet the requirements of availability unless the work is formally published. On the other hand, anyone with an urge to describe new taxa may do so ad libitum, in comparatively informal media such as blogs or newsletters, provided that they are published in a fixed format and disseminated according to the rules stipulated in the *Code*s. The rules regarding criteria of publication will undoubtedly continue to evolve as the nomenclatural transition into the electronic age moves inexorably forward.

The rule of priority may potentially be affected by alternative dates of publication. Therefore, since the introduction of online publication ahead of print a contentious issue has been whether the date of publication is the date of posting of the "early view" online version or the printed version. As long as the text and figures of the online version are in their final form, not subject to additional editing, then the date of its publication on the web stands as the date of record as far as nomenclature is concerned, even if the final printed page numbers are not yet assigned. For this reason, articles introducing new nomenclatural acts should indicate clearly the date of the "version of record" on their title page. Needless to say, posting a manuscript on a prepublication peer-review/self-promotional website such as BioRxiv does not constitute "publication," and any names so proposed are *nomina nuda* until they are actually published. For discussion of these issues, see Cranston et al. (2015), Dubois et al. (2015), and Krell (2015).

Names that are available become a permanent part of nomenclature, regardless of the quality of the taxonomic work in which they appear or whether they apply to taxa recognized by modern systematists as valid or not. To help avoid subsequent confusion, the most thorough students of historical nomenclature even keep track of unavailable published names, such as infrasubspecific quadrinomina (names for entities below the level of subspecies) and other names (in zoology) that do not meet one or more of the criteria of availability listed above when published or via subsequent modification of the *Code*.

Availability should not be confused with the concept of validity. *Availability* refers to the formal introduction of names into the literature and their resultant acceptability—in biological nomenclature—under the set of criteria specified above. *Validity* refers to the application of those names to recognized taxa, as determined under rules governing the concepts of homonymy and synonymy, which are discussed below. Every valid name must be available, but not all available names are valid.

Typification

Whereas the concepts of priority and availability were more-or-less implicit in the works of many authors from the time of Linnaeus, such was not the case with the type concept. To early authors, the most common species was often considered "typical" of a genus. Under this approach, a landmark for fixing generic limits was often recognized, but the absence of any established or formalized approach to such fixation (such as the definition of a genus by one or more synapomorphies, as encouraged by the cladistic approach) at times created great confusion. Species were often represented by a number of "typical" specimens— a "type series" encompassing males, females, and minor individual variation.

The nomenclatural type concept was clearly stated as early as 1819, by de Candolle, and recommended in the 1842 rules by Strickland, who observed, "We may obtain a great amount of fixity, in position at least, if not in the extent of our groups, by invariably selecting a type, to be permanently referred to as the standard of comparison." However, this approach was not formally codified in zoological nomenclature until the publication of the "Règles" (i.e., "Rules," the first internationally sanctioned zoological code) in 1905. The required designation of "types" produced a form of stability and order in zoological nomenclature not seen before, most particularly in the application of generic names. The mandatory designation of types in botany did not apply until January 1958.

Types are of two forms: the types of species-group names are specimens, while the types of genus-group names and family-group names are species and genera, respectively. Because the type of a genus is a species, and the type of a species is a specimen, all names are ultimately tied to a particular biological specimen that (one hopes) resides in perpetuity in a museum collection (types of fossilized specimens are not, strictly speaking, "biological," and in some historical instances, an illustration of a specimen that has been subsequently lost represents the type).

Species-group types. A type in the species-group is a specimen to which a name is attached, providing an objective criterion for establishment of usage of that name. Species-group types recognized in the *Codes* of nomenclature are called *primary types* and are of the following sorts:

- *Holotype*: A single specimen (or illustration, in botany) designated by the author of the name at the time of publication of the original description.
- *Neotype*: A specimen designated to replace the holotype (or other primary type) if the latter can be documented to have been lost or destroyed.
- *Syntypes*: A group of specimens thought to represent a given species, as designated or indicated by the author(s) of the original description. These specimens are sometimes referred to as the "type series." The term "co-type" is sometimes used in the same sense. Designation of syntypes is no longer permitted under modern nomenclatural *Codes*, but they remain in historical collections.
- *Lectotype*: One of the syntypes chosen by the original or a subsequent author(s) to function as the name bearer. Other members of a syntype series not selected as the lectotype become paralectotypes.

It is customary to deposit primary types in a recognized institution—usually a museum—dedicated to the maintenance of scientific collections. Such specimens are often given special status in museum collections because of their importance for the consistent application of names to biological entities over time. For example, the slide-mounted holotypes of aphids held by the Smithsonian Institution are secured in a fire- and waterproof gun safe.

Many other categories of so-called types will be found mentioned in the literature (see Witteveen [2016] for an illuminating history of the type concept in biology). In botany, probably most important among these is the *isotype*, a specimen collected from the same individual plant as the holotype, such as several twigs clipped from the same tree.

Paratypes are commonly designated in conjunction with the description of new species, these being specimens that were studied by the author of a new species or subspecies judged to be conspecific with the holotype, and designated by that person at the time of publication of the original description. An *allotype*—a specimen of the opposite sex from the holotype—is a kind of paratype. Although paratypes often have value to subsequent workers for reference as to how a name was applied (see *circumscription*, below in Synonymy), they have no standing in nomenclature and—as one might imagine—sometimes include misidentified specimens belonging to different species, as is also often the case with syntypes. Paratypes may serve as convenient potential candidates for designation as neotypes, in cases where holotypes are destroyed. Another common practice is to distribute paratypes (isotypes in botany, which may be twigs clipped from the same bush and so parts of a single individual) among multiple collections, to provide a consistent authoritatively identified reference to the broader community of taxonomists and to hedge against catastrophe, such as war, earthquake, or accidental fire.

Genus-group types. These types are species, comprising names, not specimens. Classical authors generally did not designate types of genera. The modern *Codes* require that for a generic or subgeneric name to become available, a type species must be designated by the author. For genera lacking a type species because they were published before institution of requirements for mandatory designation in the *Codes*, types are most commonly *fixed*, that is, assigned by one of the following:

- *Monotypy*: The genus name in question was published with only a single included species, in which case the genus would have been *monotypic* (or *monobasic*) when made available; only the single originally included species can be the type even though other species may have been subsequently added to the genus.
- *Subsequent designation*: When more than one species was originally included, the type is selected at a later time either by the original author or another person.

Family-group types. Types in this group are generic names. Family-group names are based on generic names in Linnaean nomenclature but not, as might be supposed, on the oldest genus name included in the group. Instead, priority of family-group names is based on the first use of a name at that rank, regardless of whether the genus name used is the oldest in the group or even if it is currently valid or not. A few examples may help to illustrate the point.

> In the insect suborder Heteroptera, the family Velocipedidae is based on *Velocipedes*, a junior synonym of *Scotomedes*. The angiosperm genus *Winteria* is a junior synonym of *Drimys*, yet the valid family name is Winteraceae. The nymphalid butterfly genus *Callicore* Hübner [1819] is part of the tribe Catagrammini Burmeister 1878, and the name Callicorini Orfila 1952 is considered an objective junior synonym of Catagrammini, even though *Catagramma* Boisduval 1836 is a subjective junior synonym of *Callicore*.

Family-group names are also subject to the rules of homonymy (see below), in that no two can be spelled identically, even though they may be based on different, although similar, generic names.

Homonymy

The principle of *homonymy* refers to the application of the same name to different taxa. The *Codes* state that no two names above the species-group level may be the same within either zoology or botany, although names may be duplicated between the two fields. This dictate has been applied since early on, with the greatest confusion having resulted from names with variant spellings but which nonetheless represent the same word or meaning. Currently in zoology, the rules of Latin are bent such that names that are spelled differently are treated as different even though they may be based on the same root and have the same intended meaning. The rules in botany are not so lenient: even names that are spelled differently but might cause confusion are excluded as illegitimate.

Homonyms may be of several types:

Senior homonym: The valid name on the basis of priority.

Junior homonym: The preoccupied name on the basis of priority or by ruling of the bodies governing nomenclature. A junior homonym is considered invalid and may not be used.

Primary homonym: In the species group (species, subspecies, etc.), names that are the same and were proposed in the same genus-group taxon. A junior primary homonym is invalid when proposed and must always be replaced, either by a new name or, if such exists, by a junior synonym.

Secondary homonym: If two species with the same name are placed in the same genus subsequent to their publication, they become secondary homonyms. The senior secondary homonym is the older of the two (or more) names. Junior homonyms become invalid and must be replaced. Sorting out these issues is made much simpler if the author's name and date of publication are included in citations of the name. Some of the most pernicious difficulties in establishing secondary homonymy will be encountered if the rules of Latin regarding gender agreement of adjectival species epithets are applied strictly. As an example, species that are black are often named for their coloration. The spelling of the name will vary depending on the gender of the genus: *niger* (masculine), *nigra* (feminine), *nigrum* (neuter). If two species named for their black coloration end up in the same genus, a homonym will be formed under the strict application of the rules of Latin grammar, even though the spellings were different because the genera of original placement were of different gender. Additional complications arise from orthographically incorrect constructions such as *nigrus*, whose intended meaning was obviously "black." The rules for replacement of secondary homonyms may vary depending on dates of publication and between botany and zoology. Therefore, the *Codes* should be consulted on these issues.

The following examples will help illustrate the rule of homonymy:

- *Bougainvillia* Lesson (Cnidaria) is not a homonym of *Bougainvillea* Comm. ex Juss. (Angiospermae: Nyctaginaceae) because the former is an animal and the latter is a plant.
- *Kingia* Malloch, 1921 (Insecta: Anthomyiidae) is not a homonym of *Kingia* Brown (Angiospermae: Dasypogonaceae) but is a junior homonym of *Kingia* Theobald, 1910 (Insecta: Culicidae).
- *Mononychus* Perle, Norell, Chiappe, and Clark, 1993 (Vertebrata: Dinosauria) is a junior homonym of *Mononychus* Scheuppel, 1824 (Insecta: Coleoptera).
- *Gerris* Fabricius, 1794 (Insecta) and *Gerres* Quoy and Gaimard, 1824 (Pisces) are not homonyms, despite the similar spellings. The family names based on these genera are now spelled differently, Gerridae and Gerreidae, respectively, to avoid homonymy at that level.
- *Phytocoris marmoratus* Douglas and Scott, 1869, is junior primary homonym of *Phytocoris marmoratus* Blanchard, 1852, both having been placed in the same genus at the time of their original description.
- *Phytocoris modestus* Reuter, 1908, is a junior primary homonym of *Phytocoris modestus* Blanchard, 1852, because both were originally described under the same genus, even though *modestus* Blanchard is currently placed in the genus *Polymerus*.
- *Dichrooscytus marmoratus* Van Duzee, 1910, became a junior secondary homonym of *Phytocoris marmoratus* Blanchard, 1852, on the basis of priority, when it was transferred into *Phytocoris*.

Synonymy

The concept of synonymy in nomenclature deals with the application of different names to the same taxon. As is the case with homonyms, synonyms can be of several types.

Senior synonyms: The senior name is one of two or more different names for the same taxon that is deemed valid, usually on the basis of priority, or because of its selection by the first reviser (in zoology), or by a ruling of the bodies governing nomenclature.

Junior synonyms: The junior name(s) is the one deemed to be invalid, usually on the basis of priority, or because of its designation by the first reviser, or by a ruling of the bodies governing nomenclature. Junior synonyms are available names in nomenclature, even though they are not considered valid as the name of a recognized taxon. It is therefore important to understand the distinction between availability and validity.

Objective synonyms: These are different names that by examination of nomenclature alone can be judged to refer to the same taxon. For example, any two family-group names with the same type genus, or any two generic names with the same type species, are objective synonyms; two species names based on the same type specimen would also be objective synonyms. Because of their clear-cut nature, objective synonyms usually are created only by inadvertent error, for example:

> *Saprophilus* Steubel, 1939, is a junior objective synonym of *Creophilus* Leach, 1819, both being based on the same type species, *Staphylinus maxillosus* Linnaeus, 1758. Although an unlikely eventuality, it is possible for an objective junior synonym to become the valid name of a species if its senior synonym becomes a junior homonym of another name owing to amalgamation of species with the same specific epithet into a single genus.

Subjective synonyms: These are different names whose application to the same taxon is determined by systematists applying different taxonomies (circumscriptions). For example, the names of two nominal species originally described as distinct but later treated as conspecific are subjective synonyms. This type of synonymy is common, and in some complex cases the subject of great confusion.

> *Demarata* Distant, 1884, is a junior subjective synonym of *Ceratocapsus* Reuter, 1876, because the type species of *Demarata*, *D. villosa* Distant was judged, subsequent to its description, to belong to *Ceratocapsus*.

Subjective synonymy is one of the chief sources of nomenclatural instability due to disagreement among systematists about the *circumscription* of taxa. Those systematists who prefer to recognize many separate taxa with few included members in each are sometime referred to as "splitters," while those who prefer fewer, more inclusive taxa are called "lumpers," for example:

> The genus name *Speyeria* has been used for North American argynnine butterflies, while the names *Argynnis*, *Argyronome*, *Childrena*, *Damora*, *Fabriciana*, *Mesoacidalia*, *Nephargynnis*, and *Pandoriana* have been used for Eurasian members of the same group. A morphological cladistic study by Simonsen (2006) argued that all the species in these nine genera should be viewed as members of the genus *Argynnis*. Simonsen's circumscription was empirically well supported but did not

sit well with the authors and users of the many American and European butterfly guides whose taxonomy his work rendered obsolete. A more recent molecular study by de Moya et al. (2017) supported the respective monophyly of *Argynnis*, *Fabriciana*, and *Speyeria* and resurrected the latter two genera from synonymy. The *Codes* of nomenclature make no recommendations about such decisions—there isn't even a recommendation that named groups be monophyletic (although that is certainly a desideratum for most modern systematists, as we shall argue below in Arguments for Cladistic Classifications).

The Linnaean Hierarchy

The Linnaean hierarchy (Figure 8.2) is a system in which the names associated with the nested levels in the classification denote taxonomic rank. The same result could be achieved through a system of indentation of words on the page, the degree of indentation denoting rank in the hierarchy. Indentation requires a visual presentation in order to understand rank. The Linnaean system of names does not—one need only remember the rank order of the names used to define the hierarchy. It can thus communicate rank in both written and verbal communication. A more or less universally agreed-on rule of Linnaean nomenclature is that no taxon should contain subordinate taxa of an equal or higher rank, but this rule is often bent to accommodate asymmetrical, pectinate groups above the level of families. For example, the class Aves is sister group to a clade of saurischian dinosaurs, which presents a problem if one wants to recognize Dinosauria as an order.

Because Linnaeus had no reason to believe that his system would one day accommodate millions of species, he employed a limited number of levels (ranks), but his approach nonetheless reflected the nesting of groups within groups. The systematic hierarchy in the 10th edition of the *Systema Naturae* (1758) included

Kingdom
  Class
    Order
      Genus
        Species
          Variety

As knowledge of biological diversity increased, later authors incorporated more categories above and below the level of order. McKenna and Bell (1997) discussed this issue at length with respect to the Mammalia and probably incorporated as

I. COLEOPTERA.

Elytra alas tegentia.

170. SCARABÆUS. *Antennæ* clavatæ capitulo
fiffili.
Tibiæ anticæ fæpius dentatæ.

* *Thorace cornuto.*

Hercules. 1. S. thoracis cornu incurvo maximo fubtus barbato, ca
pitis cornu recurvato: fupra dentato.
Marcgr. braf. 247. *f.* 3. *Jonft. inf. t.* 16. *f.* 1.
Olear. muf. t. 16. *f.* 1. *Pet. gaz. t.* 70. *f.* 1.
Grew. muf. 162. *Swamm. bibl. t.* 30. *f.* 2.
Ræf. fcarab. 1. *t. A. f.* 1. *inf.* 4. *p.* 45. *t.* 5. *f.* 3.
Habitat in America.

Actæon. 2. S. thorace bicorni, capitis cornu tridentato: apice bi-
fido. *Muf. L. U.*
Marcgr. braf. 246. Euena. *Olear. muf. t.* 16. *f.* 2.
Mer. fur. t. 72. *Ræf. fcar.* 1. *t. A. f.* 2.
Hoffn. pict. 1. *t.* 1. *in medio. Swamm. bibl. t.* 30. *f.* 4.
Habitat in America.

Simfon. 3. S. thorace bicorni, capitis cornu apice tantum bifurcato.
Sloan. jam. 2. *p.* 205. *t.* 237. *f.* 4. 5.
Brown. jam. 428. *t.* 43. *f.* 6. Scarabæus 4.
Habitat in America.

Atlas. 4. S. thorace tricorni: antico breviffimo, capitis cornu re-
curvato. *M. L. U.*
Marcgr. braf. 247. *f.* 1. *Olear. muf. t.* 16. *f.* 3.
Pet. gaz. t. 49. *f.* 8. *an t.* 14. *f.* 12.
Merian. furin. in titulo F. G. Swamm. bibl. t. 30. *f.* 3.
Habitat in America.

Aloëus. 5. S. thorace tricorni : intermedio longiore, capite muti-
co, elytris uniftriatis. *M. L. U.*
Ræf. inf. 2. *fcar.* 1. *t. A. f.* 6. *Pet. gaz. t.* 24. *f.* 10.

Y 5 *Habi-*

Scarabæorum *Larvæ vivunt tranquillæ fub terra; harum pleræque fimo delectantur*
& eo pafcuntur.

FIGURE 8.2. The Linnaean hierarchy as portrayed by a page from the *Systema Naturae*, including the ranks of class (Insecta), order (Coleoptera), genus (*Scarabaeus*), and species (*hercules* and so on) (Linnaeus 1758).

many levels as does any existing, formal, modern classification. Today it is common to see 10 or more taxonomic ranks in both botany and zoology.

    Kingdom
      Phylum
        Class
          Subclass
            Order
              Suborder
                Superfamily (primarily in zoology)
                  Family
                    Subfamily
                      Tribe
                        Subtribe
                          Genus
                            Species

    Botany adds the infrageneric ranks of section and series and the infrasubspecific ranks of variety and form. Zoological nomenclature recognizes the use of trinominal designations for subspecies, following the rules for species-group names, but infrasubspecific names are not recognized by the zoological *Code* and are not available if published as such.

    To facilitate the determination of rank at the family-group level (names from superfamily to subtribe), the names bear unique suffixes in zoological nomenclature and for even more inclusive taxa in botanical nomenclature, usually as follows:

Rank	*Zoology*	*Botany*
Phylum or Division	—	-phyta
		-mycota (fungi)
Subphylum or Subdivision	—	-phytina
		-mycotina (fungi)
Class	—	-phycae (algae);
		-mycotina (fungi)
		-opsida (other groups)
Subclass	—	-phycidae (algae)
		-mycetidae (fungi)
		-idae (other groups)
Order	—	-ales
Suborder	—	-ineae
Superfamily	-oidea	—
Family	-idae	-aceae

Subfamily	-inae	-oideae
Tribe	-ini	-eae
Subtribe	-ina	-inae

Formal Classifications

We have now outlined the terminology and structure of biological nomenclature as they currently exist, which tell us how to use names in conformity with the nomenclatural *Codes*. As noted above, however, the *Codes* deliberately avoid any rules or recommendations pertaining to composition of taxa or their arrangement in relation to one another. To address this fundamental task of systematics, we need to revisit the aims and principles discussed in earlier chapters.

Arguments for Cladistic Classifications

Chapters 3 through 6 of this book dealt with methods and approaches for evaluating character data and bringing them to bear as evidence for phylogenetic inference. As we discussed in Chapter 1, systematists have believed since the time of Darwin that descent with modification provides the "hidden bond" that explains the Natural System. At the same time, the empiricist philosophical tradition (Mill 1843; Gilmour 1940) held that "natural" classifications should summarize character information most efficiently. Hennig argued that there should be a direct correspondence (*isomorphy*) between a formal classification and the underlying pattern of phylogenetic diversification, as closely as it may be inferred by empirical methods. Pheneticists argued that since the true course of phylogeny is unknowable, the character data alone should be considered in building "natural" classifications.

The epistemological optimality of the empirical correspondence between natural classifications and Hennigian character transformation-based phylogenetic trees was demonstrated by Farris (1979a, 1979b) on the basis of efficiency of diagnoses. Farris emphasized that a salient attribute of an irregularly bifurcating hierarchy is the impossibility that every feature be most informative for every group considered, because some groups are subsets of other groups (Hennig's heterobathmy of synapomorphy). Farris proved formally that in this regard, a most parsimonious tree (1) allows all data to be summarized in the most succinct diagnoses of clades; (2) provides every clade in the tree with specific, identifiable synapomorphies that support it; (3) minimizes instances of homoplasy; and (4) permits the greatest number of predictions about character-state distributions between adjacent taxa. Such conclusions correspond closely to Hennig's ideas concerning the role of classifications as the general reference system for biology.

The following example illustrates the points made by Hennig and Farris. Consider the traditional classification of the Insecta:

Insecta
   Entognatha
   Ectognatha
      Apterygota
         Machilida (Archaeognatha)
         Thysanura (Zygentoma)
      Pterygota
         Odonata
         Ephemeroptera
         Neoptera

The Machilida and Thysanura were long grouped together—as above—because they lack wings (a symplesiomorphy) but nonetheless share many features not found in the more primitive entognathous insects. However, the mandibles in Machilida have a single point of articulation, while the Thysanura share with all winged insects the dicondylar structure of the mandibles. This synapomorphy is recognized in the following classification:

Insecta
   Entognatha
   Ectognatha
      Machilida (Archaeognatha)
      Dicondylia
         Thysanura (Zygentoma)
         Pterygota
            Odonata
            Ephemeroptera
            Neoptera

In this scheme the Thysanura are not grouped with the Machilida as members of "Apterygota" but rather are the sister group of winged insects on the basis of mandibular articulation and other distinctive features. Although the first classification would seem to convey information on similarity more effectively because the wingless insects are grouped together, that notion is simply an illusion. Grouping based on the complementary absence of distinctive characters allows for any arbitrary assemblage to be recognized as a "group." For example, the absence of insect wings as a group-defining character could group Machilida and Thysanura with any other organism except winged insects, such as Collembola, oak trees, and bacteria. Such an approach would require that all characteristics of all "groups" be completely enumerated in order to recognize them, an obviously inefficient system. Furthermore, such groups would be overlapping and mutually contradictory. Imagine, for example, another group of all organisms lacking a dicondylar apophysis on the mandibles, which would include Machilida and oak trees but not Thysanura. Grouping on the basis of synapomorphy (parsimonious accounting of character-state transformations rather than degree of character-state identity) produces the most efficient hierarchic representation of the data because the repetition of attributes is minimized.

Arguments Against Cladistic Classifications

We have seen that parsimony accounts for the character-state transformations most efficiently, in the form of nested patterns of synapomorphy, and provides an empirically sound basis for classifications that can be explained in evolutionary terms. However, in our quest for circumspection, we will examine arguments against cladograms as the basis of biological classifications and against the application of parsimony more generally.

For many years, the most frequently heard objections to cladistic classifications were those of the "classical evolutionary taxonomists" (e.g., Mayr 1974; Ashlock 1979), who argued that

1. the cladistic approach would require the reranking of many familiar taxa;
2. there would be too many hierarchical ranks; and
3. under Hennig's *monophyly* concept, cladistic classification schemes would discard information on similarity, an important aspect of evolution.

The pheneticists likewise argued that

1. information on similarity would be discarded;
2. there would be too many levels in the classification;
3. the cladistic classifications are generally not symmetrical, with unequal numbers of taxa in sister clades, and therefore defective; and
4. the cladistic approach would not produce true classifications because it is not based on an ultrametric and therefore does not treat rates of evolution in different lineages as equal (see Sidebar 11, Chapter 5, for further discussion of metrics).

Let us examine the substance of these complaints.

"Discarding information on similarity" is really an argument that groups based on symplesiomorphies, or combinations of synapomorphies and symplesiomorphies, should be retained. As we discussed above in the "Apterygota" example, such groups are artificial, arbitrary, and lack predictive value.

Objections to assigning new ranks to familiar taxa may have their deepest roots in the traditions of taxonomy as well as the conservative desire to preserve nomenclatural stability. Yet if taxonomic ranks are intended to connote relationship, objecting to correction of ranks or circumscriptions that incorrectly portray relationships would seem to be scientifically counterproductive and disingenuous. Alternatively, if taxonomic ranks are not intended to connote relationship, then they convey no intrinsic information at all and represent arbitrary groups formed at the whim of the taxonomist. This point was made by Gaffney (1979a) and others, who concluded that a perfectly stable classification could represent only the perpetuation of ignorance. If systematists did not modify their classifications in light of new taxonomic data, all butterflies, along with zygaenid moths and owl flies (which belong in the order Neuroptera), would still be in the Linnaean genus *Papilio* (butterflies now comprise two superfamilies), and the orangutan would still be in the genus *Homo*.

The cladistic approach to classification, reflecting the irregularly bifurcating hierarchy of the Natural System, does result in a proliferation of ranks in the Linnaean hierarchy compared with Procrustean, artificially tidy systems. Nonetheless, reranking has largely disappeared as an issue because even the most ardent supporters of Hennig's original mandate concerning ranking realized that producing a single unified hierarchical classification for all of life would involve so many ranks as to be unfeasible, cladistic classification or otherwise. In the best-case scenario, a fully resolved, completely symmetrical tree of a million taxa would have at least 20 ranks. The simple solution to this problem is to not assign a name and rank to every node. Hennig (1969) himself apparently recognized the difficulty of creating one very large classification with a consistent set of ranking conventions and consequently adopted a numbering system in his classification of the Insecta, a portion of which is shown in Figure 8.3.

The idea that a classification must be ultrametric could be dismissed as purely arbitrary, but other arguments are available. Farris (1979a, 1979b) showed that it is impossible for a clustering-level distance (ultrametric) to fit distance data better that the best-fitting path-length distance (see Sidebar 11, Chapter 5). Thus, the most parsimonious path-length interpretation will always produce the most efficient hierarchic description of the data. The fact that phenetic methods are based on clustering levels (ultrametrics) does not ipso facto mean that such methods always portray information most efficiently in a hierarchic system. In fact, that conclusion is false.

The two remaining arguments of the pheneticists have an empirical relationship to one another. Whether a classification is phenetic or cladistic in origin, given the same topological relationships of the taxa, the numbers of levels will be the same, a point made by Farris (1979a), and certainly one widely appreciated outside

1. Entognatha
    1.1. Diplura
    1.2. Ellipura
    1.2.1. Protura
    1.2.2. Collembola
2. Ectognatha
    2.1. Archaeognatha (Microcoryphia)
    2.2. Dicondylia
    2.2.1. Zygentoma
    2.2.2. Pterygota
    2.2.2.1. Palaeoptera
    2.2.2.1..1. Ephemeroptera
    2.2.2.1..2. Odonata
    2.2.2.2. Neoptera
    2.2.2.2..1. Plecoptera
    2.2.2.2..2. Paurometabola
    2.2.2.2..2.1. Embioptera
    2.2.2.2..2.2. Orthopteromorpha
    2.2.2.2..2.2..1. Blattopteriformia
    2.2.2.2..2.2..1.1. Notoptera
    2.2.2.2..2.2..1.2. Dermaptera
    2.2.2.2..2.2..1.3. Blattopteroidea
    2.2.2.2..2.2..1.3.1. Mantodea
    2.2.2.2..2.2..1.3.2. Blattodea
    2.2.2.2..2.2..2. Orthopteroidea
    2.2.2.2..2.2..2.1. Ensifera
    2.2.2.2..2.2..2.2. Caelifera
    2.2.2.2..2.2..2.3. Phasmatodea
    2.2.2.2..3. Paraneoptera
    2.2.2.2..3.1. Zoraptera
    2.2.2.2..3.2. Acercaria
    2.2.2.2..3.2..1. Psocodea
    2.2.2.2..3.2..2. Condylognatha
    2.2.2.2..3.2..2.2. Thysanoptera
    2.2.2.2..3.2..2.2. Hemiptera
    2.2.2.2..3.2..2.2.1. Heteropteroidea
    2.2.2.2..3.2..2.2.1.1. Coleorrhyncha
    2.2.2.2..3.2..2.2.1.2. Heteroptera
    2.2.2.2..3.2..2.2.2. Sternorrhyncha
    2.2.2.2..3.2..2.2.2.1. Aphidomorpha
    2.2.2.2..3.2..2.2.2.1.2. Aphidina
    2.2.2.2..3.2..2.2.2.1.2. Coccina
    2.2.2.2..3.2..2.2.2.2. Psyllomorpha
    2.2.2.2..3.2..2.2.2.2.1. Aleyrodina
    2.2.2.2..3.2..2.2.2.2.2. Psyllina
    2.2.2.2..3.2..2.2.3. Auchenorrhyncha
    2.2.2.2..3.2..2.2.3.1. Fulgoriformes
    2.2.2.2..3.2..2.2.3.2. Cicadiformes
    2.2.2.2.4. Holometabola

FIGURE 8.3. Sample of the numbered hierarchic classification of the Insecta (excluding Holometabola) from *Die Stammesgeschichte der Insekten* (Hennig 1969).

phenetic circles. However, if the symmetry of the classification is changed, then the numbers of levels may change. Whether one approach to classification produces results that are more consistently symmetrical than another would seem to be irrelevant for judging a method (see Sidebar 14, Chapter 6), unless the goal of the method was none other than to minimize the number of hierarchic levels. Such a goal has not been claimed a priori for any approach to classification in modern times (see Strickland's comments about Procrustean symmetrical systems in Sidebar 17), and its use as a criticism of cladistic results can hardly be taken seriously.

Observation reveals, nonetheless, that the results of cladistic analyses often do show pronounced asymmetries—what are frequently referred to as *pectinate* or *comblike* cladograms. As we have discussed throughout this book, cladistic classifications, pectinate or otherwise, result from applying methods designed to best accommodate all available phylogenetically informative data. Thus, the topological form of the hierarchy in no way undermines the utility of the classification, and cladistic methods do not favor one topological form over another. The classification is simply a linguistic representation that corresponds to the inferred hypothesis of relationships among the taxa.

If there is a general conclusion to be drawn from the foregoing 50-plus years of methodological discussion about biological classification, it is the erroneous nature of the assumption that grouping by overall similarity would provide a more efficient description of information on genealogical relationships than does grouping by synapomorphy. The issue of whether classifications containing only monophyletic groups discard information on similarity, as claimed by pheneticists, evolutionary taxonomists, and those enamored of "branch length" would seem to have been unequivocally resolved.

By themselves, classifications describe hypotheses about group relationships, which allow us to make predictions concerning the distribution of features unique to the contained taxa (Platnick 1978). But, as has been repeatedly emphasized, the hierarchic representation of taxonomic relationships does not by itself transmit the information on which the classification is based. That information is in the *diagnosis*—the description of synapomorphies that support a given clade. It is diagnoses, enumerations of parsimonious patterns of character-state transformation, that describe most efficiently the inferred hypothesis of genealogical relationships while at the same time conveying as efficiently as possible the information on divergence (e.g., Farris 1979a, 1979b).

Converting Cladograms to Linnaean Hierarchies

Two approaches have been proposed for converting cladograms into formal classifications. These are *subordination* and *sequencing*. The former approach was advocated by Hennig; the latter approach was apparently first codified by Nelson (1974). The results of applying these two approaches are summarized in Figure 8.4.

Subordination. This approach requires that each branching level in a cladogram receive a distinctive hierarchic designation. Subordination requires the maximum number of ranks in the Linnaean presentation of the classification but allows for exact retrieval of the branching pattern without recourse to an additional set of conventions.

Sequencing. This approach allows progressively nested sister-group relationships to be of equal rank, thereby requiring fewer ranks, particularly in rendering a pectinate scheme. However, sequencing necessitates knowing the convention by

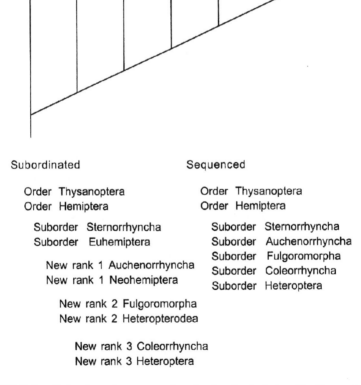

Subordinated

Order Thysanoptera
Order Hemiptera

   Suborder Sternorrhyncha
   Suborder Euhemiptera

      New rank 1 Auchenorrhyncha
      New rank 1 Neohemiptera

         New rank 2 Fulgoromorpha
         New rank 2 Heteropterodea

            New rank 3 Coleorrhyncha
            New rank 3 Heteroptera

Sequenced

Order Thysanoptera
Order Hemiptera

   Suborder Sternorrhyncha
   Suborder Auchenorrhyncha
   Suborder Fulgoromorpha
   Suborder Coleorrhyncha
   Suborder Heteroptera

FIGURE 8.4. Examples of sequenced and subordinated classifications of the Hemiptera (scale insects, cicadas, true bugs, and so on).

which the Linnaean scheme was rendered in order to retrieve the branching pattern exactly. As shown by Wiley (1979), with enough conventions, any group can be classified or reclassified without a proliferation of ranks. But those conventions must be recorded and remembered or indicated clearly on a tree diagram accompanying the classification.

Classification of fossils and taxa of uncertain position. The issue of fossil taxa and cladistics was at one time a subject of impassioned debate, primarily because of the traditional paleontological view that fossils might actually represent the ancestors of Recent taxa. We have noted in Chapter 4 that logically there is no way to test theories of ancestor-descendant relationships in a cladistic framework, for several reasons. First, the admission of ancestors to the system would allow group recognition based on the absence of characters. Second, as noted by Farris (1976b), there is no obvious way to represent ancestor-descendant relationships in a formal hierarchic classification. Note, once again, that the rejection of ancestors in cladistics is not a denial that ancestors existed but rather a recognition of the empirical and epistemological impracticality of reifying a particular fossil as such. We might then conclude that all groups—fossil and Recent—should be treated as terminal taxa and that "the main problem of classifying fossils is not the accommodation of 'ancestral groups,' but rather groups of uncertain relationship" (Nelson 1972:230). Having arrived at this conclusion, the same methods can be used for dealing with all *incertae sedis* taxa—that is, those of uncertain position—irrespective of whether those taxa are Recent or fossil.

Taxa of uncertain position often attain that distinction because they are incomplete and therefore not able to be scored for some critical characters, as is the case with many fossil specimens. In a cladogram, such taxa often form part of a trichotomy or polytomy, being related to other taxa only at the level that the available character information allows. A practical approach was advocated by Nelson (1972), whereby fossil groups to which no Recent members are assigned are preceded by a dagger (†). Taxa of uncertain position are placed at the level at which their relationships can be positively determined and labeled as *incertae sedis*. The following example will illustrate the point:

Order A
†Species A (*incertae sedis*)
  Family A
    Genus A
      Species B
      Species C
  Family B
  Species D (*incertae sedis*)

Genus B
   Species E
   Species F
Genus C
   Species G
   Species H

Species A (a fossil) is missing synapomorphies that would associate it with either Family A or Family B and can be placed with certainty only at the level of Order A. Species D exhibits synapomorphies that allow it to be associated with Family B but not either of the included genera (perhaps it is known only from a single specimen that does not exhibit the features that would allow more precise diagnosis). The species B, C, E, F, G, and H have sufficient information to place them at the generic level. Further discussion of the classification of fossil taxa can be found in the works of Farris (1976b), Patterson and Rosen (1977), and Wiley (1979).

Stem groups and crown groups. The proposals outlined above allow for the placement of both fossil and Recent taxa in a phylogenetic context. Nonetheless, the apparent desire to understand the "evolutionary history" of fossils has led to proposals that would seem to contradict the core principles of phylogenetics, namely the recognition of Recent *crown groups* (clades of extant taxa, monophyletic groups) and, more particularly, their complementary paraphyletic fossil *stem groups* (Jefferies 1979; see also Ax 1985). Paradoxically, the foundations for much of this thinking are attributed to the works of Hennig (1969, 1981) and his attempts to understand phylogenetic relationships within the Insecta.

The reasons offered for using the crown group–stem group terminology seem to be multiple. First, as in the example above, many fossil taxa are fragmentary and lack the defining attributes of groups based on morphologically complete Recent taxa. Second, irrespective of quality of preservation, extinct lineages frequently outnumber extant lineages, and their recognition as nested sister taxa of extant groups can lead to a proliferation of categorical ranks, as is well illustrated by the work of McKenna and Bell (1997) on the classification of the mammals, many lineages of which are known only as fossils. Third, in the minds of at least some authors, the absence of character information in fossils suggests their potential status as ancestral to Recent taxa. This would seem to represent an erroneous equivalence of absence of evidence with evidence of absence.

Although reference to crown and stem groups has grown dramatically in the last 20 years (Donoghue 2005), usage of the terms has been restricted almost exclusively to the paleontological literature and to efforts to define names in "phylogenetic nomenclature" (see section below). Stem groups are nearly always paraphyletic and therefore not groups at all in a phylogenetic context. The concept is thus of dubious utility and—in our view—should be abandoned to avoid confusion.

Conclusions. Representing the results of a phylogenetic analysis in a formal classification should take into account the following points:

1. Only monophyletic groups should be named.
2. Named groups should be those that need to be recognized by name (i.e., empty ranks need not be named). A good example might be the Machilida, discussed earlier in this chapter. As the sister group of all other ectognathous insects, eight ranks would be required to produce a portrayal of ranks for approximately 300 species of bristletails equivalent to that required to deal with the 1 million remaining species of ectognathous insects, namely, Dicondylia, Pterygota, Paleoptera, Paurometabola, Paraneoptera, Acercaria, Condylognatha, and the orders. In fact, the levels most commonly used are subclass Archaeognatha and order Machilida. At the other end of the scale, there is not much point in assigning a name to every node in a fully resolved cladogram for a speciose genus.
3. Traditional rankings may be retained (e.g., class Aves and class Insecta) with an implicit adoption of the sequencing convention. Hierarchic relationships are established through the naming and (sometimes) ranking of necessary inclusive groupings. The insect orders represent one of the best examples of this practice. Some of the currently recognized groupings were established by Linnaeus and have been used at the same rank level ever since. Phylogenetic studies have revealed, however, that not all insect higher groups traditionally recognized as orders merit equal rank in a phylogenetic system, as can be seen in the classification of Hennig (see Figure 8.3). However, through the use of approximately eight ranks above the level of order, the hierarchic relationships of the major insect groups can be portrayed, while at the same time retaining most of the traditionally recognized orders.

It is worth noting at this juncture that even though classification and phylogeny are viewed as having an inextricable relationship, published classifications of living organisms may not be based on explicit phylogenetic hypotheses. Over the time span encompassing three editions of this book, the desirability of basing classifications on phylogenetic analyses has become widely recognized. Our ability to achieve this goal, however, remains limited by a variety of factors, including (1) the dwindling number of practicing systematists; (2) the reality of not yet having, for many groups, data capable of resolving detailed hypotheses of relationships among organisms; and (3) the shift of focus in research and training from systematics to "phylogenetics."

Criticisms of the Linnaean System

The Linnaean hierarchy, as embodied in the *Codes* of nomenclature, has been the subject of two general categories of criticism. Both address the issue of *stability*, a caveat of the *Codes*.

First is modification of names themselves, irrespective of the taxonomic concepts attached to them. The most obvious example is the creation of a *new combination*. When a species is transferred from one genus to another and the gender of the generic names is different, if the specific name is an adjective, its ending must be changed to maintain gender agreement (in Latin, nouns can be masculine, feminine, or neuter in gender), for example:

> *Dichrooscytus speciosus* Van Duzee (a masculine genus name) became *Bolteria speciosa* (Van Duzee) in order to maintain gender agreement upon its transfer to *Bolteria* (apparently a feminine genus name).

Gender agreement has been a particular irritant to zoologists, where versions of the zoological *Code* prior to 1999 dictated the gender of many generic names, irrespective of the gender implied in their usage by the original author. The result has been that some "authorities" on nomenclature have modified many specific names for no other reason than to achieve apparent agreement between the gender of the genus—as determined by the *Code*—and all of its included species. See Turner (1967) for an entertaining critique of this practice.

The requirement of gender agreement is an impediment to stability of species-group names that could be removed by not requiring agreement between generic nouns and adjectival specific epithets. Such an approach would require deciding whether specific epithets should adopt their original termination or their current termination; the principle of priority would be served by the former. The consistent spelling of names has particular advantages for use of computer databases, which are powerful tools for the organization and management of information in biological systematics but at the same time are very myopic regarding spelling differences.

Given the clear practical advantages for stability, one might expect adherence to the rules of Latin grammar to come under more pressure for change. But, in the 1999 revision of the zoological *Code*, gender agreement was maintained despite efforts to remove its application when new combinations are created. One way to avoid the gender agreement issue with new species names is to designate them as "nouns in apposition," which have a gender of their own and always retain their original spelling regardless of the generic name to which they are attached. It should be pointed out that the availability of names is unaffected by the fact that in their original form they do not have correct Latin orthography. In such cases, bastard names have usually been treated as if they were simply an arbitrary combination of letters, a legitimate approach under the zoological *Code*. As an example, we might point once again to the erroneous use of *nigrus* as the masculine form for reference to species that are black in color.

Second, and less subject to objective resolution, is the issue of the systematic concepts attached to names. Typification brought a sense of order to nomenclature that had not been achieved previously, this "transition" occurring earlier in zoology than in botany. Yet it poses what some systematists have viewed as problems with no obvious or universally acceptable solution, especially in light of the long nomenclatural history with which biologists must contend.

At the simplest level, consider type specimens. A poorly chosen specimen for a given species may leave the identity of the taxon in permanent doubt. For example, a female holotype in a genus in which species identification is based on details of male genitalia may render comparisons between this and other species moot, if positively associated males of the female-based species cannot be found. Old, badly damaged holotypes can also be a source of ambiguity. Nonetheless, in the vast majority of cases, the existence of type specimens—in combination with well-prepared printed descriptions and figures—has greatly facilitated the uniform application of species-group names. Safeguarding type specimens in perpetuity as the anchors of biological nomenclature is one of the critical functions of museum collections.

All supraspecific categories have "nomenclatural" types. In the case of genera, a "poorly chosen" type species can cause problems with generic circumscription. In the example above, if the designated female holotype pertains to the type species of the genus, the issue of generic relationships may be complicated or unresolvable.

Another potential drawback of typification is sometimes manifested in groups with histories that predate the establishment of the *Codes*. When, upon revision or other critical examination, two long-used and well-known generic names are found to be synonymous because their type species are deemed to be congeneric, one of the generic names will be treated as invalid under the rule of priority. Such a situation would be of no import to a computer, but it has proved vexing for taxonomists and other biologists who attach particular concepts to certain names. A good example is the fondness of behavioral ecologists and conservation biologists for the generic name *Maculinea*, shown to be a subjective junior synonym of the less mellifluous *Phengaris* (see above). The last line of recourse is to petition to the international commissions for conservation of the name whose concept is more widely understood, thus setting aside the rule of priority.

If fixed types were removed from the equation, concepts could probably be made more stable by changing the type of a supraspecific taxon—from time to time—to improve conceptual agreement. This would, however, leave the field wide open for changes based on the personal choice of investigators with competing theories of generic limits and relationships (the lumpers and splitters). The choices could range from placing all species in a single genus, on one extreme, to describing a unique genus for each species, on the other—the same dire circumstances Strickland sought to avoid!

Thus, the issue boils down to whether priority and typification are the best possible solutions for maintaining a universal system of names in a situation that, under any circumstances, will always involve some compromise. At least two alternative approaches have been proposed. First was *uninominal nomenclature* (Michener 1963). The idea (using just a single name for each taxon), designed to do nothing more than promote stability of names, received little attention at the time of its introduction into the literature because of its obvious shortcomings in terms of information content (the Linnaean binomen serves the same useful function as a familial surname, allowing one to distinguish, for example, between Donald Trump and Donald Duck). Uninominal nomenclature seems to have disappeared completely from consideration. A more recent, related proposal—*phylogenetic nomenclature*—has been the subject of considerable controversy in the systematic literature over the past 25 years.

Phylogenetic Nomenclature

Starting in the late 1980s, a small group of systematists proposed that the Linnaean hierarchy should be scrapped in favor of a rankless "phylogenetic taxonomy," or what is better termed *phylogenetic nomenclature*. Their arguments are founded on the precept (introduced in Chapter 2) that Linnaean nomenclature is based on what they considered to be the incorrect ontological perspective that species are classes, rather than individuals, and that Linnaean ranks thus represent an undesirable holdover of Aristotelian essentialism (Rowe 1987; de Queiroz 1988; de Queiroz and Gauthier 1990). These claims derive from a certain philosophical viewpoint (e.g., Hull 1965; Ghiselin 1966; see Sidebar 15, Chapter 7) and can be viewed as resting largely upon a restrictive meaning of the term "definition" as denoting an "essence," with no consideration of how the meaning attached to that term may have changed over the past 2000 years or how it is employed by practicing taxonomists.

Because they are ontologically committed to taxa as lineages that exist in the world, irrespective of systematic knowledge of their characters, the advocates of phylogenetic taxonomy believe that taxa should be defined exclusively in terms of common ancestry rather than according to their attributes. For example, Lepidosauria is defined as *Sphenodon* and squamates and all saurians sharing a more recent common ancestor with them than with crocodiles and birds. This definition (more aptly, circumscription) includes not only extant taxa and known fossil taxa but also hypothetical entities. According to de Queiroz and Gauthier (1990), defining taxa this way allows evolutionary considerations to enter directly into taxonomic definitions rather than assessing them after the fact, which they view as good. Phylogenetic definitions of taxa would—in the view of its proponents—clarify the distinction between *definition* and *diagnosis.* Definitions of taxon names would become ontological statements referring to groups that are presumed to exist under the central tenet of common descent, independent of our ability to recognize them, while diagnoses would retain their standard meaning as a list of features empirically differentiating one taxon from others. Furthermore, according to de Queiroz and Gauthier (1990:313), taxa recognized in the "phylogenetic taxonomic system" would not be concepts but real things—systems deriving their existence from common ancestry relationships among their parts. Note that in order to conceive what Lepidosauria is under the above definition, you need to know a priori what *Sphenodon,* squamates, saurians, crocodiles, and birds are, not to mention exactly how they are related to one another. If the definitions of these, in turn, refer to other taxa and relationships, such definitions would seem to assume an infinite regress of a priori knowledge—rather strikingly Platonic for a system supposedly designed to be rid of essentialism.

Aside from the philosophical overcommitment to metaphysics embodied in this approach, addressed in general terms in Chapter 2, the issue of taxonomic "reality" can easily be questioned from a pragmatic perspective by considering the case of a group whose topology has been altered as a result of revised phylogenetic analysis. Advocates of phylogenetic nomenclature make little mention of this point, perhaps assuming that hypotheses of relationship in their system will be self-evidently true and neither questioned nor revised. Yet in actual practice, the "real" groups defined in phylogenetic nomenclature derive their epistemological recognition from phylogenetic analyses based on attributes thought to be diagnostic for the taxa (i.e., synapomorphies). Thus, a basic logical conflict persists in all presentations of the phylogenetic-taxonomic approach, despite the lengthy and impassioned disquisitions of its proponents. As we have noted elsewhere, the "reality" of taxonomic groups is not a self-evident property. It is, rather, the result of analysis and synthesis of observation to form successively more general and encompassing hypotheses of relationship. Those observations are represented by nothing more nor less than the lower-level theories regarding attributes (characters) of the organisms themselves. That epistemological framework, central to our attempts to acquire knowledge of genealogical relationships among organisms, will not disappear through appeals to "evolutionary considerations" by advocates of phylogenetic nomenclature. Taxa and their circumscriptions remain hypotheses, while the hierarchy of names attached to those hypotheses is anchored to individual specimens by typification. We note with some amusement that animal names defined on the basis of ancestry are excluded from zoological nomenclature under the zoological *Code* because the *Code* expressly forbids (under Article 1.3.1) the naming of hypothetical concepts—the very problem avoided by typification. We would further observe that phylogenetic nomenclature commingles names of taxa with taxonomic concepts, an approach that is also expressly forbidden by the existing *Codes* of nomenclature. These incompatibilities show that the two systems are mutually exclusive.

The claim that phylogenetic nomenclature can provide a more stable basis for our taxonomic concepts of higher taxa was addressed by Dominguez and Wheeler (1997). They pointed out that information on hierarchic relationships implicit in the ranked names of the Linnaean system would be lost in the rankless naming system of de Queiroz and Gauthier. They further observed that as hypotheses of relationships are revised, taxa named under the system of de Queiroz and Gauthier can change their level of generality radically, from being part of a group to including that same group. Thus, while the names remain stable as words, their referents can change dramatically. Nevertheless, a group of advocates has produced various drafts of a "Phylocode" (Cantino and de Queiroz 2007; de Queiroz and Cantino 2020) intended to supplant the zoological and botanical *Codes* of nomenclature by establishing a parallel system of names. This effort has met with a barrage of criticisms of its underlying philosophy and practical implications from other systematists (Nixon and Carpenter 2000; Carpenter 2003a; Keller et al. 2003; Nixon et al. 2003; Schuh 2003; Barkley et al. 2004; Rieppel 2006b). With the publication of a printed "final" version of the *Phylocode* (de Queiroz and Cantino 2020), there undoubtedly will be a reignition of this controversy (cf. Brower 2020).

Despite all the Sturm und Drang from its advocates over the past 30 years, phylogenetic nomenclature seems to have had a minimal impact on the ongoing practice of naming taxa by working systematists and to have attracted very few converts from the traditional *Codes*. Finally, we concur with Nixon et al. (2003) that if the desideratum is nomenclatural stability, then perhaps tossing out 250+ years of systematic literature and making up an entirely new nomenclature is not the best way to achieve that goal.

Aids to the Use of Nomenclature

Finding and applying names correctly in biology requires some well-organized aids. Under the currently accepted *Codes* on nomenclature, these differ between botany and zoology, and we will therefore discuss them separately.

Zoology

The ICZN maintains a website that contains links to three essential taxonomic tools. First is the zoological *Code*, which has undergone several revisions, the most recent published in 1999 (some modifications to expand and refine methods of publication have lately been proposed [International Commission on Zoological Nomenclature 2008, 2012]). For any working taxonomist, access to the most recent *Code* is indispensable, and the ICZN is to be commended for posting a free and accessible version of the current *Code* on the web. Second, the ICZN also publishes the *Bulletin of Zoological Nomenclature* online, in which petitions to and decisions of the commission appear. Petitions include proposals for setting aside the rule of priority in order to suppress long-unused names in favor of more recently published names that have been widely used and accepted by the scientific community. The commission might also be asked to decide which of two homonymous family names should remain unchanged and which should be replaced. The third resource available from the ICZN website is ZooBank, a growing electronic world register of animal names, which ultimately will replace prior published (printed) efforts.

The *Zoological Record* prepares a listing of many of the generic (and subgeneric) names newly published in zoology. These names are freely available online through the Index of Organism Names. The complete *Zoological Record* is available through institutional subscription (see also Chapter 1).

The *Nomenclator Zoologicus* (Neave 1939–1996) is now available as a searchable online listing of most generic names published in zoology from 1758–2004. Listings of species names are difficult to prepare on a comprehensive basis because of the multitude of taxa. Nonetheless, at least one work— *Index Animalium*— attempted such a universal listing for animal names published between 1758 and 1850 (Sherborne, 1902–1932). This work is valuable in discovering and attributing older names, especially those placed in large, poorly defined genera of early authors and that may now reside in several families. The *Index Animalium* is also now available online.

Botany

Botanical systematics, and its attendant nomenclature, has a somewhat more streamlined history than that in zoology. Sources such as the *Index Kewensis* (now maintained online by the International Plant Names Index, IPNI) offer a centralized source of information of a sort found only widely scattered in the zoological literature. On the other hand, the *Zoological Record* serves as a more-or-less comprehensive indexing source in zoology, whereas no equivalent publication exists in botany. The botanical *Code*, available online, not only provides an up-to-date rendering of the rules of nomenclature in botany (including mycology) but also includes a list of all "conserved" names for chlorophyll-producing organisms, fungi, and slime molds. A further online source for nomenclatural information in botany is IPNI. This online index is the result of a collaborative effort by the Royal Botanic Gardens, Kew, the Harvard University Herbaria, and the Australian National Herbarium. It provides information on names and associated bibliographical details for all seed plants, ferns, and fern allies. Traditional books, such as *Mabberley's Plant Book* (2017) remain as invaluable and readily available resources for understanding the application of botanical nomenclature at the generic level.

Digital Data Resources for Systematics

Managing information on no fewer than 2.5 million species of living organisms—and possibly 10 times that many—is a massive task. In the electronic age, two technological approaches now dominate attempts to acquire, organize, and disseminate this information. First, virtually all such efforts have database technology at their core. Second, the internet now plays a central role in making this information available.

The types of data being managed and delivered can be divided into three more-or-less discrete categories.

1. Nomenclatural data
   a. Names (including synonymy, homonymy, and nomenclatural types)
   b. Literature references
   c. Type specimen information
   d. Images
2. Specimen-based data
   a. Label data from individual specimens
   b. Specimen deposition (museum; private collection)
   c. Distributional information
   d. Biological associations, economic status, and so on
   e. Images
3. Taxon-based data
   a. Character data and analyses
   b. Images

Although these areas are intimately related to one another, historically and functionally they are distinct, and therefore each presents its own management problems and required solutions. We will discuss each category and provide examples, then discuss how they are being integrated on the internet.

Nomenclature and Taxonomy

Whereas at one time all information on biological nomenclature and taxonomy existed in printed form, it is inexorably moving to electronic media. We have already mentioned the *Nomenclator Zoologicus, Zoological Record*, and IPNI, all of which are available online and allow for rapid context-based searching. Crowd-generated online sources such as Wikispecies may be useful as an entry point into the literature, but such web-based taxonomic resources are not reliably authoritative and should not be used as primary references.

A number of national and international consortia and initiatives (e.g., the Integrated Taxonomic Information System [ITIS], Catalogue of Life [CoL+], Universal Biological Indexer and Organizer [uBio], ZooBank) have been formed to grapple with this issue of acquiring, organizing, and delivering up-to-date information on nomenclature and classification. Each embodies a slightly different primary objective and applies different methods to achieve that objective. ITIS and CoL+ (formerly Species2000) are self-contained enterprises that provide data they have organized themselves or that has been provided to them, which is delivered directly from their web servers. The ultimate goal is to provide *authority files* for the names of all taxa and to place those names in some classificatory context. In point of fact, species authorities such as CoL+, ITIS, and the Global Biodiversity Information Facility (GBIF), ECat, and Global Names Index (GNI) initiatives (Patterson et al. 2010) are databases of species names, which do not necessarily include all available names and often provide only minimal data (such as author and year of publication) and omit vital nomenclatural documentation, such as whether or not the name refers to a valid taxon. Two further problems with these sources are that they frequently share data files, so errors in one database are likely to be promulgated in others, and they generally do not engage directly with working systematists, so their nomenclature tends to be static and can become out of date (Franz and Sterner 2018).

The ZooBank initiative, which is strictly nomenclatural in nature, was organized as a way of registering names proposed in zoology. This initiative on the part of the ICZN will potentially add an improved sense of order to the introduction of new names and allow for the assignment of life science identifiers (LSIDs; a restricted example of the global unique identifier [GUID] concept) to new names in zoology.

UBio is a web "aggregator" of information otherwise available on the internet, but because of the location of that information on myriad sites it is not directly interpretable through the results assembled by web search engines. UBio uses nomenclature as a core organizing principle. Through the application of sophisticated software approaches it is capable of taking nomenclatural data—and other attendant biological information—and rendering it in a much more meaningful format than a search engine can. The UBio approach relies on computer algorithms to parse nomenclatural data but sometimes with limited success; as for example, in the case of homonyms such as Archaea, which is a name applied at a high level to basal unicellular lineages and to a genus of spiders.

Many freestanding sources that deliver information on nomenclature and classification for particular taxa are now available on the internet. Online catalogs also rely on databases (usually relational) to store and deliver information. Although this approach may appear as a "black box" to the user, it offers the power to search and organize information in a way that is not possible with "flat" HTML pages, such as the Catalog of Sphecidae (Pulawski 2008).

Possibly the largest online nomenclatural database is Systema Dipterorum (Pape and Evenhuis 2019), which delivers information on nomenclature and classification for well over 100,000 species of true flies and the literature pertaining to those names. Hymenoptera Online (2019) delivers information for a large number of groups within the Hymenoptera, most of the data being supplied by specialists in the field rather than being compiled directly by the creator of the database. The World Spider Catalog (2020) delivers information on more than 48,000 spider species and allows extensive searching as well as access to relevant literature. Species File is an online utility for the preparation of web-based systematic catalogs, all presently dealing with nonholometabolous insects such as Orthoptera and Hemiptera.

Another example is the On-line Systematic Catalog of Plant Bugs (Schuh 2002–2013), which delivers detailed nomenclatural and other information on more than 11,000 valid species of Miridae (Insecta: Heteroptera) via a database that makes connections between the names, the literature, and information on host associations in the group. This web application allows for searching via categories of information (e.g., genus, species, author, host, geographical distribution) or "drilling down" through a taxonomic hierarchy, the latter approach helping to ameliorate the "black-box" effect. A simplified representation of the database schema is shown in Figure 8.5.

Within the Vertebrata, Eschmeyer's Catalog of Fishes (Fricke et al. 2020) is a valuable authoritative online source for the largest monophyletic subgroup. Other online resources for Vertebrata include Amphibian Species of the World (Frost, 2020) and Mammal Species of the World (Wilson and Reeder, 2020).

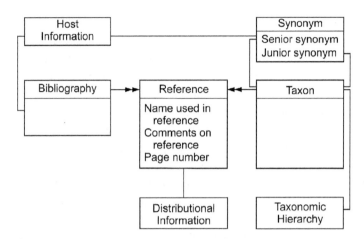

FIGURE 8.5. Relational database model for managing bibliographic and nomenclatural information, simplified from the application of Schuh (2002–2013).

The resources enumerated here are just a sample, and the number of databases and sites devoted to nomenclature increases daily—see, for example, iDigBio, Darwin Core, TDWG, and Plazi. Readers interested in particular taxa are invited to consult major research libraries or search the internet for additional online resources.

Specimen-Based Data

The creation of specimen-level databases in electronic form is a practice that goes back at least three decades and that has seen a progression of development directly associated with the increased availability of powerful database software and the widespread movement of information management and transmission to the internet. All of these databases were originally freestanding products used within their own institutions, usually as collection management tools, and were essentially an electronic replication of ledger books or card files. Progressively they have moved to serve a much broader purpose, allowing for the assembly of specimen data across collections and the mapping and analysis of distributions, with the consequence that large quantities of data are now available to the general and scientific public over the internet (e.g., Symbiota, Gries et al. 2014)

The Costa Rican Instituto Nacional de Biodiversdad (INBio) was one of the first institutions to apply database technology to the acquisition and maintenance of systematic collections from the day of its founding. Those efforts involved the attachment of barcode labels to all specimens acquired by the institute and the entry of all specimen data into the database at the time the specimens came from the field and were mounted and labeled. Barcode labels—not to be confused with DNA barcodes (see Chapter 7)—provide each specimen with a unique identifying number that links it to searchable electronic records. This "prospective" approach offers opportunities for future users of the collection to digitally mine its data and perhaps discover patterns that would be onerous to derive by retrospective data capture.

Australia is distinguished for having the greatest achievements in national specimen-database coordination, particularly in botany through Australia's Virtual Herbarium information system (Australasian Virtual Herbarium 2020). Information on Australian plant taxa is available through a portal that allows for assembly of data from 22 major herbaria in Australia and New Zealand. Not only are the specimen holdings of these herbaria largely available in database form, localities for a very large number of the specimens have been "georeferenced," allowing for the creation of distributional maps in real time on the internet.

The Australian initiative is seen in a global form in the Global Biodiversity Information Facility (GBIF), which is a confederation of member institutions around the world, each of which brings data to bear on the issue of assembling biodiversity information. The GBIF was conceived as a clearinghouse for information as opposed to being a primary data provider. All information made available through the GBIF internet portal comes from its member institutions and organizations. At the core of organizing that information is biological nomenclature, but the data being delivered are derived almost entirely from specimens residing in collections housed in museums around the world. The track record of ongoing success of this initiative has met with mixed reviews.

Within the United States, several consortia were formed on the basis of taxonomic affiliation rather than on the basis of political geography: Mammal Networked Information System (MANIS), Fish Net 2, ORNIS, and HerpNet. These were consolidated into a single digital portal for all vertebrates, VertNet, in 2014. This large consortium shares features in common with Australia's Virtual Herbarium, collectivizing the search for specimen data across a range of collections.

While most of the data referred to above was assembled over a local-area network, or even on a stand-alone basis, data capture itself has now also moved to the internet. The National Science Foundation-funded Planetary Biodiversity Inventories project, overseen by Randall T. Schuh and Gerasimos Cassis, implemented what has become a more-or-less "standard" relational data model for the capture of specimen data, the core elements of which are shown in Figure 8.6. Although the structure of the data used by Schuh and Cassis is not unique, they forged a new approach to data capture. Because members of their team of investigators were located on three continents in five institutions, they chose to develop

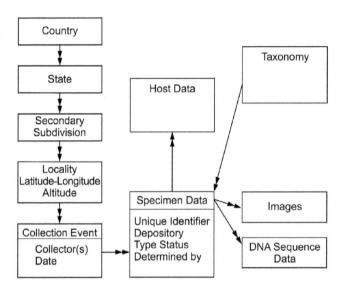

FIGURE 8.6. A relational database model for managing specimen data, simplified from the Arthropod Easy Capture application of Schuh et al. (2004–2014).

a web interface that allowed all investigators to enter data into a centralized database in real time, rather than capture data on a stand-alone basis and integrate it at a later date, either in a single database or through a web-aggregator such as GBIF or iDigBio. This approach allowed them to enter data from the site of any collection with a high-speed internet connection and further allowed for the application of centralized georeferencing and database management, aspects that provided more precise control over data quality.

Databases of vertebrate collections are—like those for plants—relatively straightforward from a logistical point of view because the specimens are large, often have individual unique identifying numbers associated with them, and the absolute number of specimens is manageable. Electronic databases exist for many vertebrate collections; a substantial number of specimens in many botanical collections are also databased.

Insects, and some other nonvertebrate groups, present distinct challenges when creating specimen databases because of the very large numbers of specimens and because individual specimens are seldom labeled so that they can be uniquely identified. Barcoding, widely used in plant-specimen databases, has been utilized to a limited degree for insects and may offer a solution to providing the unique specimen identifiers required for accurate specimen tracking. For example, INBio was one of the first institutions to use small barcode labels as unique specimen identifiers. The Arthropod Easy Capture project shown in

Figure 8.6 adopted "matrix codes," which are more compact and therefore well suited to use on small specimens but nonetheless allow for the encoding of a very large amount of information.

As specimen data accumulate in web-based archives, it becomes increasingly important for researchers to ensure that their voucher specimens—the individual organisms studied or remaining parts thereof—are uniquely labeled and archived as well. Morphologists have always been reasonably conscientious about this, but it has been a problem with DNA-based studies. Preservation of voucher material ensures that future researchers will be able to verify identifications of specimens examined. It is also critical to ensure that disassociated parts, such as genitalic slide mounts or DNA extractions, are labeled with unique identifiers that allow their reassociation with the voucher specimen(s).

Taxon-Based Data

There are a number of public databases with professional managers/curators that provide archival storage and access to data generated by individual researchers. One of the largest and most familiar of these is National Center for Biological Information (NCBI) GenBank, run by the US National Institutes of Health. GenBank stores DNA sequences and data associated with them, one of the few taxonomic databasing efforts deemed significant enough, due to its medical applications, to warrant dedicated facilities and permanent staff. The sequences in GenBank are derived from a wide variety of taxa, resulting from various molecular systematic and population genetic studies. With the growth of genomic data sets, the amount of data in GenBank has increased exponentially, and managing it has become a daunting task. Sequences of genes of interest can be accessed via a semiphylogenetic hierarchical classification scheme. Although NCBI invests great effort in ensuring that submitted sequences are properly annotated, there seems to be little to no comparable concern about specimen vouchering, and GenBank relies on the (sometimes faulty) taxonomic expertise of its contributors to provide accurate identifications of the organisms from which the sequence data were derived. Many journals now require submission of DNA data to GenBank as a criterion for publication.

The huge and still rapidly growing Barcodes of Life Database (BoLD) relies on an operational "majority rule" approach to attach names to its mitochondrial sequences. Sequences are grouped phenetically into barcode index numbers (BINs) based on percent similarity, and when several that have been identified as the same species by collaborating systematists, all sequences in that BIN de facto correspond to the assigned identification. If the same name is associated with several BINs, this suggests hypothetical sibling species, as in the aforementioned

Astraptes fulgerator example. Conversely, if multiple names fall into the same BIN, this implies either misidentification or synonymy.

Several additional database initiatives with a systematic focus, some sponsored by the US National Science Foundation, help to bring systematic data to the internet. MorphBank is a searchable repository of digital images that represent various taxa and their characters that have been employed in phylogenetic analyses. TreeBase (Piel et al. 2007) is a database composed of executable data sets from published systematic studies. Both of these rely on individual systematists to submit their data, which is posted on the web under a creative commons license. The Tree of Life website (Maddison and Schultz, 2007) represented an effort to summarize hypotheses of the hierarchy of life in a single tree. The individual clades were the responsibility of groups of volunteer specialists, and therefore the Tree of Life contains pages representing only a fraction of biological diversity. Unfortunately, when funding for the project ended, submission of new data was discontinued, and the site now exists only as a static snapshot of phylogenetic hypotheses contributed prior to 2013. The Encyclopedia of Life (Parr et al. 2014) is large-scale initiative that has expressed the ambitious and somewhat hubristic goal to present a web page for every species. DiscoverLife packages and provides data and images on a large and diverse array of plant and animal taxa, using authority files, interactive keys, species pages, and real-time mapping.

There are also many specialized websites and databases related to applied aspects of nomenclature and distribution, coordinated by the Commonwealth Agricultural Bureaux International (known by its acronym, CABI), the US Department of Agriculture, and other national and international entities. Because the correct name of an organism allows researchers and regulatory managers to associate a specimen discovered, for example, in a crate of imported fruit, with the biological literature relating to the taxon to which the organism belongs, nomenclature provides the critical link that allows correct determination of pest status, distribution, and other factors that influence quarantine decisions. The impacts of identifications of pests, pathogens, noxious weed seeds, disease vectors, and the like can be immense and may result in billion-dollar eradication programs for invasive agricultural or forestry pests such as the Asian longhorned beetle (*Anoplophora glabripennis*, Cerambycidae) or the light brown apple moth (*Epiphyas postvittana*, Tortricidae). We like to think of this sort of work as "regulatory biogeography." It is in such biosecurity activities that systematics is transformed from what some might view as a rather esoteric pastime into critical, high-stakes work that is eminently practical and economically consequential.

Our efforts to revise and update this chapter over the years lead us to a concluding observation: the proliferation, conglomeration, and occasional expiration of web-based biodiversity resources offers sometimes frustrating challenges to users, as well as to those endeavoring to document them. Projects are frequently subsumed or rendered redundant by other projects, and there appear to be many independent research groups attempting to address similar problems, sometimes at cross purposes. An aspirational goal for the future is a global, permanently established, regularly updated, well-funded, and authoritative one-stop shop for biodiversity informatics. The benefits to society would certainly outweigh the costs of establishing an institution to manage such data. Unfortunately, this does not appear to be a public priority at the present time.

SUGGESTED READINGS

Blackwelder, R.E. 1967. Taxonomy: A text and reference book. New York: John Wiley and Sons. [A now somewhat dated but informative review of the relationship between nomenclature and taxonomy]

Borror, D.J. 1960. Dictionary of word roots and combining forms. Palo Alto, CA: Mayfield Publishing. [An aid to the formation of scientific names]

Brown, R.W. 1956. Composition of scientific words. Published by the author. Reprinted by Smithsonian Books, 2000. [A useful compendium of Latin used in biological nomenclature]

Crowson, R.A. 1970. Classification and biology. Chicago: Aldine Publishing. [A pointed disquisition on the *Codes* of nomenclature and their history, as well as a readable and cogent essay on biological classification]

Farris, J.S. 1979. The information content of the phylogenetic system. Systematic Zoology 28:483–519. [A sometimes technical discussion of how and why hierarchic classifications store and transmit information]

Graham, C.H., S. Ferrier, F. Huettman, C. Moritz, and A.T. Peterson. 2004. New developments in museum-based bioinformatics and applications in biodiversity analysis. Trends in Ecology and Evolution 19: 497–503.

Knapp, S., G. Lamas, E.N. Lughadha, and G. Novarino. 2004. Stability or stasis in the names of organisms: the evolving codes of nomenclature. Philosophical Transactions of the Royal Society of London B, Biological Sciences 359:611–622.

Stearn, W.T. 1992. Botanical latin. History, grammar, syntax, terminology, and vocabulary, 4th ed., rev. Portland, OR: Timber Press. [A valuable aid to the formation of scientific names]

Wiley, E.O. 1979. An annotated Linnaean hierarchy, with comments on natural taxa and competing systems. Systematic Zoology 28:308–337. [A detailed discussion of classificatory conventions]

Witteveen, J. 2016. Suppressing synonymy with a homonym: the emergence of the nomenclatural type concept in nineteenth century natural history. Journal of the History of Biology 49:135–189. [A thorough and engaging history of the evolution of the type concept]

THE INTEGRATION OF PHYLOGENETICS, HISTORICAL BIOGEOGRAPHY, AND HOST-PARASITE COEVOLUTION

In Chapter 6, we saw how consensus techniques can be used as a way of summarizing information shared among multiple most-parsimonious cladograms and for forming classifications. We will now examine the use of this and other approaches for understanding historical biogeographic relationships and patterns of associations between parasites and their hosts.

This chapter deals primarily with analyses that are often placed under the heading of *cospeciation* or *codivergence*, situations in which hosts and their parasites appear to have intimate, long-standing historical connections and in which speciation in the host may result in speciation in the parasite. This type of association appears to obtain for many internal parasites and for certain external parasites, such as lice. In addition, or on the other hand, there is a whole class of host associations, such as those between herbivorous insects and their food plants, in which the relationship of the parasite and the host generally does not show such long-term fidelity but involves many apparent host shifts. These latter situations are often referred to under the more liberal heading of *coevolution* and will be addressed in Chapter 10.

Historical Biogeography

Background

Whereas methods for discovering the ordered distribution of attributes among organisms are now well established, the same cannot necessarily be said for techniques used to analyze recurrent patterns of geographic distribution. Nelson (1978a:269) noted, "Biogeography is a strange discipline. In general, there are no institutes of biogeography; There are no departments of it. There are no professional biogeographers—no professors of it, no curators of it." We will review very briefly the history and nature of biogeographic inquiry from the late 1850s forward, paying particular attention to the relationship of cladistics and biotic distributions. Reviews of the pre-Darwinian history of biogeography are given by Nelson (1978a) and Nelson and Platnick (1981: chapter 5).

The term *biogeography* has had many meanings attached to it. We will treat it as having two connotations: (1) the study of biotic distributions and the ecological processes that influence them in the short term, such as temperature, rainfall, and seasonality, a field often referred to as "geographical ecology" or "ecological geography"; and (2) the study of biotic distributions resulting from longer-term historical factors, an area of inquiry usually called "historical biogeography." Treatises on the subject of biotic distributions have not always made clear these distinctions, and indeed some authors have deliberately blurred them (Endler 1982). Generally speaking, ecological biogeography relates to the physical or biotic constraints of the environment in where species *can* live, while historical biogeography relates to the particular contingencies that explain where they actually *do* live. Thus, cacti and hummingbirds are native to the New World, penguins and *Araucaria* trees occur only in the Southern Hemisphere. The human-facilitated range expansions of exotic invasive species, such as gypsy moths or starlings, and the decimation of human populations by introduced pathogens like small pox or bubonic plague, provide clear indications that there are other parameters in addition to appropriate ecological conditions that regulate the native distributions of taxa. Biogeography as used in the following pages means *historical biogeography*—the patterns of, and explanations for, the often seemingly enigmatic distributions of taxa.

The study of historical biogeography traditionally has been heavily influenced by the state of knowledge of geology and geologic processes. Thus, when the orthodoxy of academic geology did not fully embrace the possibility of continental movements (as late as the 1960s in some quarters), biogeographers often attributed distributional *disjunctions* solely to the *dispersal* powers of the organisms involved. When plate tectonics became the accepted explanation for orogeny, ocean formation, and other feature changes on the earth's surface, more biogeographers began to couch their explanations of organismal distributions to mesh with the new geological paradigm.

Some biogeographers have proposed distributional theories independent of influence from other fields of inquiry. For example, Darwin's contemporary, Joseph Dalton Hooker, one of the most influential botanists of his time,

was, and remained, an ardent supporter of the idea that the present-day biotas of New Zealand, Australia, and South America achieved their current distribution via previous land connections among these land masses. In the same
vein, Alfred Russel Wallace (1860) argued for the former connection of land
masses as a way of explaining his observations on the distributions of animals
in the Malay Archipelago. Whereas Hooker remained true to what he believed
to be the most obvious explanation of the data, Wallace soon capitulated to the
Darwinian—dispersalist—point of view, which was virulently against modification of geological features as a method for explaining observed biotic distributions (Fichman 1977). We should note that both of these distributional patterns
are currently explained, at least in part, by the former connection of land masses:
in the case of Hooker's continents, the explanation lies with Mesozoic plate tectonics (Humphries and Parenti 1999); for Wallace's archipelago, an explanation
is changes of sea level as recently as the Pleistocene (2.6 million years ago to the
human epoch), which periodically exposed land connections from the shallow
seas between the Malay Peninsula, Borneo, Sumatra, and Java and between Australia, Tasmania, and New Guinea (Voris 2000).

The stabilist view of geography adopted by Darwin and many of his contemporaries derived from their aversion to the "fanciful" creations such as land
bridges invoked by Hooker and others to explain modern-day continental biotic
disjunctions involving tremendous distances. They believed dispersal of the
organisms themselves, across the oceans or circuitously over the contiguous land
masses, offered a more reasonable explanation for these observed patterns of
distribution. One can appreciate what seemed like a parsimonious perspective,
prior to the discovery of an alternative naturalistic explanatory theory. Like the
Natural System itself, the evident order of geographical distributions was a puz-
zling empirical pattern that begged a material explanation for these nineteenth-
century scientists.

Such an explanation was soon proposed. After reading the early twentieth-
century expositions of Alfred Wegener (e.g., Wegener 1966) on continental drift,
you will likely conclude that the data—even nearly 100 years ago—spoke clearly
and strongly for the concept of continental movements. For example, Wegener's
arguments included geodetic data about the movement of Greenland relative
to Europe; the excellent fits of the continental outlines of North America and
Europe and of South America and Africa, the latter as shown by the Paleozoic
geological similarities of the Cape Province of South Africa and the mountains
in Buenos Aires Province, Argentina; and the known distribution of evidence
for Permo-Carboniferous glaciation on the southern continents showing the
"invalidity of [continental] permanence theory." Wegener also found substantial
corroboration of his theory in the distributional patterns of Recent plants and

animals. In short, there seemed to Wegener to be a massive body of geological, paleontological, and biogeographical corroborative evidence for the theory of continental movement as opposed to the then-current theories of a contracting Earth, subsiding land bridges, or stable continents.

Wegener's views were received enthusiastically by some, including the South African geologist-paleontologist Alexander du Toit. However, influential Americans, including mammalogist William Diller Matthew, geologist-paleontologist George Gaylord Simpson, and entomologist-biogeographer Philip J. Darlington were resolute in rejecting Wegener's ideas as a way of explaining biotic distributions. This, in the case of Darlington and Simpson, even after the acquisition of modern geophysical evidence for continental movements in the middle of the twentieth century. An analogy may be drawn with the rejection of Darwin's theory of evolution by the powerful Victorian vertebrate anatomists Louis Agassiz and Richard Owen.

One might wonder whether the evidence of biotic distributions has a legitimate voice of its own, or if it must always be tempered with the more "tangible" knowledge of earth processes derived from studies in geology. History suggests that patterns of distribution of plants and animals do have a story to tell, and those who have disregarded the biotic evidence have often failed to appreciate the ultimate strength of the data provided by organisms themselves. One need only contrast the conclusions of Darlington (1965) and Brundin (1966) with regard to the biotas of the far Southern Hemisphere. Darlington rejected cladistics and also resisted accepting continental drift as an explanation of the distributions of those biotas. He "felt certain" that all plants and animals have reached New Zealand across the water, at least during the Tertiary and probably before that. Darlington drew this conclusion even for those many plant groups with distributions restricted to the south-temperate portions of the southern continents and with clear connections of phylogenetic affinity among those land masses. In a great leap of logic, he then concluded (Darlington, 1965:107): "If these plants and all other plants and animals that have reached New Zealand crossed water gaps, no land connections are needed anywhere across the southern end of the world to explain the distribution of far-southern terrestrial life." This type of fallacious argument is called a petitio principii, in which the assertion that the premise is true is the "evidence" that supports the conclusion.

In an early application of Hennig's phylogenetic principles, Brundin (1966) demonstrated repeated patterns of intercontinental connection within multiple lineages of chironomid midges in the far south (see Figure 9.4). He attributed those phylogenetic connections to former land connections, including Antarctica, which he explained via continental drift. In short, whereas Darlington clung to the prevailing systematic and geological dogma of his formative years and embraced ad hoc explanations of patterns easily accommodated with the bolder tools of plate tectonics and cladistics, Brundin accepted evidence before preconceived theory and treated concordant patterns of geographic distribution parsimoniously, as the result of common underlying causes. Subsequent authors (e.g., Cranston et al. 2010) have corroborated Brundin's results.

Having discovered that there are meaningfully consistent and repeated geographical patterns to the distribution of plants and animals, how might we analyze them? The early history of this enterprise was largely descriptive and is probably best characterized in the work of the British ornithologist Philip Lutley Sclater. Sclater's mark on the field was made in his paper, "On the General Geographical Distribution of the Members of the Class Aves" (Sclater 1858), in which he proposed six biogeographic "regions," which are employed today nearly in their original form under the names Nearctic, Neotropical, Palearctic, Ethiopian, Oriental, and Australian regions. Sclater's division of the world's biota corresponded roughly to the major continental land masses, with the exception of the Oriental region (Regio Indica) (Figure 9.1) and was soon accepted by Wallace in his works on biogeography (Wallace 1876) and by many others. Sclater's regions represented recognizable geographic features that showed major biotic differences among themselves and provided an organizational scheme for cataloging the world's biological diversity into relatively distinct regional floras and faunas. It would be some time before botanist Leon Croizat and dipterist Willi Hennig recognized that patterns from distributional data might be subject to new and

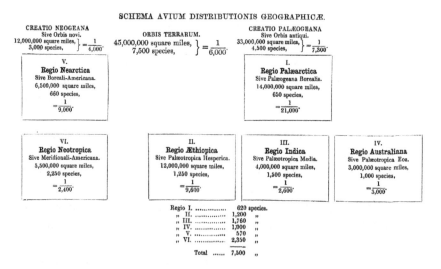

FIGURE 9.1.   Sclater's (1858) scheme of the biotic regions, based primarily on the study of birds.

improved means of analysis, irrespective of the geological orthodoxy. The works of both authors proved formidable to comprehend.

Croizat's works are prolix in the extreme, achieving a level of verbosity possibly unequalled in the biological literature. His major ideas are most mercifully summarized in *Space, Time, Form: The Biological Synthesis* (Croizat 1962). Croizat said, in reference to common patterns of animal and plant distributions across the continents, that "nature forever repeats," an observation that he believed demanded explanation. He dubbed his approach "panbiogeography," the core activity of which was "track analysis." Croizat's data on relationships and distributions of taxa came from monographs and revisions, many of them from works that had nothing to do with the Euphorbiaceae, the group of plants on which Croizat himself had specialized. Croizat drew tracks to connect disjunct areas possessing related organisms, such as members of the angiosperm family Proteaceae, as shown in Figure 9.2. When multiple groups of organisms displayed the same disjunct distributions, Croizat termed the phenomenon "vicariance." Its explanation was geologic history. (Note that the term *vicariance* has evolved in the more recent literature to indicate the process of subdivision of ancestral populations by extrinsic barriers—a mechanism of allopatric speciation.)

Croizat's approach has been applied in its original form by only a few workers, notably Robin C. Craw (e.g., Craw 1982, 1983; Craw et al. 1999), although the seeming reality of many of the repetitive distributions identified by Croizat has been documented by many other authors. An example comes from the work of Weston and Crisp (1994) on the "waratahs," a subgroup of Proteaceae (Figure 9.3), which, when analyzed in detail, portray a pattern of distribution very similar to that depicted by Croizat.

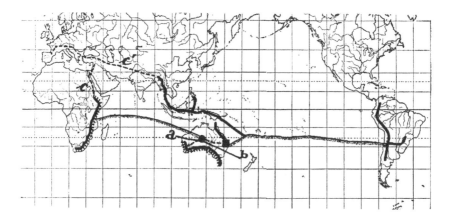

FIGURE 9.2. Tracks indicating the distribution of the Proteaceae (Personioideae and Grevilloideae) (from Croizat 1962: figure 42, p. 169).

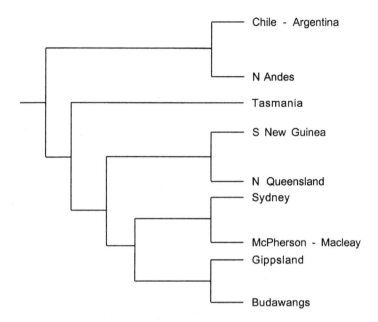

FIGURE 9.3. Intercontinental connections between eastern Australia and montane and southern South America as shown by the waratahs, a subgroup of Proteaceae (from Weston and Crisp 1994).

Hennig's work on biogeography (1960, 1966) was described by Ashlock (1974) as "at once brilliant and badly conceived [, and] its title and organization are such that few but dipterists would be attracted." Hennig's major empirical contribution to biogeographic analysis (Hennig 1960) was an extension of his methodological work on phylogenetic relationships, addressing the Diptera of New Zealand as an example problem. Hennig's approach relied heavily on the *progression rule*, the idea that if an organism has migrated to new areas, developing new characters as its distribution expands, the progress of migration is marked by sequentially more-derived characters. This rule assumes that organisms in the ancestral area of origin undergo little or no change, a questionable assumption at best and one that would seem to reject clocklike evolutionary patterns (see Chapter 11).

Hennig's biogeographic theory, despite what are now viewed as its flaws, is notable because it established a direct connection between phylogenetic analysis and the study of biotic distributions. Nonetheless, it might have remained obscure, even in translation, had it not been for the monographic work of the Swedish entomologist Lars Brundin (1966), who applied both Hennig's phylogenetic and biogeographic methodologies to his studies of chironomid midges from the southern continents. Brundin showed that there were repetitive distributional patterns among different clades of midges of that part of the world (Figure 9.4) and that—in vindication of the views of Hooker with regard to plant

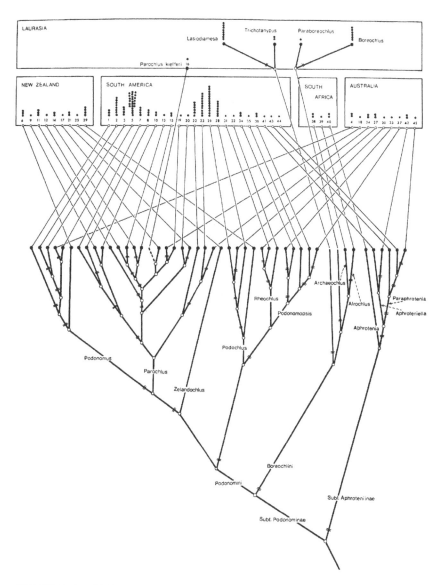

FIGURE 9.4. Disjunct distributions on southern continents as evidenced by midges of the family Chironomidae (Brundin after Nelson and Ladiges 1996: figure 2; courtesy of the American Museum of Natural History). Note the repeated pattern of occurrence in New Zealand, South America, and Australia in separate groups of midges.

distributions—the midge distributions implied the connection of the southern continents. Furthermore, in Brundin's view, the patterns supported Hennig's progression rule. The significance of Brundin's findings was widely recognized. The appearance of his work coincided felicitously with the spreading consensus that continental drift was a scientifically credible theory, as well as with the publication of a review article and English translation of Hennig's book-length exposition of his theory of phylogenetic systematics, making phylogenetic methodology accessible to a much broader audience (Hennig 1965, 1966).

At the American Museum of Natural History in New York, Donn Rosen's empirical studies (1975, 1978) and the theoretical advances of Gareth Nelson and Norman Platnick (Platnick and Nelson 1978; Nelson and Platnick 1981; Platnick 1981b) followed directly from the seminal works of Hennig and Brundin. The approach advocated by these authors has come to be called *cladistic biogeography*, as by Chris Humphries and Lynne Parenti (1999), and for a time dominated the historical study of plant and animal distributions as practiced by systematists. The analytic precision of cladistic biogeographic approaches has been refined through the appearance of computer algorithms to solve the sometimes extremely complex problems, discussed below, in Cladistic Biogeographic Methods. Concomitantly, alternative approaches to the analysis of historical biogeographic patterns have sprung up, most being placed under the banner of "event-based" approaches. Lastly, the vigor with which systematists and biogeographers have pursued the search for congruent distributions during the last 30 to 40 years has been challenged by some who argue that the types of patterns once championed by Brundin can be just as satisfactorily explained without the influence of plate tectonics and other long-term events of earth history. In the twenty-first century, vicariant biogeographical hypotheses have been largely deemphasized in favor of ad hoc, mainly dispersalist geographical origin scenarios for individual clades based on inferred optimizations of hypothetical ancestral areas. We will now examine the premises and methods of each of these approaches. Some useful terminology is listed in Sidebar 18.

Sidebar 18
Biogeographic Terminology

The following terms are widely used in the discussion of cladistic biogeography. Clarification of their intended meaning will help the reader to comprehend the remainder of the discussion in this chapter.

component   a (monophyletic) group of taxa (areas) connected at a node on a cladogram
general area cladogram   a cladogram showing the resolution of area relationships among a variety of taxon-area cladograms
missing taxon   a taxon absent from one of the areas in one of the area cladograms being compared, the result being noncorrespondence of distributions across taxa

redundant (paralogous) distributions   the sequential or multiple occurrence of the same area on an area cladogram; a phenomenon observed to occur frequently among the basal taxa (areas) on cladograms (Nelson and Ladiges1996)

taxon-area cladogram   (or **area cladogram**) a cladogram of taxa for which the distributions of the taxa have been substituted for the taxa themselves

term   a terminal taxon (area) on a cladogram

widespread taxon   a taxon that occurs in more than one area of endemism, as areas of endemism are interpreted on the basis of the distributions of other taxa

Cladistic Biogeographic Methods

As with systematic methods themselves, parsimony-minded researchers have endeavored to strip unjustified assumptions from analytical methods in historical biogeography, with the intent of discovering patterns in the data without presupposition of untestable scenarios and invocation of ad hoc hypotheses. Among historical assumptions that are no longer deemed valid are the following:

Center of origin: This long-held biogeographic dictum postulates a priori that taxa arise in one identifiable area and disperse from there eventually to achieve their present distribution. This is certainly a possible scenario (such as, for example, the origin of *Homo sapiens* in Africa and our subsequent geographical spread throughout the world), but it needs to be demonstrated empirically rather than assumed as a model at the outset of a biogeographical study.

Progression rule: A corollary to the center of origin concept, this Hennigian tenet treats all distributions as having polarity, with the oldest taxa inhabiting the ancestral area, while more recently derived taxa occur progressively farther from the center of origin. As in the previous case, such a pattern could be empirically demonstrated but should not be presumed as a general expectation.

Ad hoc theories of dispersal: This is the means traditionally invoked to explain the present distributions of all taxa, that is, to expand their range beyond the identifiable center of origin. Since almost any organism has a life stage that can walk, swim, fly, or be borne on currents of wind or water, dispersal is by its nature an ad hoc hypothesis that can be (and has been) used to accommodate nearly any pattern of distribution. While it is indisputable that some taxa, such as terrestrial inhabitants of oceanic islands, have achieved their current distributions by dispersal, such an explanation has no predictive value for the distributions of other taxa. Unlike vicariance, which is a parsimonious common-cause explanatory hypothesis epistemologically parallel to the concept of homology, dispersal can be viewed as the "homoplasy" of biogeography and should be invoked only to explain those aspects of distribution that do not conform to common patterns explicable by geological events and other dynamic forces of earth history that have potential to affect the distributions of entire biotas.

These postulates were largely replaced with the view that shared biogeographic patterns are the result of historical range fragmentation over time, a concept to which the term *vicariance* is now generally applied. Dispersal is then invoked to explain patterns that do not conform parsimoniously to the general pattern, such as widespread taxa. Somewhat paradoxically, the fragmentation of ranges, through orogeny, continental drift, or other means, conforms closely to the concept of allopatric speciation under which most modern precladistic biogeography was envisioned by noncladistic systematists—using the dispersal model. Wiley and Lieberman (2011) used the term *geodispersal* to describe a sort of reverse vicariance, a common-cause explanation in which geological phenomena such as the emergence of the Isthmus of Panama or the Bering land bridge allow concurrent range expansions of a variety of terrestrial taxa. Such patterns have previously been described as biotic dispersal (Wiley 1981).

The data of modern historical biogeography are *taxa* and *areas of endemism*. An area of endemism "can be defined by the congruent distributional limits of two or more species" (Platnick 1991). Taxa represent the analog of characters in phylogenetic analysis; areas of endemism are the analog of taxa.

What distinguishes nearly all current synthetic biogeographic methods from past practice is the necessity of having (multiple) explicit phylogenetic hypotheses of relationships for taxa as the initial data. The method, as first proposed by Rosen (1975, 1978), involves replacing taxon names on a cladogram with the areas of endemism where the taxa occur, forming what are known as *area cladograms* or *taxon-area cladograms*. Thus, a hierarchic scheme of areas is revealed (see Figure 9.5). If all taxa have responded to earth history in a like manner, this will be reflected in their common patterns of distribution, a single scheme of area interrelationships—a general area cladogram—will be the result, and no further analysis is required. Observation suggests, however, that such is seldom the case, for a variety of reasons which might include the following:

1. Taxa may be of different ages and therefore have undergone geographical differentiation in response to different vicariant phenomena.
2. Different taxa may have responded to the same phenomenon in different ways.
3. Theories concerning areas of endemism may be in error because of subsequent dispersal, extinction, poor sampling, or the confluence of formerly separated areas.

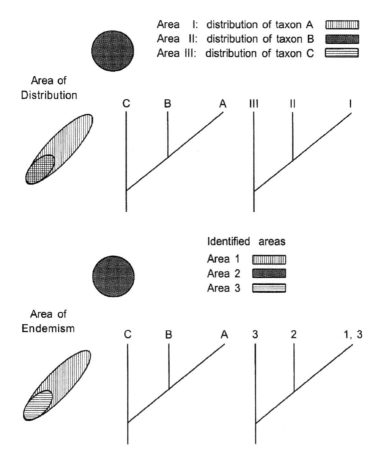

Area   I: distribution of taxon A
Area   II: distribution of taxon B
Area   III: distribution of taxon C

Area of
Distribution

C   B   A   III   II   I

Identified areas
Area 1
Area 2
Area 3

Area of
Endemism

C   B   A   3   2   1, 3

FIGURE 9.5.   Determining areas of endemism as potentially different from areas of distribution, according to the method of Axelius (1991). The upper portion of the figure shows distributions coded as areas. The bottom portion of the figure shows areas of endemism coded to recognize overlapping areas as distinct for all taxa.

The empirical observation necessary for identifying areas of endemism is simply the known geographic distribution of a taxon, and the test of the theory—that other taxa occur there—is usually much less clear-cut than is the case with morphological characters (see Harold and Mooi 1994; Humphries and Parenti 1999). Nonetheless, there are guidelines, possibly even methods, for recognizing areas of endemism. Platnick (1991) observed that we might always wish to pursue biogeographic studies using taxa with the smallest ranges and the largest numbers of species for the area under study. One of his favored examples of distributional oversimplification is the treatment of cool-temperate southern South

America as a single area, even though many groups of spiders, and some other chelicerate arthropods such as Opiliones, show extreme diversity and highly localized distributions within this geographic region. Again, we emphasize that just as there are clades nested within clades, so may there be areas of endemism nested within areas of endemism.

The treatment of the known distributions of species as de facto areas of endemism was criticized by Axelius (1991) regarding cases where the distributions of species are overlapping. He offered the solution shown in Figure 9.5. The argument for this approach lies in the observation that the overlap in the distributions of taxa A and B must be due to dispersal of one or both of the species. Therefore, the portion of the distribution of taxon A overlapping with the distribution of taxon B should be coded separately from the nonoverlapping portion. Otherwise, the resulting area cladograms will not be completely informative with regard to the areas occupied by the taxa. In cases of partial overlap, Axelius recommended coding the overlapping portion as a separate area, in anticipation of determining which part of the distribution was the result of dispersal (note that this approach assumes allopatric speciation).

Whereas implementation of the above-stated criteria is admittedly a somewhat idiosyncratic process, Claudia Szumik, Pablo Goloboff, and colleagues (Szumik et al. 2002, 2019; Szumik and Goloboff 2004, 2015; Goloboff 2005, 2016; Arias et al. 2011) have developed explicit criteria and produced computer programs that aid in the recognition of areas of endemism. Their approach proceeds by first dividing the area occupied by the species under study into a grid, then choosing sets of cells from the grid to which collections of species are restricted. They have successively refined the technique by applying a more complex set of criteria for whether a species is present or not, thus allowing for the recognition of areas based on patchier distributions. The size of the grid cells can be varied to accommodate the density of sampling. Success in recognizing areas of endemism through the use of these algorithms ultimately depends on multiple taxa occupying the same area and the refinement of the grid size to fit the particular situation.

If each taxon's area cladogram implies the same pattern, then the result is straightforward. But what if they disagree? The core problem of cladistic biogeography then becomes, "How do we combine nonidentical area cladograms for different groups to produce a summary of the information common to them?" The process proceeds as follows (Nelson and Platnick 1981; Page 1990a; Morrone and Carpenter 1994):

1. *Construct taxon-area cladograms from taxon cladograms.* Areas of endemism are determined for all taxa, the widespread taxa being coded as occupying more than one area of endemism. The taxon cladogram is then labeled with the areas of endemism as shown in Figure 9.6.

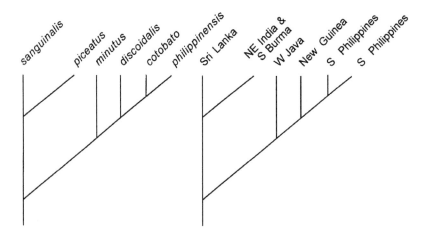

FIGURE 9.6. Taxon cladogram (Stonedahl 1988) and corresponding area cladogram (Schuh and Stonedahl 1986).

2. *Convert taxon-area cladograms into resolved area cladograms.* As originally described in detail by Nelson and Platnick (1981) in their exposition of "component analysis," any given area cladogram may imply a few to many "resolved area cladograms," primarily because of the occurrence of widespread taxa. Missing areas will result in some resolved area cladograms being different but will generally not increase the numbers of resolved area cladograms.

Four approaches have been proposed for solving this problem, representing a set of assumptions—with arcanely uninformative names—about how to deal with widespread taxa and missing areas. The solutions range from very restrictive to quite liberal. An alternative solution for paralogous (redundant) distributions was offered by Nelson and Ladiges (1996) and will be discussed below, in Subtree Analysis and Area Paralogy. The attributes of the four assumptions can be summarized as follows (modified from Page 1990a):

Approach	*Missing areas*	*Widespread taxa*
Brooks parsimony analysis:	Uninformative	Sister areas
Assumption 0:	Primitively absent	Sister areas
Assumption 1:	Uninformative	Paraphyletic
Assumption 2:	Uninformative	Float all but one occurrence

Brooks parsimony analysis (BPA; Wiley 1988) derives directly from methods proposed by Brooks (1981) for the analysis of host-parasite data, where additive binary coding methods are used to prepare a matrix of data from area cladograms, which can then be analyzed using a standard parsimony program. The BPA and assumption 0 (Zandee and Roos 1987) use similar methods for forming resolved area cladograms. They both treat the areas occupied by widespread taxa as monophyletic; that is, the areas that combine to form a widespread distribution are more closely related to one another than to any other areas occupied by the taxa under consideration. This aspect of BPA and assumption 0 makes them the most restrictive of the four. The BPA and assumption 0 differ only in their treatment of missing areas, which are viewed as uninformative by BPA and primitively absent under assumption 0.

Assumption 1 (Nelson and Platnick 1981) is less restrictive, in that area relationships for widespread taxa can be either monophyletic or paraphyletic with respect to the taxon inhabiting them (Figure 9.7). Thus, the widespread taxon may be informative in combination with other taxon-area cladograms if the components of the cladogram with the widespread taxon are combinable with the components in a taxon-area cladogram where the widespread distribution is resolved. The taxon with the widespread distribution will never be recognized as representing more than one taxon.

Assumption 2 (Nelson and Platnick 1981) is least restrictive with regard to the treatment of widespread taxa, and in the minds of some is the most reasonable interpretation of many biogeographic data sets. A widespread taxon on a single cladogram comprises a component containing unresolved area relationships, but when considered in concert with other cladograms the unresolved area relationships may become informative (Humphries and Parenti 1986). Under this interpretation, some—but not all—of the areas of endemism comprising the distribution of widespread taxa may "float" on the general area cladogram. Thus, what were uninterpretable area relationships on the single taxon-area cladogram become resolved (Figure 9.7). The taxon with the widespread distribution may in the future be interpreted as representing more than one taxon, potentially allowing selection among alternative hypotheses of area relationship.

As pointed out by Page (1990a), BPA and assumptions 0, 1, and 2 need not be the only assumptions one might make about the resolution of distributional information in order to prepare resolved area cladograms. For example, all possible combinations of ways of treating missing areas and widespread taxa could be invoked. The lack of a means for selecting among these alternative criteria that may imply different patterns has been a confounding source of ambiguity in the interpretation of complex biogeographic distributions.

3. *Prepare general area cladograms.* Determining what information is
shared in common across taxa can be achieved in a variety of ways. Both

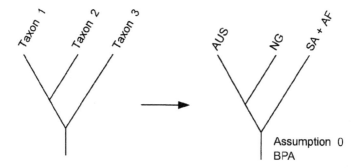

Assumption 1

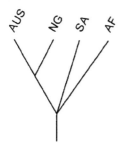

Assumption 2

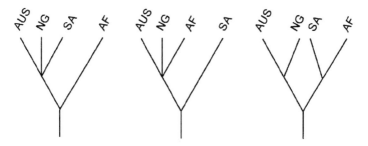

FIGURE 9.7. Taxon-area cladogram and resolved area cladograms under assumption 1 and assumption 2. AUS = Australia; NG = New Guinea; SA = South America; AF = Africa. Assumption 1 does not require that SA + AF be monophyletic, as would be the case with Brooks parsimony analysis (BPA) and assumption 0, but it does not specify how either one might be related to the remaining areas. Assumption 2 specifies those possible area relationships (from Humphries and Parenti 1986).

parsimony and consensus techniques have been used for finding general area cladograms; some of the possible approaches have been described by Page (1988, 1989, 1990a). As pointed out by Morrone and Carpenter (1994) in their comparison of automated methods, two areas still caused confusion: (1) the choice between parsimony and consensus is not clear because neither method produced results that Morrone and Carpenter found conclusive; and (2) the available software could solve only problems involving a small number of taxa and areas.

Subtree Analysis and Area Paralogy

Area *paralogy* (redundant distributions of taxa among areas) has been identified by Nelson and Ladiges (1996) as the source of apparent inconsistency in biogeographic data, as seen, for example, in the results of Morrone and Carpenter (1994). Basically, if a clade is older than the areas it inhabits, and several of its subordinate components exhibit congruent distributions across the same areas, these redundant patterns can be collapsed to yield a single hypothesis of area relationships (Figure 9.8). The removal of paralogy, through what Nelson

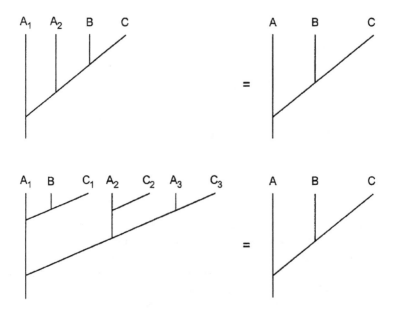

FIGURE 9.8. Two examples of taxon-area cladograms with paralogous distributions, before and after subtree analysis. Subscripts represent the repetitive occurrence of different taxa in the same area (from Nelson and Ladiges 1996).

and Ladiges termed *subtree analysis*, appears to offer improved results. Improved, in the sense used here, implies (1) the discovery of greater congruence among taxon-area cladograms, (2) the ability to solve problems involving more than 10 taxa and areas (including widespread distributions), and (3) the ability to analyze these larger data sets in a reasonable period of time.

Nelson and Ladiges (1996) coded matrices for their data in the form of components, following the approach outlined by Nelson and Platnick (1981), and as three-area statements (a biogeographical application of three-taxon statements; see Chapter 5). The results were analyzed using the venerable Hennig86 and PAUP parsimony programs available at the time the paper was published. The unsupported nodes produced by the programs were deleted through a node-by-node examination of the resultant general area cladograms. The dissection of complex area relationships into subtrees (some examples of which are shown in Figure 9.8) and the analysis of the data with a parsimony program are responsible for the efficiency with which the results are produced.

Nelson and Ladiges (1996) demonstrated the utility of their approach by reanalyzing most of the data sets used by Morrone and Carpenter, which allows direct comparability between the results of the methods applied by the two sets of investigators. They also analyzed several large data sets, including those of Brundin (1966), Schuh and Stonedahl (1986), and Mayden (1988). Whereas Morrone and Carpenter often found multiple solutions, sometimes thousands, Nelson and Ladiges found one to a few solutions for the same data sets. The latter authors concluded that removal of paralogy from area cladograms allows for solutions to complex problems and that the number of possible solutions is greatly reduced. The "assumptions" described above are no longer needed with subtree analysis, as ambiguity among distributions is removed by other means.

Criticism of Cladistic Biogeography

Cladistic biogeography is by its very nature concerned with forming connections among biotas with disjunct distributions. Robin Craw (1982, 1983) criticized "cladistic" biogeography for being preoccupied with fragmentation and therefore being incapable of providing a reliable analysis of the biotas of composite areas. Dan Polhemus (1996:63), without citing Craw, echoed the same sentiments, suggesting, "Cladistic and vicariance biogeographers . . . developed methods that due to their reliance on dichotomously branching diagrams were best suited to the depiction of faunal patterns arising from continental fragmentation . . . these methods have been notably less successful when applied to areas of continental convergence . . . Present component and parsimony analyses that rely solely on dichotomous branching . . . are clearly insufficient tools for such investigations." Both authors suggested that a single land area appearing in more than one pattern would represent failure of the method, presumably because of its being based on dichotomy and fragmentation. They used New Zealand, New Caledonia, and New Guinea as examples.

The complaints of Craw were interpreted by Platnick and Nelson (1984:331) as suggesting that "the geographic pattern exhibited by the preponderance of evidence would be accepted as real, and that conflicting real patterns shown by other taxa would be rejected as unreal (i.e., the biogeographical analog of homoplasy, due to chance dispersal)." They indicated that no such assumptions had ever been made under the method, and that indeed, in the case of geologically composite areas, incongruence between general patterns would be expected.

Schuh and Stonedahl (1986) found New Guinea appearing in two places on a taxon-area cladogram, in one case relating New Guinea to tropical Asia and in the other case to continental Australia. Polhemus (1996) used this as an example of "the futility of attempting to search for a single area cladogram that will accurately represent the faunal relationships of an area in which arc collisions and subsequent composite terrains have been dominant influences." Nonetheless, Schuh and Stonedahl made it clear that the method would force the conclusion that New Guinea was of hybrid origin, although they did not try to resolve the problem of areas of endemism within New Guinea.

Event-Based Approaches

Panbiogeographers are not the only critics of cladistic biogeographic methods, as articulated by Rosen, Nelson, Platnick, Humphries, Parenti, and others. Foremost among the alternate-school critics are Fredrik Ronquist and his colleagues, primarily Isabel Sanmartin. Their main criticisms are that cladistic biogeography (1) assumes that there is a single historical pattern of differentiation among areas and that the method is incapable of accounting for co-occurring lineages with different biogeographic histories, (2) assumes that the ancestral distributions of taxa are widespread, allowing only for subsequent fragmentation with no account of how the widespread distribution was achieved in the first place, (3) does not incorporate information on divergence times, (4) does not allow for statistical tests to assess the significance of historical biogeographic hypotheses, and (5) requires the a posteriori interpretation of events that led to observed distributions (see, e.g., Ronquist1997; Sanmartin and Ronquist 2004; Sanmartin 2007; Ronquist and Sanmartin 2011). Their recommended solutions to these perceived problems come in the form of what are usually called *event-based methods* of analysis, or—somewhat presumptuously—"phylogenetic biogeography."

Event-based methods in biogeography (and host-parasite coevolution) are the analog of maximum likelihood or Bayesian analysis in phylogenetics, in that they incorporate models as part of the analytical protocol. As stated by Sanmartin (2007:139), "Event-based biogeographic methods rely on explicit models with states (distributions of terminals) and transition events between states (biogeographic processes)." (See also Ree et al. 2005; Ree and Smith 2008). Different models assign different costs to different events. Although proponents of these approaches advance the results as objective historical narratives, it cannot be demonstrated that any of these scenarios are realistic portrayals of what actually happened in any particular situation.

Ronquist (1996, 1997) offered a description and algorithmic implementation of *dispersal-vicariance analysis* (DIVA). Like other event-based approaches, DIVA allows for the assignment of differential costs to different types of events based on the "likelihood" of their occurrence. Like cladistic biogeography, all analyses are performed using a cladogram of taxon-area relationships for the group under study. The method does not require, however, a general hypothesis of area relationships. Vicariance (allopatric speciation) is assumed to be the primary driving force for diversification across a geographic range and, along with duplication (Sanmartin 2007), is assigned a cost of 0. Dispersal events, which are assigned a cost of 1, are the result of adding a unit area to the distribution of a taxon; extinction events, which are also assigned a cost of 1, are the result of removing a unit area from the distribution of a taxon. Analyses under DIVA incorporate the use of a three-dimensional matrix for the assignment of differential costs to separate event types. The cost under the DIVA model is computed for a single taxon-area cladogram. Yu et al. (2010, 2015) introduced modifications to account for uncertainty (S-DIVA, RASP). A recent implementation of these approaches examining patterns of distribution in a group of southern South American lizards was performed by Hibbard et al. (2018).

Parsimony-based tree fitting (Page 1994; Ronquist 2002, 2003), like DIVA, assigns costs to events. Some studies have suggested that optimal cost assignments are 0.01 for vicariance and duplication, 1 for extinction, and 2 for dispersal (Ronquist 2003; Sanmartin, 2007). Unlike DIVA, however, this is a two-step process. Costs are first computed for a set of taxa with an observed distribution. This involves optimization of distributional states on a cladogram under a given set of costs. The topology of that optimization is then fitted to a "general area cladogram", the tree-fitting part of the process. The topology of the general area cladogram might be inferred from geology, mirrored from other taxa, or chosen from all possible topologies for a given set of areas. Whatever the source, the more areas involved, the more complex the solution.

The significance of the results in event-based reconstructions is established by comparing the observed costs of a given reconstruction with costs derived from permuted area relationships, where areas and/or taxa have no assumed historical association. This process is analogous to the calculation of bootstrap values in phylogenetic analyses.

Sanmartin (2007) has suggested, on the basis of her own empirical studies, that each of the two event-based methods described above has its own particular utility in historical biogeographic reconstructions. She argued that DIVA is most appropriate when the biogeographic history is reticulate, such as in the Northern Hemisphere, where a hypothesis of unique vicariant events does a poor job of explaining observed distributions. Sanmartin argued that tree-fitting methods, on the other hand, are more appropriate when there is a strong vicariant pattern, such as in the analysis of classic Gondwana distributions, of the type first dealt with in a cladistic context by Brundin (1966).

Sanmartin and Ronquist (2004) performed a large-scale analysis of Southern Hemisphere distributions based on a data set including a large number of plant and animal distributions. These authors used the program TreeFitter 1.3 (Ronquist, 2002) to perform their analyses. Their results suggested that animals show a stronger signal of vicariance than do plants and that that signal is correlated with geological scenarios relating to the breakup of Gondwana (Africa (New Zealand (southern South America, Australia))). Their results further suggested a lesser role for vicariance in the distribution of southern plant taxa, with the major pattern being (southern South America (Australia, New Zealand)). Because of the event-based nature of their analysis, they invoked dispersal across Antarctica as a significant factor in determining observed animal distributions.

The idea that plants have dispersed in the Southern Hemisphere has gained some traction beyond the work of Sanmartin and Ronquist. Winkworth et al. (2002) reviewed a series of analyses that suggest relatively recent colonization and diversification by plant groups in New Zealand. Others, including Cook and Crisp (2005), have argued that the problem is with flaws in method and that because of their shortcomings, tree-based (systematic) methods of analysis should be rejected in favor of process/likelihood-based approaches. One can detect a desire to return to the pre-Brundin world in the writings of McGlone (2005), who argued for the primacy of dispersal as an explanation in a commentary piece entitled "Goodbye Gondwana." Similar dispersalist sentiments are expressed by Alan de Queiroz (2014) in the trade book, *The Monkey's Voyage*.

Many of these critiques are fueled by molecular clock-based age estimates that appear to be incompatible with vicariant mechanisms that might explain the current distributions of the taxa. For example, Sauquet et al. (2012) attacked the classic vicariant hypothesis for southern beeches (*Nothofagus*) proposed by Humphries (1981), and Rowe et al. (2010) invoke trans-Atlantic rafting by rodents from Africa to South America. These examples notwithstanding, phylogenetic studies continue to offer evidence for the strong connection between patterns of continental fragmentation and the topological relationships of taxa with little or no dispersal capability, such as the cyphophthalmid Opiliones (Arachnida) (Boyer et al. 2007). One of the most steadfast defenders of vicariance is Michael Heads, who elaborated his critique of molecular biogeography in a 2012 treatise. The bottom line is that dispersal scenarios can "explain" any pattern of distribution, but such narratives are intrinsically ad hoc and do not provide common explanations of more general patterns and are thus deficient in their scientific character.

Host-Parasite Coevolution

The observation that hosts and their parasites have evolved in concert is long standing and was discussed extensively by Hennig (1966) under the heading of "Fahrenholz's rule." The development of improved techniques for determining to what degree "coevolution" has actually occurred, however, has gone hand-in-hand with the development of cladistic biogeographic methods.

Hennig (1966:112) believed that the relationships of parasites contained much information of great value to understanding the phylogenetic systematics of host taxa, but he lamented the unsatisfactory state of its theoretical basis. He asserted, "Even the most extreme advocates of the thesis that the phylogeny of the parasites usually parallels the phylogeny of the host (in the sense of Fahrenholz's rule) do not assume that the parallelism is so close that every process of speciation in the one corresponds to a process of speciation in the other" (Hennig, 1966:111). This statement clearly suggests a parallel with historical biogeography: just as taxa can be vicariant passengers of the evolving geographical features they inhabit, so can parasites become isolated by vicariant (species splitting) events of their host taxa.

Host-parasite comparisons benefit from the fact that both hosts and parasites exhibit patterns of relationships that can be inferred from intrinsic character data (i.e., both are subject to phylogenetic analysis in their own right). Thus the problem of "homologous taxa" does not arise as it does with areas of endemism. The similarities between host-parasite relationships and biotic distributions might be listed as follows:

1. The hosts are equivalent to the "areas," and speciation events in the hosts act as vicariant events splitting apart populations of the parasite taxa and providing "allopatric" conditions favorable to divergence.

2. Some hosts may have no parasites, the equivalent of certain taxa being absent from certain areas.
3. Some parasite species may occur on more than one host, the equivalent of widespread distributions.
4. Some hosts have more than one parasite species, the equivalent of redundant distributions.

Daniel Brooks (1981) explicitly treated the analysis of phylogenetic relationships of multiple parasites on a single group of hosts as a way of establishing (testing) the relationships of the hosts themselves. His approach used the parasite relationships as character data and assumed that when multiple parasites were evaluated, signal would overwhelm noise and a corroborated answer would result, showing a correct host cladogram based only on cladograms of parasites. Most subsequent authors have not followed Brooks's approach and have avoided a priori assumptions about coevolution. Rather, they have more frequently used consensus techniques of the types described in Chapter 6 to compare cladograms of hosts and parasites as a way of determining to what degree parasites have actually evolved in parallel with their hosts.

As a first example of such congruence-based studies, let us examine the work of Brian Farrell and Charles Mitter (1990) on leaf beetles of the Lamiales-feeding genus *Phyllobrotica* (Coleoptera: Chrysomelidae) and their hosts. These authors constructed a cladogram for the 14 species of *Phyllobrotica* and two outgroup taxa. They also constructed a scheme of relationships for the hosts based on the best estimates available in the literature. They then compared the topologies of the host and parasite cladograms, excluding from the comparison those five beetle species for which no host information was available. Farrell and Mitter calculated the consensus and the Adams compromise for the two cladograms. They considered the Adams's result superior to the strict consensus for comparisons of this type because it allowed that, "both parallel phylogenesis and host transfer may contribute to a given set of insect/plant interactions." In the Adams's result, six of the eight possible groupings were resolved, suggesting a high degree of parallel diversification between the lamialean hosts and chrysomelid parasites.

As a second example, let us consider work on coevolution between pocket gophers and their louse parasites as originally conducted by Mark Hafner and Steven Nadler (1988, 1990) (Figure 9.9). The data were reanalyzed by Roderic Page (1990b, 1994), who included substantial observations on methods for seeking congruence between host and parasite phylogenies. The objects of this study were eight species of pocket gophers in the genera *Geomys*, *Orthogeomys*, and *Thomomys* (Rodentia: Geomyidae). They are the hosts of 10 species of chewing lice belonging to the genera *Geomydoecus* and *Thomomydoecus* (Insecta:

Gophers
(hosts)

Lice
(parasites)

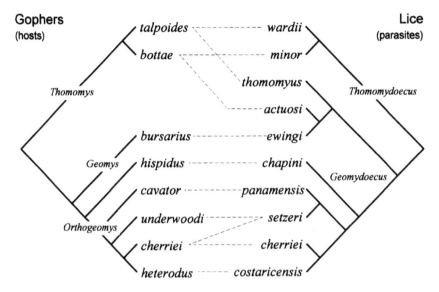

FIGURE 9.9. Phylogenetic relationships of pocket gophers and those of their louse parasites compared. Dotted lines indicate relationships of the hosts and their parasites (from Page 1994: figure 9).

Phthiraptera). Hafner and Nadler identified six components common to the two phylogenies and postulated three dispersal events to explain the remaining louse distributions. Page reanalyzed their data to determine whether or not they had identified the maximum number of cospeciation events. He found that they had but concluded that the anomalous distributions attributed to dispersal events could be explained alternatively as the result of sampling error and extinction. Under further analysis, Page concluded that rather than two parsimonious scenarios for the coevolution of relationships for the pocket gophers and their lice there were actually six, but that the maximum number of cospeciation events was always six, and the differences among the schemes concerned alternative explanations for incongruence between the lice and their hosts.

Some further examples of phylogenetic coevolutionary studies include termites and attine ants and their respective fungal symbionts (Mueller et al. 1998; Aanen et al. 2002); birds and lice (Johnson and Clayton 2003; Smith et al. 2004); histricognath rodents and their pinworms (Hugot 2003), and evolution of mimicry in *Heliconius* butterflies (Brower 1996). In the last case, the paradigm of parallel evolution of aposematic wing patterns has been challenged to some extent by an alternative hypothesis of adaptive introgression (Gilbert 2003; Zhang et al. 2016).

More recently, the advent of environmental genomics and the study of microbiomes has led to a number of coevolutionary revelations. For example,

Moeller et al. (2016) documented a high degree of taxonomic congruence between phylogenetic trees for several species of gut-inhabiting bacteria and their hominid hosts. The ever-expanding literature on evolutionary patterns of bacterial gut biomes in the Heteroptera: Trichophora (Insecta) hosts were reviewed by Schuh and Weirauch (2020). On the other hand, event-based narratives have spilled over from biogeographical research to coevolutionary studies as well. Nylin et al. (2017) emphasized the importance of colonization of novel host plants by herbivorous insects as an analogy to a dispersalist biogeographical scenarios.

SUGGESTED READINGS

Crisci, J.V., L. Katinaa, and P. Posadas. 2003. Historical biogeography, an introduction. Cambridge, MA: Harvard University Press. [A review of methods and approaches in historical biogeography]

de Queiroz, A. 2014. The monkey's voyage: How improbable journeys shaped the history of life. New York: Basic Books. [A pro-dispersal trade book]

Ebach, M.C., and R.S. Tangney, eds. 2008. Biogeography in a changing world. Boca Raton, FL: CRC Press. [A collection of papers describing relatively recent approaches in biogeography]

Heads, M. 2012. Molecular panbiogeography of the tropics. Berkeley: University of California Press. [A critique of the exuberant rejection of vicariance biogeography with molecular data]

Humphries, C.J., and L.R. Parenti. 1999. Cladistic biogeography: Interpreting patterns of plant and animal distributions. Oxford: Oxford University Press. [A concise but thorough treatment of background and method]

Lomolino, M.V., D.F. Sax, and J.H. Brown, eds. 2004. Foundations of biogeography. Classic papers with commentaries. Chicago: University of Chicago Press. [A broad cross section of classic papers in biogeography]

Morrone, J.J., and J.V. Crisci. 1995. Historical biogeography: introduction to methods. Annual Review of Ecology and Systematics 26:373–401. [A review of methods in historical biogeography]

Nelson, G., and N. Platnick. 1981. Systematics and biogeography: Cladistics and vicariance. New York: Columbia University Press. [Historical review and detailed discussion of biogeographic methods]

Nelson, G., and D.E. Rosen, eds. 1981. Vicariance biogeography: A critique. New York: Columbia University Press. [A collection of symposium papers, including discussion, dealing broadly with historical biogeography]

Page, R.D.M., ed. 2003. Tangled trees: Phylogeny, cospeciation, and coevolution. Chicago: University of Chicago Press. [A collection of papers addressing relatively current issues in coevolution]

Parenti, L.R., and M.C. Ebach. 2009. Comparative biogeography: Discovering and classifying biogeographical patterns of a dynamic earth. Berkeley: University of California Press. [An updated, slightly less "cladistic" edition of Humphries and Parenti (1999)]

Sanmartin, I., and F. Ronquist. 2004. Southern Hemisphere biogeography inferred from event-based models: plant versus animal patterns. Systematic Biology 53:216–243. [A large-scale study of Southern Hemisphere distributions using event-based methods]

EVALUATING HYPOTHETICAL SCENARIOS OF EVOLUTION, ECOLOGY AND ADAPTATION

In Chapter 9 we examined issues that fit broadly under the umbrella of coevolution but more specifically within the restrictive subset of phenomena often referred to as cospeciation. The methods for testing theories of cospeciation are not well suited to test coevolutionary theories of ecological association, adaptation, and more loosely constrained patterns of host association. It is, nonetheless, desirable to evaluate such theories in a rigorous historical context. Indeed, the ability to use the results of cladistic analyses to evaluate ecological and adaptational theories represents a truly powerful application of the method. This area of inquiry has become a standard approach in contemporary phylogenetic research, to the extent that one's chances of publishing a study in a high-profile journal may be diminished if the tree is not couched as a test of such a hypothesis. Within the cladistic framework, two interrelated approaches to evaluating adaptational hypotheses have been proposed—mapping and optimization. We will review examples of the application of each.

This chapter describes methodological approaches that are best suited to what we refer to in this volume as *extrinsic data*, but which also apply to the optimization of heritable traits that were not part of an analysis. Our rationale is based on the desire to provide an independent test of theories about the evolution or association of attributes within individual lineages when there is no straightforward way to produce a hierarchic scheme for those attributes. For example, what criterion of homology does one apply to ecological characters such as "entomophagous," "herbivorous," "ectoparasitic," and "sanguivorous on vertebrate blood"? Likewise, how does one homologize host associations—and particularly character adjacencies—in phytophagous insects when a single monophyletic group may feed on such disparate plant groups as Fabaceae, Fagaceae, Ericaceae, Salicaceae, and Sapotaceae (Schuh 2006)?

No doubt, our approach is not the only one that might be taken. Indeed, Miller and Wenzel (1995) saw no problem with coding ecological and other "extrinsic" characters alongside intrinsic features and described in detail methods and justifications for doing so, as was also the case advocated by Kluge (1989), whose "total evidence" concept was discussed in Chapter 6. On the other hand, although including such data in a character analysis may well offer a broader sample of evidence, the use of the resulting phylogenetic hypothesis as an assessment of the nature and sequence of "ecological" or other changes with a clade cannot be viewed as independent of the analysis itself and therefore offers a somewhat tautological test of such theories. Regardless of whether extrinsic characters are included in the phylogenetic analysis or not, it is clear that support for patterns implied by those characters needs to come from other congruent characters to provide independent corroboration of implied scenarios.

Information on geographic distributions is certainly "extrinsic," and the nature of state associations between areas of endemism cannot be defined through the use of homology criteria specified in Chapters 3 and 4. Nonetheless, the analysis of distribution patterns still seems to fit under the coevolutionary model because the comparisons are being made among multiple hierarchic schemes, all of which are derived from cladistic analyses, the subject matter of Chapter 9. The situations examined in the present chapter, however, all involve the distribution of traits within single lineages rather than comparisons among lineages.

Evolutionary Scenarios and Their Tests

The acquisition of novel attributes and associations over evolutionary time has long fascinated biologists. Views on the how, when, and why of such acquisitions have often been expressed in the form of scenarios. For example, arachnologists long held the view that the orb web evolved twice from a cobweblike structure because of its superior prey-catching ability (Coddington 1988). Although such statements might be viewed as hypotheses to be tested, they have frequently been propounded as if testing were irrelevant or as if their truth were self-evident. Whatever the intention of the authors of such claims, many such scenarios are subject to test via knowledge derived from systematic studies. In other words, by testing an evolutionary hypothesis in a systematic framework, we might come to some understanding of whether an idea represents mere assertion or whether available data actually support it. Such tests have provided a fruitful demonstration of the relevance of systematics to areas of biology such as ecology and animal behavior, where it has been ignored in the past.

Corroboration, or testing, of evolutionary scenarios can be divided into two broad approaches: (1) *cladistic methods*, which provide a way of understanding the sequence of ecological and adaptive changes relative to the branching pattern of the taxa that exhibit them; or (2) *statistical techniques*, which are used to test theories of adaptation with uncorroborated cladograms or nonphylogenetic classifications uncritically employed as standards of comparison (Crisp 1994). We will examine the former approach in some detail. The latter approach, whose base of inquiry is rooted in ecology rather than in systematics, has been promoted under the banner of "the comparative method" by Harvey and Pagel (1991) and Martins (1996) and has burgeoned into a distinct and, in our view, rather confused discipline (cf. Swenson 2019). These authors generally view phylogenetic relationship as a source of bias or a nuisance parameter to be factored out of the study of adaptation. That perspective will not be covered here.

Mapping

Once a cladogram is available, the distribution of attributes among the terminal taxa can be examined and understood explicitly. Any character included in the cladistic analysis can, of course, be traced on the cladogram to reveal inferred positions of character-state transformation, patterns of shared ancestry of alternate states, or homoplastic occurrence of independent gains and/or reversals. Certain morphological attributes, for example loss of legs in various squamate lineages, may be revealed as homoplastic features that have multiple origins. In addition, ecological associations and other extrinsic characteristics not used in the phylogenetic analysis can be "mapped" onto a cladogram as a way of understanding the number of origins, the sequence of origin, and the hypothetical ancestral conditions of those features as well. The procedure is analogous to the method used for biogeographic analysis described in Chapter 9.

We remind the reader, as discussed in Chapter 6, that the best supported phylogenetic hypotheses are those derived from simultaneous analysis of all relevant evidence (but as noted, not including extrinsic characters). We therefore emphasize the difference between the approach described here and the hubristic view of certain boosters of molecular systematics who espouse the notion that "phylogenies" should be based on molecular data and that morphology, due to its supposed fickle adaptive nature or difficult-to-interpret patterns of homology, should be mapped onto those trees (see, e.g., Gould 1985; Hedges and Maxson 1996; Scotland et al. 2003; Avise 2006) Such an approach strikes us as fundamentally wrongheaded, for at least three reasons: (1) the legitimacy of molecular systematics is derived from its almost universal corroboration of hypotheses of relationships based on morphology; (2) relying solely on DNA means that many taxa—fossils and those extant groups unsampled for DNA, for example—cannot be included in the analysis; (3) there are strong logical arguments for analyzing all relevant data simultaneously, as we have discussed in Chapters 3 and 6; and (4) the idea that molecular data are less homoplastic than morphological data is manifestly false, as is the idea that homoplastic data do not contain information on synapomorphy (Källersjö et al. 1999).

Mapping of the adaptive consequences of life-history strategies was used by Mitter et al. (1988) to test the widely invoked—but seldom tested—hypothesis that "diversification is accelerated by adoption of a new way of life, i.e., movement into a new adaptive zone." Using an independent contrasts approach, Mitter et al. mapped the feeding habits of major clades of insects onto cladograms of those groups and compared the numbers of known species in adjacent clades. Their results showed that in 11 out of 13 comparisons, the phytophagous clade was substantially more diverse than its nonphytophagous sister group, suggesting that evolution of phytophagy promotes diversification in insects.

Some monophyletic groups of phytophages are known to maintain strict associations with a particular host group. This is not the case with many groups, however, nor does it obtain at higher levels of relationship. Because there is no test available to determine homology and state order for such extrinsic characters, mapping is the only way to understand the sequence of change. Miller (1987) used mapping to demonstrate that patterns of larval food-plant association in the swallowtail butterflies of the family Papilionidae were often not tightly constrained at the tribal or generic level (Figure 10.1). Nonetheless, on the basis of the patterns of evidence, he concluded that host switching in the Papilionidae is probably constrained by plant chemical compounds.

Mapping was used at the species level by Futuyma and McCafferty (1990) in their examination of host relationships of the chrysomelid beetle genus *Ophraella* and its association with members of the plant family Asteraceae. Although the beetles are restricted to one plant family, critical analysis showed no clear patterns of cospeciation between the beetles and the eight genera of Asteraceae that serve as their hosts. The authors concluded that host shifts by the beetles postdate the divergence of the host plants. Similar studies have been performed for several butterfly groups (Brower 1997; Weingartner et al. 2006; Penz 2007).

Mapping has also been used to analyze the historical origins of traits that are at least in part genetically determined. As one possible example, Nils Møller Andersen (1997) studied the evolution of flightlessness and wing polymorphism in insects generally and the family Gerridae of the suborder Heteroptera in particular. Within the Gerridae, these features are determined by a combination of

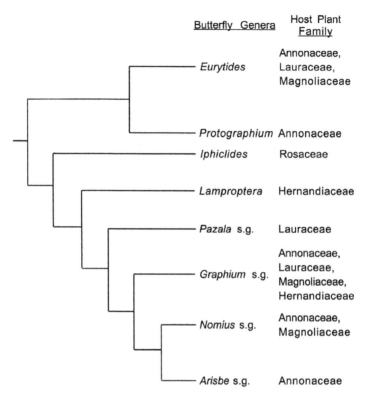

FIGURE 10.1. Cladograms comparing phylogenetic relationships among genera of the tribe Graphiini (Papilionidae) and host plant associations within the group (Miller 1987). Comparison shows not only that different butterfly lineages feed on the same host group but also that some butterfly lineages feed on multiple host groups. Thus, we must assume multiple colonization events or losses of association with some plant groups.

genetic and environmental factors. Andersen found that patterns were difficult to discern at the level of insect orders, probably because the nature of the condition is difficult to define. Within the Gerridae, however, his analysis produced the "unexpected" conclusion that—contrary to popular conception—the ancestral condition for the wings is not always monomorphic macroptery (fully developed functional wings). Andersen's analysis revealed that within the Gerridae, wing dimorphism (fully winged forms and those with reduced wings and flight musculature) is primitive, with monomorphic macroptery being a secondary acquisition. As can be seen from Figure 10.2, Andersen's analysis permits determination of the ancestral condition for nodes on the cladogram but on a less rigorous basis than would be the case under the "optimization" approach described below.

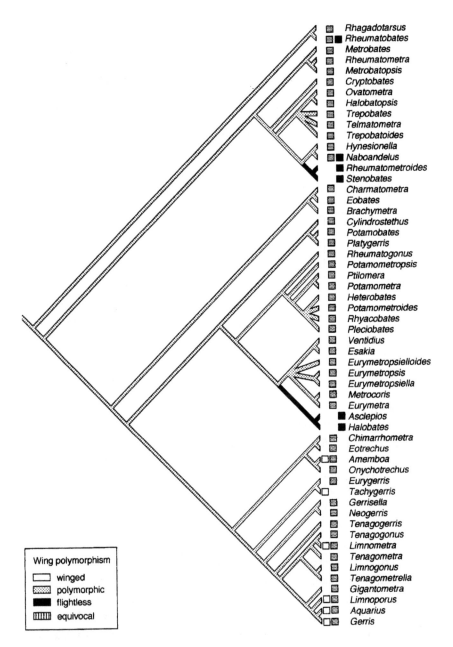

FIGURE 10.2. Mapping flightlessness on a cladogram of generic relationships within the Gerridae (Andersen 1997:98), showing the widespread and apparently primitive nature of wing polymorphism and the relative rarity of occurrence and fixity of flightlessness and permanent macroptery (fully developed functional wings).

In an analogous study, Whiting et al. (2003) showed that wings in stick insects appear to have re-evolved from wingless ancestors at least four times within the order. They mapped the presence or absence of wings on a cladogram based on nuclear ribosomal RNA and histone gene sequences. Subsequent controversy over the optimization of wing losses versus gains is a good example of the effect of imposing asymmetrical character-state transformation costs (Trueman et al. 2004; Whiting and Whiting 2004).

In another case, John Wenzel (1993) examined congruence between behavior and morphology. His study group comprised 23 of the 28 generic-level taxa of the paper wasp subfamily Polistinae. Wenzel treated the attributes of nest architecture as behavioral characters. Although we have argued that behavioral characters can be treated as intrinsic information and therefore used as part of a "total evidence" analysis, Wenzel chose to construct a cladogram solely on the basis of behavioral traits to examine the degree to which those attributes produced a scheme topologically congruent with one based on morphology. His comparison of the two schemes is shown in Figure 10.3, revealing a substantial degree of congruence. He also optimized his "architectural" data directly onto the morphology-based cladogram of Carpenter (1991), a process that revealed a consistency index of 56 percent for the architectural data, compared with 48 percent for the morphological data on which the cladogram was originally based. Wimberger and de Queiroz (1996) surveyed a variety of studies employing behavioral characters and showed that patterns of homoplasy in behavioral versus morphological characters like those discovered by Wenzel are common.

Some groups of Australian papilionoid legumes are bird pollinated, whereas others are pollinated by bees. Michael Crisp (1994) used cladistic methods to determine the origins of bird pollination among them. The groundwork for understanding bird pollination was laid in revisionary works by Crisp and Weston in their analyses of relationships among the 30 genera belonging to the endemic Australian papilionoid tribe Mirbelieae. On the basis of floral morphology, Crisp concluded that bird pollination had arisen independently at least five times within the Mirbelieae. Toon et al. (2014) conducted a more detailed study of the genus *Gastrolobium*. As shown in Figure 10.4, even within this one genus, bird pollination had arisen three times in this primitively bee-pollinated group. In all cases, Crisp and colleagues inferred phylogenetic trees for the legumes based on only characters not associated with bird pollination and retrieved the same results, offering evidence for the independence of his conclusion concerning the number of origins of bird pollination.

A prevalent use of the comparative mapping approach is in efforts to assess and explain inferred differences in DNA sequence divergence rates among various lineages. As we note in Chapter 11, there are a variety of factors hypothesized

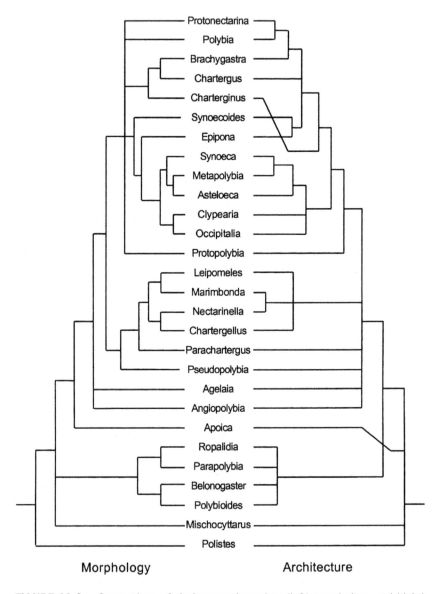

Morphology   Architecture

FIGURE 10.3. Comparison of cladograms based on (left) morphology and (right) nest architecture for the paper wasp subfamily Polistinae (from Wenzel 1993).

to contribute to variability in evolutionary rates, including metabolic rate and generation time. Bromham (2016) reviewed several of these. In one example, the rate of molecular evolution in parasitic plants was inferred, based on comparisons of multiple parasitic lineages with their nonparasitic relatives, to be faster than that in autotrophic plants. This could be explained by the parasitic habit or by some confounding correlated factor, such as plant stature (parasitic plants

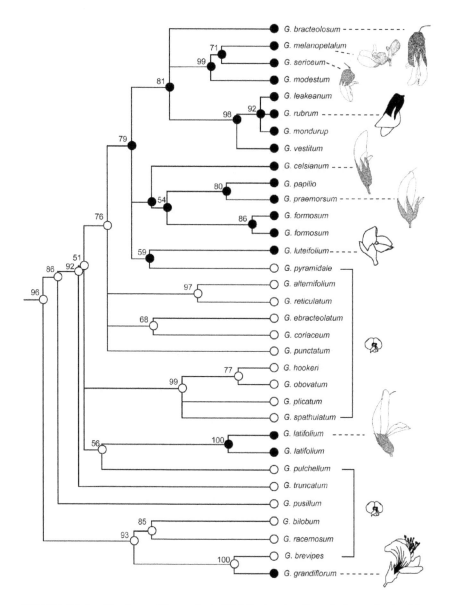

FIGURE 10.4. The tree of Toon et al. (2014) showing three independent origins of bird pollination in the Australian papilionoid legume genus *Gastrolobium*. Black dots indicate presence and hypothetical ancestral states for bird pollination in the group.

tend to be small, and lineages of smaller plants also apparently have faster rates of molecular evolution than larger ones (Lanfear et al. 2013).

Optimization of Hypothetical Ancestral Character States

In Chapter 5, we discussed character optimization in the context of discovering most parsimonious (or otherwise optimal) trees. As introduced in Chapter 4, the term *optimization* is also employed to refer to the inference of the state occurring at internal nodes in a cladogram, which can be of interest in the development of evolutionary and coevolutionary scenarios. Most of the time, the inferred states of internal nodes are unambiguous on a given tree: if terminal sister taxa share the same state, then the most parsimonious state for their hypothetical ancestor is the state they share, based on the interpretation that a single origin of the feature in the common ancestor is more parsimonious than independent origins in each of the daughter lineages. Ambiguity among alternate equally parsimonious optimizations arises when adjacent terminals or groups of terminals exhibit different states. You have probably seen little pie charts on the internal nodes of published phylogenetic trees, that pictorially represent uncertainty about inferred ancestral states. Farris (1970) proposed—and Swofford and Maddison (1987, 1992) elaborated—a method for calculating internal states that has been implemented in software such as MacClade (Maddison and Maddison 1992) and WinClada (Nixon 2000).

The basic method for assigning character states to internal nodes with nonadditive data is to start at the terminals and work toward the base of the tree, using the state of the immediately distal nodes to assign a state to progressively more basal ones. If the distal nodes differ in their states, then the basal node is designated as ambiguous for the alternate states. If one distal node is ambiguous, but one of its possible states is shared by its sister node, then the basal node is resolved as the shared state. The "down-pass" phase is complete when the root of the tree is reached. Ambiguous nodes may often be resolved by reversing the process and performing an "up-pass," using the shared state criterion.

As discussed in Chapter 4, there are times when alternate equally parsimonious optimizations imply different states for internal nodes. Swofford and Maddison (1987) described two scenarios, accelerated and delayed transformation (*ACCTRAN* and *DELTRAN*), that characterize two ends of the spectrum of possible transformation scenarios. Under accelerated transformation, state changes are hypothesized to take place as close to the root of the tree as possible, which tends to favor early gains of features with subsequent reversals. Delayed transformation, by contrast, hypothesizes transformations as occurring closer to the tips of the tree, thereby implying independent gains of features. De Pinna (1991) made the intuitive evolutionary plausibility argument that losing a feature is easier than

gaining one, and that therefore ACCTRAN should, in general, be favored. Agnarsson and Miller (2008) have questioned the universality of this assumption.

An example is helpful to illustrate these concepts. Many members of the Vespidae (hornets, yellowjackets, and potter wasps) are eusocial, with a single female
acting as the reproductive queen and the remainder of females as nonrepro-
ductive workers. A long-standing question in this group is whether eusociality
evolved more than one time. Based on a phylogenetic analysis of morphological
characters, James Carpenter (1988b) hypothesized a single origin of eusociality
in Vespidae. Carpenter (1989) optimized hypothesized changes in vespid social
behavior on morphology-based cladograms of the groups in question. His procedure was first to map observed behaviors on terminal taxa and then determine
the optimal character-state set for the hypothetical ancestors of those groups.
The approach allowed for critical examination of the order of the proposed
stages in the evolution of social behavior, a theory that was derived outside of a
phylogenetic context. A portion of Carpenter's example is shown in Figure 10.5.

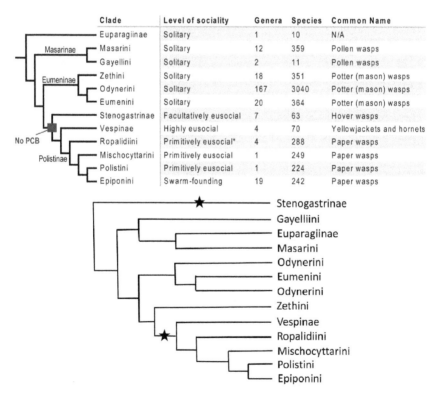

Clade	Level of sociality	Genera	Species	Common Name
Euparagiinae	Solitary	1	10	N/A
Masarini	Solitary	12	359	Pollen wasps
Gayellini	Solitary	2	11	Pollen wasps
Zethini	Solitary	18	351	Potter (mason) wasps
Odynerini	Solitary	167	3040	Potter (mason) wasps
Eumenini	Solitary	20	364	Potter (mason) wasps
Stenogastrinae	Facultatively eusocial	7	63	Hover wasps
Vespinae	Highly eusocial	4	70	Yellowjackets and hornets
Ropalidiini	Primitively eusocial*	4	288	Paper wasps
Mischocyttarini	Primitively eusocial	1	249	Paper wasps
Polistini	Primitively eusocial	1	224	Paper wasps
Epiponini	Swarm-founding	19	242	Paper wasps

FIGURE 10.5. (Top) Hypothesis, based on morphology, indicating a single
origin of eusociality among vespid wasps (star) (after Carpenter 1989). (Bottom)
Phylogenomic tree implying two origins of eusociality (stars) in Vespidae. Both
figures adapted from Piekarski et al. (2018).

Other authors, using rather modest molecular data sets, hypothesized a diphyletic origin for eusociality (e.g., Schmitz and Moritz 1998; Hines et al. 2007), but Carpenter steadfastly defended his interpretation of the pattern with critical reanalyses of the data that continued to lend support to a monophyletic origin (Carpenter 2003b; Pickett and Carpenter 2010). However, in the face of overwhelming phylogenomic evidence accumulated over the past decade, Carpenter has, after 30 years, capitulated, expressing support for the diphyletic hypothesis (Piekarski et al. 2018). The moral of this story is that any adaptive scenario, no matter how venerable and seemingly well substantiated, remains subject to test and falsification.

We have discussed the issue of the evolution of host associations in phytophagous insects on several occasions. Schuh (2006) used optimization techniques to evaluate patterns of host shifting in the *Phymatopsallus* group of genera (Heteroptera: Miridae) from the North American Southwest. He first calculated the most parsimonious tree for a group of 32 taxa. He then coded the host data as a single unordered multistate character and determined the most parsimonious distribution of those data on the morphological cladogram. The results are shown in Figure 10.6.

At a more inclusive level, Weirauch et al. (2019), based on a combined phylogenetic analysis of morphological and ribosomal RNA gene data, examined the evolution of habitat and feeding preferences across the entire Heteroptera. These extrinsic attributes were mapped onto substantially congruent parsimony, dynamic homology, and likelihood trees, using both parsimony and likelihood character optimization approaches. According to this hypothesis (or these hypotheses), true bugs evolved an aquatic lifestyle three separate times and phytophagy evolved twice, from a plesiomorphic predaceous condition.

The reader may raise an eyebrow at the revelation that, after our steadfast advocacy of parsimony and eschewal of employing multiple methods, that this paper coauthored by Schuh adopted a *syncretic* approach and even selected the maximum likelihood tree over the most parsimonious tree as its preferred topology for mapping the ecological characters (in this case, because the likelihood tree provided greater resolution and thus a framework for a more precise classification scheme). We note that Brower has participated in similarly syncretic projects (e.g., Wahlberg et al. 2009; Toussaint et al. 2018). As Brower (2018a) observed:

> There are diverse reasons/excuses for syncretism. Sometimes, there may be a theoretical question that warrants comparison of results from multiple methods, such as testing the success of some novel approach against traditional approaches. Sometimes, a cladist may be engaged in a collaborative empirical project with systematists who hold alternative

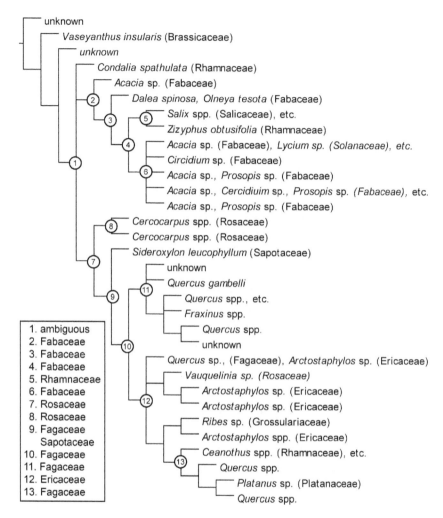

FIGURE 10.6. The optimization of host associations on a morphology-based cladogram, with the host condition indicated for each of the nodes (adapted from Schuh 2006).

views, and to keep the peace, multiple analytical approaches may be presented as a compromise. Sometimes, poorly trained systematists possess limited familiarity with the theoretical bases of the various methods they choose to employ, and perhaps imagine that getting a similar result from multiple methods might increase corroboration (a false notion— see Brower 2000[a]). Sometimes, systematists may fear that expressing a preference for one method or another will alienate readers (or reviewers) from an opposing camp, and conversely, sometimes, imperious editors may demand that multiple analytical approaches be employed. And some systematists appear to try all the methods and select the one that gives them the tree they were hoping for (sometimes called "taxonomic congruence" e.g., Wheeler et al. 2017).

Although such compromises expose cladists to accusations of hypocrisy, from an optimistic perspective, we can nevertheless be comforted by the fact that since the taxonomic congruence between parsimony and model-based analyses found in the Weirauch et al. (2019) study is apparently a general pattern (Rindal and Brower 2011), colleagues with differing methodological preferences can at least agree, in most cases, on the interpretation of the biological results.

The mapping and optimization approaches described in this chapter may seem less precise or appear to deal with sloppier data than the coevolutionary approach described in Chapter 9. The diminished analytical complexity does not impugn the scientific value of the results, however. The patterns we observe among terminals, and the states we can infer for internal nodes, offer substantial predictive power about the habits and associations for members of groups yet to be collected or observed. This predictive power is the same as for attributes whose existence derives directly from inheritance.

Tests of Adaptational Hypotheses

From the selectionist view of evolutionary biology come theories predicated on the idea that natural selection for superior adaptations acts as the force driving evolutionary change. Many such theories have not been subjected to rigorous testing but exist more in the realm of conjecture, and there are alternative explanations for the distribution of features among organisms that include phylogenetic constraint (Lewontin 1978; Wanntorp 1983). For example, for many years it was hypothesized that the enormous antlers of the Irish elk were the result of orthogenesis or runaway sexual selection. Gould (1974) showed that the large antlers could be explained simply as a result of an allometric growth relationship that holds among the Cervidae: small deer have disproportionately small antlers, while large deer have disproportionately large antlers. Thus, there is no need to invoke adaptive scenarios to explain the large antlers of the Irish elk.

Jonathan Coddington (1988) proposed an approach utilizing cladistic methods to test adaptational hypotheses. The requirements of his approach, portrayed in Figure 10.7, demand that cladograms be inferred for the group being studied on the basis of attributes other than those involved in the hypothesis of adaptation. Also included is the necessity of a specified relationship between form

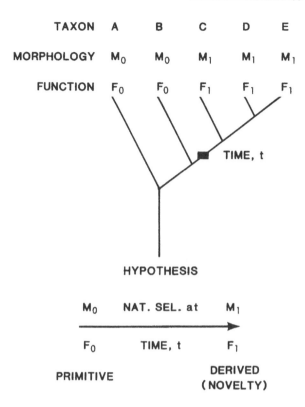

FIGURE 10.7. The approach of Coddington (1988: figure 1) for testing hypotheses of evolution through adaptation (M = morphology; F = function; t = time; subscripts indicate direction of hypothesized change 0 → 1). Selection is inferred to have acted at time t to produce the derived, presumably adaptive, synapomorphic state M1.

(morphology, M) and associated adaptation (function, F), as well as the stipulation of transformation from primitive (ancestral) to derived. One of the hypotheses Coddington tested was that mentioned at the beginning of this chapter: that orb webs evolved twice from cobwebs. Based on cladograms for orb-web- and cobweb-weaving spiders, Coddington showed that orb webs are ancient and had evolved only once and that cobwebs are advanced relative to orb webs, not vice versa. Coddington's exclusion of the "adaptive" character under test from the analysis has been criticized from the "total evidence" perspective (e.g., Deleporte 1993; Wenzel and Carpenter 1994; Zrzavý 1997; Kluge 2005a; see Chapter 6). However, even if the character had been included in the analysis, it is doubtful that a single character with two states would change the resultant topology. Indeed, the traditional "two gains" and Coddington's "gain and subsequent loss" scenarios are alternative optimizations of equally parsimonious distributions of the character, each requiring two steps, and so, at least in that case, including or excluding the character has no impact on the resultant hypothesis of relationships.

The mapping and optimization approaches described above allow us to structure our understanding of extrinsic attributes in light of the best available theories of phylogenetic relationships. These extrinsic data may or may not have been construed under some prior theory of relationships for the taxa exhibiting them. Nonetheless, the types of adaptational theories discussed by Coddington imply theories of relationships themselves. Such theories will survive testing only if they conform to the requirements specified by Coddington, and also are congruent with the independently derived phylogenetic hypothesis.

There is, of course, a vast noncladistic literature addressing the comparative method. Pertinent reviews may be found in O'Meara (2012), Garamszegi (2014), and Swenson (2019). A premise of this body of work seems to be that phylogenetic relationships are a nuisance parameter that needs to be discounted before the statistical verification of the adaptive process under study may be achieved. Indeed, synapomorphies may be seen as a material impediment to this work, as expressed by the following revealing quotations (abridged for succinctness) from Uyeda et al. (2018:1091):

> Their singular and unreplicated nature seems incompatible with models that we typically use to describe change over time, such as Brownian motion . . . or the Mk model . . . It is these sorts of idiosyncratic and unreplicated events that we often think of when we think of the need to consider phylogeny in analyses of comparative data. And this is not an abstract concern; a wide breadth of macroevolutionary data suggest that abrupt shifts and discontinuities have been a major feature of life on Earth . . . But as recent controversies in phylogenetic comparative biology have highlighted, our current methods . . . are not designed to deal with such dynamics.

This statement embodies the problematical, operationalist perspective of statistical approaches to phylogenetics, discussed in Chapter 5, that unruly data that are not easily accommodated by simplistic model parameters represent an inconvenience to the comparative method. And what is the fix? Even more complicated statistics: "one way to refine our inferences is to re-imagine phylogenies as probabilistic graphical models. . . ."

As empiricists, we think the opposite: synapomorphies are the evidentiary bases that allow the inference of evolutionary history. A hypothesis of relationships is the result; propounding adaptive reasons for homoplasy, or the absence of homoplasy, is a separate task (Brower et al. 1996). Whether statistical inference can confirm or reject the occurrence of unique historical events depends, we surmise, upon one's prior commitment to the endeavor. "The more you believe it, the truer it is." A final observation we might make is that when a researcher simply describes the distribution of some attribute on a tree, be it an evolutionary trend in some ecological association of interest or a dispersalist biogeographical scenario, it does not constitute a test of a hypothesis (Fitzhugh 2016). It is merely an ad hoc hypothesis to be tested.

SUGGESTED READINGS

Brooks, D.R., and D.A. McLennan. 1991. Phylogeny, ecology, and behavior: A research program in comparative biology. Chicago: University of Chicago Press. [A relatively strong introduction to cladistics, with analytic tools for ecology and adaptation, biased toward the use of parsimony rather than congruence]

Cadotte, M.W., and T.J. Davies. 2016. Phylogenies in ecology: A guide to concepts and methods. Princeton, NJ: Princeton University Press. [A recent textbook on the subject]

Eggleton, P., and R. Vane-Wright, eds. 1994. Phylogenetics and ecology. Linnaean Society Symposium Series 17. London: Academic Press. [Papers on the application of cladistics in understanding ecological issues]

Grandcolas, P., ed. 1997. The origin of biodiversity in insects: phylogenetic tests of evolutionary scenarios. Mémoires du Muséum national d'Histoire naturelle 173. [Discussions of methods and empirical tests of evolutionary scenarios]

Harvey, P.H., and M.D. Pagel. 1991. The comparative method in evolutionary biology. Oxford: Oxford University Press. [A relatively weak treatment of cladistics, with analytic tools tending toward the statistical]

Lewontin, R.C. 1978. Adaptation. Scientific American 239:213–230. [Not systematics, but a classic paper discussing the problems of adaptation and adaptationism]

Miller, J.S., and J.W. Wenzel. 1995. Ecological characters and phylogeny. Annual Review of Entomology 40:389–415. [A review of relationship and use of ecological data in phylogenetic analysis]

Rose, M.R., and G.V. Lauder, eds. 1996. Adaptation. San Diego: Academic Press. [A collection of reviews and case studies from a variety or perspectives]

Swenson, N.G. 2019. Phylogenetic ecology: A history, critique and remodelling. Chicago, University of Chicago Press. [A thoughtful, if perhaps overly optimistic, review and critique of a field that has deviated substantially from systematics]

UNDERSTANDING MOLECULAR CLOCKS AND TIME TREES

Background and Basics

In 1962, biochemists Émile Zuckerkandl and Linus Pauling introduced the concept of the *molecular clock*. Although laden with numerous process assumptions that, as we shall see, may or may not be true (or knowable), the idea is appealingly straightforward: if amino acid substitutions in proteins occurred at a relatively steady pace that were more or less constant both over time and along each of the branches of a diverging evolutionary tree, then the number of substitutions would be directly related to the time since the taxa in question diverged from one another. If an independent time calibration could be associated with a node on that tree, then it could be compared with the amount of difference among the proteins and the resultant inferred rate of x substitutions per million years could be employed to extrapolate the ages of other uncalibrated divergence points. Thus, for example, if we inferred from fossil evidence that species A and species B diverged four million years ago and we observed eight amino acid differences between their protein sequences, then the estimated rate of protein evolution, assuming clocklike regularity, would be one substitution per lineage per million years (once they split from one another, each lineage is evolving independently, so each one would accumulate its own substitutions after the divergence date). If there were another species, C, for which we had no fossil evidence, that exhibited four amino acid differences from B, and if this clocklike pattern were generally true, then we could infer that C is the sister taxon to B, and that the two diverged two million years ago. This scenario is diagrammed in Figure 11.1.

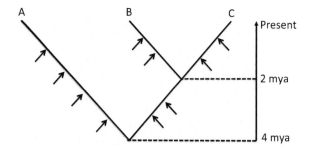

FIGURE 11.1. An idealized chronogram for three taxa with character-state transformations indicated by the arrows (MYA = million years ago).

A simple corollary of the clock hypothesis is that the number of sequence changes from the root to the tip of each branch on the tree should be (more or less) the same. This property is referred to as ultrametricity (see Sidebar 11, Chapter 5). Thus, in Figure 11.1, there are four inferred changes on the branch leading from the root to A, there are two changes between the root and the node where B and C diverged, and two changes on branches leading from that node to B and C, respectively. The rate constancy assumption of the molecular clock hypothesis may be tested, even in the absence of independent calibration points, by the relative rate test (Fitch 1976), the expectation that the number of substitutions between A and B should be equal to the number of substitutions between A and C. If this is not true, then the implied rate differs between the branches leading to B and to C (the path from A to the B + C node is the same for both), and the rate constancy assumption is falsified.

In order to provide the background for understanding the principles discussed in this chapter, we must stray away somewhat from the pattern-based realm of systematics to consider the nuts and bolts of molecular evolution and the concept of evolutionary neutrality. Amino acids make up proteins, proteins are part of the phenotype of an organism and, therefore, proteins are potentially under selection to a greater or lesser degree, depending on their function in the organism's internal economy. Generally, the primary structure of a protein, the string of amino acids assembled by the ribosome, folds up into a three-dimensional structure. Some parts of this folded molecule form the active sites that interact with other molecules to carry out the life of the cell and other parts form the framework that provides the necessary structure for those active sites to perform their interactions. The more important a particular section of a protein is to the protein's function, the more likely its sequence is to be constrained by selection—that is, mutations to the DNA encoding a protein that change its amino acid sequence in active sites are likely deleterious to the organism and therefore not passed on to subsequent generations. Selection on such sites is referred to as purifying selection and tends to slow down inferred rates of evolutionary change.

Given such constraints, there almost certainly will be variability in the rate at which different amino acids evolve in a protein. The same is true for rates of evolution among different proteins, taken as an overall average of the rates of change among their individual amino acids. Neither variation of rates within an individual protein, nor among diverse proteins, would affect molecular clock estimates because whatever the rate for a given protein may be, as long as it is constant through time and among lineages, it may be used to infer node ages, once calibrated with independent evidence from the fossil record or elsewhere. On the other hand, if rates have changed through time between the root and the tips of the tree, or vary from one branch to another, then all bets are off.

In the 1960s, evidence seemed to support clocklike evolutionary patterns of proteins, and the cracking of the genetic code led to the realization that the DNA that encoded them was perhaps even more likely to evolve in a largely clocklike manner. Motoo Kimura (1968) proposed the neutral theory of molecular evolution as an explanation for these patterns: unlike proteins themselves, protein-coding DNA sequences exhibit variability that seems to be selectively neutral due to the redundancy of the genetic code. (As readers should be aware, ribosomes read three nucleotides per amino acid to translate the genetic message into a protein.) There are 64 codons and only 23 amino acids. In many instances, whichever of the four nucleotides occupies the third position in a codon (some first positions as well), the codon is translated into the same amino acid. Therefore, nucleotide substitutions in these positions are not expressed in the phenotype and so are invisible to selection. The same was thought to be true for the large swaths of eukaryotic DNA sequence (e.g., introns, flanking regions) that had no known function and were referred to as "junk DNA." Kimura's protégé, Tomoko Ohta, refined her mentor's theory with a "nearly neutral" addendum that explained apparent neutrality even for DNA changes that affect amino acid sequences and are therefore subject to selection. She calculated that in smaller populations, stochastic evolutionary forces like genetic drift would overwhelm selection against mildly deleterious mutations (Ohta 1973). The iconoclastic suggestion that most of the genome was not subject to selection, changing because of mutational pressure and stochastic fluctuations alone, and therefore evolving under "non-Darwinian" rules, caused intense controversy among population geneticists of the time.

Prior to the advent of fast and easy DNA sequencing, a number of methods that took DNA neutrality as a given were developed, including DNA–DNA hybridization, already mentioned in Chapter 3. The molecular clock-dependent premise of that method was that the more distantly related two taxa were, the more differences there would be between their genomes on average, resulting in fewer hydrogen bonds holding together the nucleotides of two strands of hetero-duplexed homologous DNA from two different species. This could be measured, at least in theory, by the reduced melting temperature at which the strands would separate from one another. A relatively large amount of DNA from one species, the "driver," sheared into short pieces, was melted into single strands and mixed with a small quantity of similarly sheared, radioactively labeled DNA from another species, the "tracer." When cooled, the DNA would reanneal into double-stranded molecules, and the large difference in abundance of driver and tracer molecules meant that most of the reannealed pairs would be unlabeled driver-driver homoduplexes, but the radioactive tracer fragments would almost always stick to a complementary strand of the homologous sequence of the driver rather than to another tracer copy. The DNA was heated back up, and the melting temperature at which half the tracer molecules disassociated themselves from the driver DNA was recorded and compared with the melting point from a homo-duplexed control with driver and tracer DNA from the same species. The larger the difference in the observed melting temperatures between homoduplexed and heteroduplexed samples, the greater the inferred divergence between the taxa being compared. DNA–DNA hybridization was technically difficult, hard to control experimentally, and subject to asymmetries in divergence estimates when reciprocal comparisons of the same two taxa were performed (not to mention that the method represents a sort of ultimate phenetic measure, reducing the similarities and differences between the taxa compared to a single numerical distance value). Owing to these practical and conceptual challenges, DNA–DNA hybridization was largely abandoned by the mid-1980s (its last gasp was the iconoclastic "tapestry" of bird phylogeny published by Sibley and Ahlquist in 1990).

While some genes appeared to have evolved in a clocklike manner among some taxa for some periods of evolutionary history, by the 1980s it became clear that the molecular clock was not a general phenomenon. Wu and Li (1985) and Britten (1986) offered some of the earliest critical reviews of the accumulating data, showing significant inferred rate variability among lineages. To save the clock hypothesis, several ad hoc sources of variation were proposed: an inverse correlation of molecular evolutionary rates with generation times of organisms or cell lineages, a positive correlation of rates with metabolic rate, and/or rate variability due to historical differences in population size (the last, being impossible to measure, is a favorite refuge of irrefutability for population geneticists when the data do not fit their preferred model). Ayala (1999) reviewed patterns that falsified these explanations, since even when examined in the same taxa (thereby controlling for generation time, metabolic rate, and historical population size) rates vary among genes in different directions in different groups, with one gene apparently speeding up in one lineage versus another, and another slowing down. His conclusion was that "only fluctuating and unpredictable natural selection can account for these erratic patterns of molecular evolution"—a clear refutation of the clock hypothesis.

But a good story is hard to quell, and with the advent of the polymerase chain reaction and increasingly automated DNA amplification and sequencing in the late 1980s, the abundance of DNA sequences offered a great temptation to continue to seek clocklike patterns in molecular data. Particularly among closely related taxa, such as those examined in phylogeographic studies (see Chapter 7), most variability might be expected to occur in relatively unconstrained third-codon-position silent sites. At least in the initial stages of divergence, most observed sequence differences might represent unique, selectively neutral mutational events that could accumulate in a reasonably clocklike manner. Brower (1994b) used this as a rationale to propose an arthropod mitochondrial DNA clock, with the explicit caveat that it should not be extrapolated to infer ages of clades above some three million years, beyond which the linearity of nucleotide substitutions with time might be expected to be dampened by selective constraint (Brown 1983). Others felt no qualms in extending molecular clock estimates to infer ages of much older and more inclusive taxa, including vertebrates and metazoans inferred to have diverged tens or hundreds of millions of years ago (Hedges et al. 1996; Wray et al. 1996). Once this Pandora's box was opened, the allure of molecular clock narratives became too strong for many researchers to resist. The past two decades have seen a vast proliferation of time-calibrated evolutionary scenarios. Let us now examine some of the assumptions that underlie modern molecular clock estimates.

Calibrating Ages of Nodes on Phylogenetic Trees

The names of many geological periods, such as the Jurassic and the Cretaceous, are based on famous sedimentary rock strata. These were recognized, and their relative ages were well understood, by the mid-1800s. Geologists also knew, based on Charles Lyell's principle of *uniformitarianism*, that the earth must be old, but they did not know how old. Best estimates ranged to some millions of years—orders of magnitude older than the age established by the Sacred Chronology of James Ussher (1650), which deduced from the pedigrees of biblical patriarchs that the Earth was created at 9 a.m. on Sunday, October 23, 4004 BC. Lord Ernest Rutherford proposed the idea of radiometric dating in 1905, but it was not until the 1950s that estimates of the absolute ages of rock strata, and the fossils embedded within them, could be estimated with a degree of precision and accuracy.

Radioactive isotopes undergo spontaneous decay into other isotopes, according to a stochastic but measurable "half-life" (the average time it takes for half of the atoms in a sample to decay; in the second half-life, half of the remaining atoms decay, and so on, leading to an asymptotic diminution of the quantity of the parent material and corresponding increase of the quantity of daughter material). The stochastically clocklike decay of radioactive isotopes trapped within a rock's crystalline matrix provides a metric of the age of that rock by comparison of the ratio of parent to daughter isotopes (assuming that there were no daughter isotopes present when the rock was formed). Different parent isotopes have different half-lives, and therefore different ranges of effectiveness for date inference. One of the most familiar, ^{14}C ("carbon fourteen"), has a half-life of 5730 years and is useful mainly for dating archaeological remains that are less than 50,000 years old. By contrast, ^{40}K, which decays to ^{40}Ar, has a half-life of 1.3 billion years and has been used to estimate the age of the Earth at 4.57 billion years—a date that was derived from rocks brought back from the Moon! (Because of the Moon's lack of either an atmosphere or mobile tectonic plates, its surface is less prone to erosion, weathering, and subduction, and so rocks apparently dating from shortly after the separation of the ancient, molten moon from the Earth are accessible on its surface. The oldest discovered rocks on Earth are estimated to be ~3.8 billion years old.)

As discussed in Chapter 5, cladistic methods treat fossils the same as any other taxon with regard to their epistemological status as semaphoronts providing character evidence to populate a data matrix, and they appear as terminals on a cladogram, not nestled among its internal nodes. However, unlike extant taxa, fossils have the additional characteristic of being old. Of course, this does not mean that they are "ancestors" or that they necessarily bear ancestral character states, but it does mean that fossils can provide information about the ages of the clades to which they belong, if a reliable age can be inferred for them. In particular, a 10-million-year-old fossil that exhibits the synapomorphies of clade X tells us that clade X is at least 10 million years old. Tomorrow, we might discover a 15-million-year-old fossil that belongs to the same clade, and that would tell us that the clade is at least 15 million years old. Fossils thus provide hypotheses of *minimum ages* (Norell 1992)—a one-tailed age estimate truncated toward the recent end of the time scale. Benton and Donoghue (2007) have argued that "soft" maximum ages may be inferred by the absence of fossils of a particular lineage below a certain geological stratum when other taxa present in their paleoecological communities persist uninterrupted in the stratigraphic column. The inclusion or exclusion of such calibrations can have major effects on model-based age estimates (cf. Hedges et al. 2018; Morris et al. 2018). However, given the adage that absence of evidence does not equal evidence of absence, the proposition of plausible soft maximum ages is a tenuous one, at best.

There are also a number of uncertainties intrinsic to inferring the age of the fossil itself. Fossils are generally discovered in layers of sedimentary rock. First of all, there is some stochasticity in dates inferred from isotope ratios, meaning that a radiometric date estimate represents the mean of a distribution with some associated variance. Further, the ages of many sedimentary layers may be reported as multi-million-year ranges based on inferred ages of surrounding igneous rocks, rather than as precise point estimates for the particular rock layer where the fossil was deposited, and so a fossil's age may also be expressed as that range. In such a case, since a fossil only provides a minimum age for its clade, the conservative choice is to report the younger end of the range as its age. For example, claiming that a clade is at least 100 million years old is more temporally inclusive than claiming it is at least 150 million years old.

A second potential source of uncertainty is that many fossils are incomplete and do not exhibit all the necessary character states to assign them definitively as sister group to a particular clade. The more fragmentary a fossil is, the more features are likely to be unobservable, and the more general the group with which it may be unambiguously associated is likely to be. De Jong (2017) has eloquently described the challenges of using fossils to infer the ages of clades of butterflies, which, like other insects, do not generally preserve well and have a concomitantly sparse fossil record. His review showed that despite the historical assignment of fossils by their describers to rather narrowly circumscribed taxa (sometimes even extant genera), in many instances they bear only the apomorphies that allow them to be placed at the family level (note that the absence of apomorphies in a fossil is uninformative with regard to its placement on a cladogram). The ability to say that a family is at least x years old is a much less specific hypothesis than being able to say the same of a particular genus. Another effect of this ambiguity is that many vaguely classifiable fossils may provide redundant age estimates for the same node on the tree of the group to which it belongs. Thus, while there might be dozens of fossil specimens with associated date estimates belonging to a taxon, they might provide only one or a few independent calibration ages for a molecular clock hypothesis (cf. Särkinen et al. 2013).

Another geological source of evidence that may be employed to calibrate divergence dates on a phylogenetic tree is a vicariant event (Kodandaramaiah 2011; Ho et al. 2015; see also Chapter 9). Phenomena such as the elevation of mountain ranges, closure of waterways, and shifting of river drainage patterns may bisect the distributions of widespread species, interrupting gene flow and allowing each population fragment to pursue an independent evolutionary path.

In principle, vicariance theory has the advantage of providing a common cause explanation for congruent distributions of multiple pairs of sister taxa. However, vicariant events are often not instantaneous and may affect different taxa in different ways. For example, species confined to deep water might be isolated much sooner by the gradual emergence of the Isthmus of Panama than species able to live in the shallows (Vermeij 1993; Marko 2002). And even if a vicariant event did have a particular and rapid time of occurrence, such as a stream capture, it may be difficult to date precisely when that event took place, just like events in the fossil record.

A third means to infer node ages on a tree is to employ "*secondary calibrations*." If we think of the age of a fossil or a vicariant event that relates to one or more taxa under study as a "primary" calibration because it is derived from evidence independent of molecular clock assumptions, a secondary calibration is one that is derived from a prior molecular clock estimate from analysis of a different data set. This sort of extrapolatory inference is common in situations in which calibrations from the fossil record or other sorts of evidence are rare or unavailable. For example, Garzón-Orduña et al. (2015) compared ages inferred from two sets of secondary calibrations for the Ithomiini, a group of Neotropical butterflies. One set of calibrations was derived from fossil-based node ages inferred in a higher-level analysis of the butterfly family Nymphalidae (Wahlberg et al. 2009), and the second set was obtained from a fossil-calibrated tree of Solanaceae (Särkinen et al. 2013), the larval host plant family of many ithomiine taxa, on the assumption that the butterfly clades could not be older than their larval host plants. In each case, node ages on the source tree were inferred using molecular clocks, and inferred ages for butterfly clades, or the ages of their obligate larval host plants, were plugged in as assumptions in the ithomiine analysis. Obviously, in situations such as this, all assumptions and uncertainties explicit or implicit to the source analyses carry over to ages inferred in the secondarily calibrated tree and, indeed, are likely to be compounded in potentially unpredictable ways that may greatly decrease precision and skew accuracy.

Models, Assumptions, and the Resurgence of Phenetics in Time Tree Calibration

As we have discussed in Chapter 5, the belief that phylogenetic inference is empowered by statistical models has become widespread, and as in other realms of systematics, efforts to improve and refine molecular clock estimates with elaborate likelihood or Bayesian techniques have burgeoned in the last two decades (Sauquet 2013). While complicated algorithms that generate dates with confidence limits appear to add a patina of quantitative sophistication, accuracy, and precision to the inference of clade ages, the devil, as usual, is in the details. Whenever one encounters such estimates, one must always bear in mind that the accuracy, precision, and/or confidence offered by these efforts is meaningful only if the model is correct, or nearly so. With a different, equally defensible model, one might obtain radically different estimates of the ages of taxa (Duchene et al. 2014; Bromham 2019). Such wide variability in results throws into doubt the accuracy, and even more so the precision, of clock calibrations. Here, we qualitatively outline some of the models and assumptions that underlie these methods.

Sanderson (1997) proposed a likelihood approach to infer ages of divergence from DNA data using a method called nonparametric rate smoothing (NPRS). The premise of this approach is that even if there is no "universal clock" (steady rate of molecular evolution across the entire Tree of Life, such that the tree is ultrametric and every distance from the root to each tip is equivalent), there might be localized autocorrelation of rates of molecular evolution in less inclusive parts of the tree. That is to say, closely related taxa could tend to have similar molecular evolutionary rates (a not unreasonable, parsimony-based assumption), and if this were true, then different parts of a tree could be anchored with local calibration points that assume local clocklike evolution in nearby clades but allow potentially rather different evolutionary rates on more remote branches. Based on this assumption, rate smoothing averages out differences between sister clades that might be evident from an empirical comparison like a relative rate test. Analogous rate-smoothing approaches are implemented in the popular Bayesian dating programs MCMCtree (Yang 2007) and BEAST (Drummond and Rambaut 2007; Bouckaert et al. 2014).

Sanderson (1997) and his followers have repeatedly emphasized a contrast between "clock-based estimates" and NPRS, as though the latter is not based on assumptions about clocklike evolutionary rates. That is, of course, untrue. The power of NPRS to infer the ages of nodes on a tree, other than those immediately associated with a calibration date from a fossil or another of the sources of evidence discussed above, is entirely dependent on local clocklike rates, as dictated by one or another probabilistic model of evolutionary change. Otherwise, all that may be concluded is that nodes closer to the tips than calibrated nodes are at most as young, or younger than, those nodes, and nodes closer to the root are at least as old, or older.

We would argue that the phylogenetic position of a fossil and its age are two distinct and independent inferences. Since very early on (cf. Patterson 1981), the cladistic treatment of fossils has deliberately excluded the age of a specimen from the consideration of its phylogenetic relationships. Ronquist et al. (2012) rejected the cladistic perspective, offering what they referred to as a "total-evidence approach to dating with fossils," in which the phylogenetic positions and ages of fossil taxa are both treated as variables to be estimated simultaneously. While this sounds like a broad-minded extension of Kluge's (1989) views (see Chapter 6), it seems to us to conflate background knowledge (the phylogenetic hypothesis) with the hypothesis being tested (the inferred age of nodes in said tree) in a manner that undermines the transparency of empirical observation and adds unrealistic and burdensome additional assumptions to the inferential procedure. Ronquist et al. began by pointing out the obvious: that for a given phylogenetic hypothesis, "we do not know any of the nodes with absolute certainty," and that "younger fossils from the same group do not provide any additional information on the minimum age of a calibrated node." Their Bayesian model provides posterior subjective probabilities for their preferred topology and allows the ages of redundant fossils to become "relevant" by waving away the logical requirement that a fossil represents only a minimum age for its group: "the age of each fossil [was] fixed to the estimated age of the bed from which it was retrieved." They viewed uncertainty associated with that estimate as "negligible compared to other sources of error in the analysis." Furthermore, we may deduce that if the presence of redundant fossils influences clade age estimates, then so might the absence of fossils elsewhere in the tree. Therefore, Ronquist et al.'s approach makes large and almost certainly unrealistic assumptions about the evenness and sufficiency of sampling of fossil taxa. This sort of casual dismissal of empirical facts and glossing over of fundamental assumptions exemplifies the statistical chicanery that enables these methods to achieve supposed "greater precision and accuracy."

The reader will recall that phenetic methods cluster taxa based on overall similarity, and we find a clear reversion to phenetic reasoning in Ronquist et al. (2012:994): "The more similar a fossil is to the inferred morphology of an ancestor in the extant tree, and the more complete it is, the more it will influence the dating of the extant tree." In other words, the more symplesiomorphies the fossil exhibits, the greater its significance, because "all dating using fossils is based to some extent on the assumption that morphological similarity indicates temporal proximity." This is the opposite of cladistic reasoning (cf. de Jong 2017), under which, as we have repeatedly stated, only synapomorphies provide evidence of grouping.

Other clock modelers overtly advocate the syncretic aims voiced long ago by evolutionary taxonomists. Consider, for example, the following statements. "A phylogenetic analysis of species has two goals: to infer the evolutionary relationships and the amount of divergence among species," and "phylogeny has two components, the splitting of evolutionary lines, and the subsequent evolutionary changes of the split lines." The former is from Heath et al. (2014), the latter is from Mayr (1982:230). As we discussed in Chapter 5, to the extent that a phylo-genetic hypothesis is adulterated by the joint consideration of synapomorphies, autapomorphies, and their complementary symplesiomorphies, the resulting tree may be dismissed as a confection of incompatible evidence and methods representing the arbitrary belief or opinion of its authors, rather than a philo-sophically coherent empirical inference.

Although it is universally acknowledged that the inferred ages of fossils are the sole source of direct empirical minimum age estimates for clades, the advo-cates of model-based approaches often exhibit profound disdain for the quality of this evidence. For example, O'Reilly et al. (2015) said, "the molecular clock is the only viable means of establishing an accurate timescale for Life on Earth, but it remains reliant on a capricious fossil record for calibration." To them, it seems, the filter of models is able to extract more information from the evidence: "the wide range and flexibility of probability distributions has allowed the accu-rate incorporation of uncertainty into fossil calibrations" (whatever "accurate uncertainty" might be). But later in the same paper, they argued, "establishing the nature of a probability density function spanning minimum and maximum constraints has little justification beyond gut-feeling. Unfortunately, arbitrary choices between competing parameters have an almost overwhelming impact on divergence-time estimates. Finally, the node calibrations specified by users are invariably transformed in the establishment of the joint time prior, to the extent that they sometimes bear little relation to the original fossil evidence."

Others have gone so far as to suggest that the dating of geological events should be informed by molecular clock estimates. Baker et al. (2014) proposed a methodology called "geogenomics," in which genetic data should be used to infer the ages of geological phenomena, such as the emergence of the Isthmus of Panama. One might charitably describe such endeavors as a form of reciprocal illumination, or less so as a rather literal instance of pulling one's self up by one's bootstraps. Either way, using molecular clocks to calibrate earth history would seem to beg the question of the sources and reliabilities of the calibrations for those clocks.

Take Home Messages

The inference of clade ages has become a major area of applied statistical research over the past two decades. Concern regarding its limitations, such as we have articulated above, is largely drowned out amidst the flood of new models and applications, and it is now difficult to publish a phylogenetic hypothesis that does not include some sort of time tree analysis and accompanying narrative scenario about the temporal evolutionary history of the group. However, unbridled enthusiasm for these techniques does not render them immune from criticism. Just as it is a naive mistake to view a bootstrap value as a measure of support for the metaphysical existence of a clade (see Chapter 6), so is it foolish to jump to conclusions about rates of diversification and such matters without possessing a clear understanding of the myriad assumptions that underlie a given clade age estimate. Here is a summary of the issues addressed in this chapter:

1. Evidence does not support a universal molecular clock.
2. Evidence might or might not support "local" clocklike evolution among closely related taxa over relatively short time spans.
3. Although absolute minimum ages for clades may be inferred from fossils, from biogeographical patterns, or extrapolated from secondary calibrations, such age estimates are subject to potentially significant error due to vagaries of geological dating as well as ambiguities of fossil identity.
4. Relaxed clock models offer precision and accuracy only to the extent that their assumptions legitimately represent reality. Whether that is true is debatable, at best.
5. Owing to all of their intrinsic embedded errors and assumptions, naive employment of these methods to spin evolutionary narratives about the diversification of one's favorite clade is to a great extent an exercise in historical fiction.
6. The test of a time tree hypothesis is to discover new fossil evidence that corroborates or falsifies it. Evidence permits inference; inference does not create evidence.

SUGGESTED READINGS

Bromham, L. 2019. Six impossible things before breakfast: assumptions, models, and belief in molecular dating. Trends in Ecology and Evolution 34:474–486. [A circumspect, if rather rosy, review of assumptions in model-based time tree analyses]

Donoghue, P.C.J., and M.P. Smith, eds. 2004. Telling the evolutionary time: molecular clocks and the fossil record. London: Taylor and Francis. [A now somewhat dated collection of essays from a Systematics Association symposium]

Parham, J.F., P.C.J. Donoghue, C.J. Bell, T.D. Calway, J.J. Head, P.A. Holroyd, J.G. Inoue, et al. 2012. Best practices for justifying fossil calibrations. Systematic Biology 61:346–359. [A useful review, for the justification minded]

Sauquet, H. 2013. A practical guide to molecular dating. Comptes Rendue Palevol 12:355–367. [An introduction to dating models]

12

BIODIVERSITY AND CONSERVATION

Increasing awareness of the global destruction, diminishment, degradation, and fragmentation of natural habitats in terrestrial, aquatic, and marine environments has brought the study of "biodiversity" to a new level of intensity. *Biodiversity*, as the term is currently used, has many meanings, and its study ranges broadly across biology. There are, however, aspects of biodiversity that are strictly systematic, including (1) recognition and enumeration of the world's biota, and (2) inference of historical relations—both genealogical and geographical—among members of the biota. These types of knowledge can be used directly to inform our efforts for staving off continuing extinction at the hand of the human species. It is these areas that are the subject of this chapter.

Recognition and Enumeration

Between 1740 and 1767, Linnaeus, the eminent eighteenth-century Swedish naturalist, took upon himself the task of describing the world's biota, at least as it was available to him for study. Given the immense diversity of tiny and unobtrusive organisms, such as soil mites, fungi, and bacteria, we now appreciate that a complete enumeration of all living species may never be achieved, no matter the number of workers performing the task. Nonetheless, we should not abandon the effort because much of importance remains to be learned. We might mention, among other possibilities, the following practical areas impacted by improved knowledge of the world's biota: the functioning of ecosystems, human health, agriculture and other aspects of food production, discovery of natural products, and conservation decision making. Since the second edition of this book was published, exploding areas of research have been environmental genomics and the discovery and characterization of the complex ecosystem of microorganisms that inhabits the human body. For example, a "human" is composed of more bacterial cells than eukaryotic cells, and disruption of this microbial community has been implicated as a cause for various intestinal and respiratory ailments. Discovery and characterization of these organisms is the work of systematics, which, to turn the phrase of G. Evelyn Hutchinson (1965), sets the stage for understanding the ecological interactions among the players.

Estimates for the total number of species of living organisms range from 2.5 million to about 30 million, depending on who is doing the calculations and what criteria they apply. Whatever the number, one thing is certain: large numbers of species remain unstudied (see Sidebar 19), even at the most basic level of describing and naming them so as to formally recognize their existence. To these we can add, at least in principle, the fossil remains of millions of additional species that existed in the past.

Sidebar 19
The Taxonomic Impediment

Some have lamented the "taxonomic impediment" (Ramsay 1986; Kitching 1993; Hoagland 1996; Godfray 2002; Giangrande 2003), a perceived lack of systematic expertise to address the challenges of discovering and describing the diversity of life. Systematists have called for and secured increased support for their work (Rodman and Cody 2003) and proposed various schemes to automate aspects of discovery (Weeks and Gaston 1997; Gaston and O'Neill 2004; Godfray 2007; La Salle et al. 2009; Cao et al. 2016; DNA Barcoding, below). Others have pushed back against technological solutions as superficial stopgaps (de Carvalho et al. 2007). Technology can certainly improve the efficiency of systematists, but it cannot replace them. And, of course, neither can phylogeneticists, bioinformaticians, and their ilk.

Certain aspects of current knowledge of the world's biota may in some ways seem counterintuitive. For example, the insect fauna of Europe is extremely well known. The description of a new European species might be considered an important event. By comparison, the insect fauna of North America is not nearly so well known. This may seem a surprise, given the potentially immense resources available for science in the United States and Canada. Yet, large numbers of species remain undescribed, and some groups, particularly in the western United States, have never been studied in detail.

Even with the existing limitations of knowledge of the North American fauna, the temperate Northern Hemisphere is extremely well known compared with the tropics and the temperate Southern Hemisphere. The biota of Australia may be least well known, but other areas are in desperate need of study as well. One of the ways we can estimate the undescribed diversity of a region is by comparing numbers of described and undescribed species in collections. Of course, this presupposes that we have adequately collected what is "out there," which in many instances is not the case.

Some examples may provide perspective on the scale of the problem. Within the true bug family Miridae, 2000 species have been described from North America north of Mexico (Henry and Wheeler 1988; Schuh 2002–2013). One might surmise that when the North American fauna is "completely" known, that number might increase by as much as 15 percent. Australia, with a comparable land area, has a currently described mirid fauna of only a few hundred species (Cassis and Gross 1995; Schuh 2002–2013). The authors of *The Insects of Australia* (Carver et al. 1991) estimated the number at 600, although this number is certainly much too small as well, judging from specimens available in collections and from the large numbers of taxa described in recent publications (e.g., Schwartz et al. 2018; Symonds and Cassis 2018, among many others). One might conclude that, whereas approximately 15 percent of the North American species remains to be described, fewer than 50 percent of the Australian species has been described.

In the butterflies, arguably the best-known group of insects and among the best-known nontetrapod organisms in general, there are huge numbers of species awaiting description. A comprehensive catalog of Neotropical butterflies (Lamas 2004) indicated hundreds of "n. spp." taxa recognized as distinct but not yet formally described, particularly among the hairstreaks (Lycaenidae) and satyrines (Nymphalidae). As these are described, still more are revealed as new (e.g., Pyrcz et al. 2018). There is little doubt that many additional species remain to be discovered in the field.

DNA barcoding (see Chapter 7 and further discussion below) has provided another avenue for exploring the realm of cryptic or sibling species (taxa that are morphologically similar but genetically distinct). Paul Hebert and colleagues (2004) purported to reveal "ten species in one" by examining hundreds of individuals of the skipper butterfly "species" *Astraptes fulgerator* from a single site in Costa Rica. Comparable numbers could be repeated for many groups of insects, other arthropods, and other nonvertebrate animals.

The situation for most groups of insects might also be compared with that for flowering plants. In North America the discovery of new species of angiosperms would probably be considered newsworthy. There are several floras for North America and often multiple floras for given "regions" within North America. By comparison, Australia and South America as yet have no complete printed flora, and large numbers of species of flowering plants remain undescribed. Nonetheless, the online Australasian Virtual Herbarium provides remarkable online access to herbarium specimen records for the Australian flora, the completeness of which is unmatched elsewhere in the world.

Recent Approaches: The Changing Face of Biodiversity Knowledge

DNA Barcoding

One approach to documenting "biodiversity" is an ambitious program initiated by Hebert and colleagues (Hebert et al. 2003a, 2003b) to obtain a 640 base-pair segment of the mitochondrial cytochrome oxidase I gene (mtDNA COI) from as many animals as possible and to use these data in the development of a standard for identifying unknown samples (other gene regions are employed for plants and fungi). This gene has been employed by molecular systematists and population geneticists in the study of insect systematics since the early 1990s (e.g., Bogdanowicz et al. 1993; Brower 1994a, 1994b; Sperling et al. 1994). The innovative aspects of the barcoding initiative is simply the identification of a pair of polymerase chain reaction primers that allow amplification of the same gene region from a wide range of organisms, the industrial scale at which the project has been undertaken, and the shift from hypothesis-driven studies of particular taxa to an omnivorous effort to document "life" with a common-denominator molecule. If barcoding were successful, it could change the way undetermined specimens are identified, from the traditional consultation of experts in museums to operational, automated queries of a sequence database (often by means of phenetic algorithms).

Needless to say, many museum scientists have perceived barcoding, or at least the rhetoric with which it has been advocated, as a threat to systematics and have criticized it accordingly (Sperling 2004; Will and Rubinoff 2004; Ebach and Holdrege 2005; Prendini 2005; Wheeler 2005; Will et al. 2005). Others have pointed out with empirical case studies that the mtDNA COI region is not necessarily isomorphic with species boundaries recognized by alternative means to the degree that its proponents have claimed (Meyer and Paulay 2005; Brower 2006b; Meier et al. 2006; Elias et al. 2007). Still others have raised the profiles of their traditional systematic research programs by forging collaborations with Hebert and other prominent boosters of the barcode endeavor (e. g., Janzen et al. 2009). Another application of barcode primers in Next Generation Sequencing is metabarcoding, in which the sequences of multiple species may be obtained from a combined heterogeneous sample., such as a shovelful of soil. This technique holds promise to reveal countless heretofore nondescript taxa. Debate over the role of the rapidly expanding DNA barcode database is likely to continue as resources for systematic research become scarcer. See Chapter 7 for further discussion.

Planetary Biodiversity Inventories and Dimensions of Biodiversity

Beginning in 1993 the US National Science Foundation (NSF) began funding large-scale collaborative projects under the heading of Planetary Biodiversity Inventories (PBI). The program had a strongly systematic focus, with the aim of documenting monophyletic groups of organisms with global distributions. Such studies have long been undertaken by systematists, in the form of comprehensive taxonomic revisions and world systematic catalogs, but seldom have resources been available to examine in detail groups comprising 1000–5000 species. Nor, one might argue, was the technology at hand to make available in a comprehensible format the vast amount of information that inevitably results from such an effort. Eight PBI projects were funded, dealing with groups including catfishes, plant bugs, parasitic wasps, goblin spiders, parasitic tapeworms, slime molds, tomatoes and potatoes, and spurges; legacy websites for the projects can be found online.

In 2010, NSF replaced the PBI program with Dimensions of Biodiversity, another program to support documentation of global patterns of biotic diversity, this time with a more system-oriented approach, examining interactions of taxa in an ecological or evolutionary context. Dozens of collaborative projects were funded, many with international partners. The products of these efforts are still forthcoming (National Science Foundation 2019).

Encyclopedia of Life: Beyond Web Integrators

There now are several well-known, large-scale efforts to bring nomenclature, biological classification, and specimen data to the internet, including Catalog of Life, Integrated Taxonomic Information System (ITIS), and Global Biodiversity Information Facility (GBIF), among others. What has eluded most efforts up to now is the funding necessary to assemble—on the web and on a truly grand scale—the kind of information being generated by the PBI projects mentioned above, and other comparable efforts, for all of life. This has been the goal of the

Encyclopedia of Life (EoL). This nonprofit, foundation-funded effort is grandiose in its ambitions but remains a somewhat fragmentary and error-rife work-in-progress at the time of this writing. The primary objective, nonetheless, is to develop web tools that will allow for the assembly of *species pages* for every taxon of living things.

The data included in "species pages" have long been the core of systematic revisionary studies. In the context of printed publications, they have usually consisted of information on nomenclature (a "synonymy"), a description, an enumeration of specimens examined, illustrations (including distribution maps), and commentary on other relevant biological information. When coupled with an appropriate classification, such internet-based pages have the potential to offer a roadmap of knowledge of the world's biota. Under this vision, not only can they include the information traditionally included in printed monographs, but they can be continually updated, assembling a wealth of information in real time. They can also include pointers to other relevant information and greatly broaden the context in which information is accessed. Like all other efforts to assemble species-level information, the success of the EoL will to a great degree depend on the success of its recruitment of expert providers of content from the biological research community. The general challenge of such efforts is that they require continuous curatorial maintenance, just like physical collections, and when the funding dries up, the project often becomes moribund. A case study of this regrettable phenomenon follows.

Tree of Life

Whereas the EoL, DiscoverLife, and other similar websites are primarily species based, the Tree of Life (ToL) looked at biodiversity information from a phylogenetic perspective. Thus, rather than starting at the tips of the branches in the phylogenetic tree, the ToL focused on the internal nodes of the tree, offering an interactive phylogenetic framework to browse the entire diversity of life. Because phylogenetic hypotheses—beyond the level of classifications with limited hierarchic content—were available for only a limited number of higher taxa, the ToL lacked detail in many areas of its tree. Some clades, however, were represented in significant detail, including crayfishes and butterflies (largely depending on contributions of individual participating researchers). The ToL, like the EoL, depended on both the cooperation of the broader systematics community for the contribution and assembly of relevant information and on grant support to maintain its content curation, software engineering, and server capacity. In 2010, NSF funding ended, and this project fizzled out. Although the ToL site is still accessible, the pattern of relationships represented therein has not been expanded or updated since that time. Indeed, it seems that the only two continuously supported and maintained, aspirationally comprehensive classification resources on the internet are the classification of DNA sequences in GenBank and Wikispecies, a taxonomic component of Wikipedia. While gene annotations and other aspects of sequence characterization are meticulously managed by GenBank staff, the consistency and accuracy of the taxonomy employed is passively managed by contributing systematists who happen to report errors. The crowdsourcing model of Wikispecies entails that data are contributed by interested volunteers, and in theory anyone can correct (or introduce) taxonomic mistakes.

Whereas much discussion of biodiversity has focused on species diversity and richness, genealogical issues are nonetheless important, and it is in this area that the results of phylogenetic analyses and their assembly (through sites like the ToL) can gain broader recognition, as we will discuss in the next section.

Historical Relations

Genealogical Patterns

Without the relatively complete enumeration of species, knowledge of phylogenetic relationships will always be limited. Yet, it is inference of the hierarchical pattern of relationships that provides systematists with their most powerful tools. To believe that empirically supported hypotheses of relationships exist for most groups of plants and animals would be naive: for many taxa, as noted above, we barely comprehend the scope of the problem. Furthermore, there is no obvious correlation between our knowledge of the species diversity of a group and the level of knowledge concerning the phylogenetic relationships of those species and the more inclusive groups to which they belong.

For example, the species-level diversity of mammals and birds is quite well documented. It is relatively rare that new species are described, and most of the intraspecific variants have been named as well. The dramatic fluctuations in "numbers of species" of birds and mammals described in Chapter 7 are the result of changes in circumscription of taxa and their elevation to, or demotion from, the species rank rather than to recognition of previously unknown taxa. However, phylogenetic relationships among these taxa are not as clearly and uncontroversially resolved as one might expect. We might say that ornithologists were distracted by the "new systematics" for most of half a century, during which time they devoted their efforts to examining population-level variation within birds, almost to the exclusion of studying higher-level relationships (but see Sibley and Ahlquist [1990] on DNA hybridization). Only recently have large-scale phylogenetic hypotheses for relationships among higher groups of birds begun to appear (e.g., Hackett et al. 2008; Jarvis et al. 2014; Prum et al, 2015; Kimball et al. 2019). Relationships among mammals may be better known; at the very minimum, an integrated view of the structure of relationships at the generic level and above was portrayed in detail by McKenna and Bell (1997), and there is an ongoing controversy regarding conflicting patterns of evidence supporting relationships among mammalian orders (e.g., D'Erchia et al. 1996; Gatesy et al. 1999b; Lin et al. 2002; O'Leary et al. 2013; Springer and Gatesy 2016; Tarver et al. 2016; Beck and Baillie, 2018).

Conservation biologists have expressed dismay at the lack of stability of species concepts and resultant nomenclature (Mace 2004; Frankham et al. 2012; see also Sidebar 16, Chapter 7). Some have called for regulatory oversight to rein in the perceived caprices of systematists, who they view as incapable of reaching consensus on the numbers and delimitations of species, or even criteria for defining what a species is (Garnett and Christidis 2017). Although we doubt that any systematist intentionally strives to create nomenclatural instability, the "metaphysical quagmire" of species concepts, (Chapter 7) inevitably leads to disagreement among researchers with different data, conceptual frameworks, and research agendas. Conservation laws work most effectively when protections or proscriptions are clearly and unambiguously delimited, but nature is not tidy like that. Further complicating matters, the US Endangered Species Act of 1973 offers the following definition: "The term 'species' includes any subspecies of fish or wild-life or plants, and any distinct population segment of any species or vertebrate fish or wildlife which interbreeds when mature" (https://www.fws .gov/international/pdf/esa.pdf). Aside from apparent grammatical errors, this biological species concept-based definition presents all the metaphysical challenges raised in Chapter 7 regarding delimitations and criteria for identifying what might constitute a "subspecies" or a "distinct population segment."

A pertinent case study is the federally endangered Florida panther, *Felis concolor coryi*, which is (or was) considered a biogeographically and ecologically differentiated taxon of the more widespread mountain lion, which ranges across North America into South America. O'Brien et al (1990) demonstrated that members of the extant population of Florida panthers, although inbred, were not genetically distinct from the remainder of the species, perhaps due to interbreeding between native individuals and imported escapees from Miami zoos. Subsequently, the population was "restored" by introducing members of another subspecies from Texas (Johnson et al. 2010). What, then, is a "Florida panther"? Evidently, a panther that lives in Florida.

The above examples represent only a glimpse into where things stand. Species as we currently understand them come and go epistemologically, under the kaleidoscope of alternative species concepts. At the same time, new species of many taxa are described on a daily basis, and new techniques such as environ-mental genomics provide evidence of untold microdiversity in habitats such as soil, freshwater, and marine environments. What seems clear is that without a substantially greater expenditure of effort, many species will never be recognized as existing, let alone be studied at the level of articulating and agreeing upon pat-terns of phylogenetic relationship.

Geographical Patterns

Knowledge of historical biogeographic relationships is dependent on phylogenetic knowledge. The significance of the relationship between well-supported genealog-ical hypotheses and historical biogeography at a global level may be appreciated by examining a critique by Platnick (1992). Platnick noted that traditionally theories of historical biogeographic relationships among animals have been dominated by studies of vertebrates. This domination can be portrayed by observing that a one-time widely used textbook of "zoogeography" by the entomologist Philip J. Dar-lington (1957) used vertebrates as its sole exemplar organisms. Platnick further noted that studies of zoogeography have been heavily weighted toward the North-ern Hemisphere, creating what he referred to as the "boreal megafaunal bias." He emphasized that it is not just the tropics that contain large numbers of taxa com-pared with the Northern Hemisphere—as popular conception would have it—but that the southern continents in general deserve much greater recognition as centers of great biotic diversity and, therefore, biogeographic importance. On the latter point, he stressed that areas of endemism in the far south often seem to be much smaller than those in the Northern Hemisphere (cf. Millar et al. 2017; Apo-daca et al. 2019). The latter was, after all, repeatedly scraped clean of its biota by ice sheets during the Pleistocene, as recently as 14,000 years ago. So Nearctic and Palearctic taxa have had relatively little time to specialize and differentiate in situ, at least in the more northerly parts of their ranges, a situation apparently quite different from that experienced on the southern continents.

We might profitably extend Platnick's discussion to the issue of regional phy-logenetic diversity. But first, let us augment some of Platnick's numbers. We can easily agree that the numbers of species in some areas of the Southern Hemi-sphere are virtually unparalleled anywhere in the world. For example, the South African flora is made up of more than 20,000 species of flowering plants (Russell 1985). Western Australia should also be cited in this context, with an estimated 4000 species of flowering plants occurring in the Southwestern Botanical Prov-ince alone (Corrick et al. 1996), those species making up at least 20 percent of the Australian angiosperms but occupying roughly 4 percent of the continental land area. By comparison, North America probably has no more than 9000 species of flowering plants, with no single area of the continent showing anywhere near such high diversity.

We might also wish to consider Southern Hemisphere diversity in plant-feeding insects. Data compiled by Zimmerman (1991–1994) for the weevils (Insecta: Coleoptera: Curculionidae) of Australia indicated an estimated 6000–8000 species in 1000 genera (with a very large proportion remaining to be described), whereas the North American fauna, by comparison, comprises 2500 species in 375 genera. Another example can be found in the work of Slater on the monocot-feeding lygaeoid family Blissidae (Heteroptera). The South African fauna comprises 66 species placed in 11 genera, these feeding on the monocot families Poaceae, Cyperaceae, Haemodoraceae, Juncaceae, and Restionaceae (Slater and Wilcox 1973). The North American fauna, by comparison, contains only 26 species in 2 genera, these all feeding on the Poaceae (Slater 1964). Yet, the land area of South Africa is less than 13 percent of that of North America. All of these observations offer support for the idea of higher diversity and smaller areas of endemism in the southern continents.

Southern floristic connections in groups such as the Proteaceae were well-known in the time of Hooker and Darwin. However, it was the Swedish entomologist Brundin (1966) who first showed through the use of phylogenetic methods the detailed nature of the massive connections existing between the southern continents, exclusive of the northern biota. Whereas the type of pattern exemplified by Brundin's examples from the midges of the family Chironomidae involved the cool temperate portions of the Southern Hemisphere—presumably including Antarctica—tropical patterns also exist. For example, the plant bug tribe Pilophorini has been shown (Schuh 1991, Schuh and Menard 2011) to demonstrate biotic connections between tropical South America, tropical Africa, tropical Asia, and Australia in its relatively more basal lineages, whereas only the relatively most-derived lineages occur in the Northern Hemisphere, albeit with a large number of species.

Discrete distributions and large numbers of clades often distinguish the biotas of the southern continents. By contrast the fauna of the Holarctic is often much more homogeneous, with many of the genera and species being relatively widespread. Some species, such as *Ursus arctos* (the brown and grizzly bears, and arguably polar bears, as well) and *Biston betularia* (the peppered moth of industrial melanism fame) have Holarctic distributions. In a review of biogeographic relationships within the Holarctic, Enghoff (1995) divided the Northern Hemisphere into only four areas: eastern and western Nearctic and eastern and western Palearctic. No matter how poorly this division reflects the actual number of areas of endemism, it nonetheless graphically reflects the degree to which many taxa have widespread distributions.

Conservation

How, then, can systematic, phylogenetic, and biogeographic information be used in conservation decision-making?

Recognizing Hot Spots

Biodiversity *hot spots* were brought into the popular and scientific consciousness by Norman Myers (1988). He argued that the greatest threat to extinction of the earth's biota was in the tropical forests, and he identified 10 areas that he viewed as particularly susceptible. Subsequently, the number of hot spots has been expanded greatly, and the concept has received a formal definition. Indeed, hot spots have been adopted by Conservation International, the influential nongovernmental conservation organization, which features a website, Biodiversity Hotspots. They are the subject of recent volumes, which update the concept and the nature of the currently recognized areas (Mittermeier et al. 2005; Zachos and Habel 2011).

The concept and significance of hot spots is straightforwardly synthesized in the abstract of an article by Myers et al. (2000) as follows:

> Conservationists are far from able to assist all species under threat, if only for lack of funding. This places a premium on priorities: how can we support the most species at the least cost? One way is to identify "biodiversity hotspots" where exceptional concentrations of endemic species are undergoing exceptional loss of habitat. As many as 44% of all species of vascular plants and 35% of all species in four vertebrate groups are confined to 25 hotspots comprising only 1.4% of the land surface of the Earth. This opens the way for a "silver bullet" strategy on the part of conservation planners, focusing on these hotspots in proportion to their share of the world's species at risk.

Whatever the strengths or weaknesses of "hotspot science," there can be little doubt that the basic information at play comes from the field of systematics. Knowledge of species, their very recognition, has long been the core activity of taxonomy. Knowledge of species distributions, except possibly in the case of the charismatic megafauna, is likewise developed through the works of systematists. Furthermore, knowledge of rarity, range diminution, or extinction of a given species is fundamentally dependent on records of the distribution and abundance of the species in the past. For most taxa, such data are most readily and reliably available from museum collections, in the form of label transcriptions from individual specimens.

Phylogenetic Information

Two nearly diametrically opposite approaches have been suggested for using phylogenetic information in conservation decision making, with a number of variants in between (e.g., Nixon and Wheeler 1992).

Protect radiating lineages. Under this approach, lineages that appear to be radiating (speciating) on a dynamic basis would be those targeted for protection because they are the ones that would "create future biodiversity" (Erwin 1991). The extreme application of this approach would be to treat lineages that are not radiating and therefore "doomed to go extinct" as expendable and unworthy of efforts at protection. Of course, the time scale and taxonomic scope of such questions is quite subjective. Under this criterion, a family including only five or six species might be viewed as not particularly diverse and thus of little conservation interest—Hominidae!

Protect the most lineages. Under the "opposite" approach, absolute numbers of species per lineage would be deemed less important than the diversity of the lineages that those species comprise (Vane-Wright et al. 1991). Thus, ancient and species-poor phyletic lines are deemed more important than recent and species-rich lineages. Once again, such arguments are always tempered subjectively by the charisma of the taxa in question; otherwise we would devote most of our conservation efforts to preserving lineages of bacteria. See Winter et al (2013) and Kling et al. (2018) for recent reviews of these issues.

Examples. An application of the latter approach can be appreciated by using the case of the tuatara of New Zealand, the sole surviving species (either one or two) of an otherwise long-extinct diapsid "reptilian" lineage (see May 1990). If two species are actually represented, as taxonomic studies suggest, then it might be worth a concerted effort to protect them both because of their phylogenetic "uniqueness." If, on the other hand, such a conservation question involved one of many species of murine rodents (the largest group of mammals), the answer certainly would not be couched in terms of phylogenetic uniqueness, although it might be treated as an example of affording protection to a radiating lineage.

On a more global scale, Jetz and Pyron (2018) used a massive phylogenetic hypothesis for virtually all extant amphibians (7238 species) to highlight evolutionary distinctness they viewed as being of particular conservation concern. They noted a relatively weak correlation between imperilment and evolutionary distinctness but made the argument that having a phylogenetic framework to assess conservation needs provides a sound foundation for conservation decisions.

Biogeographic Information

Conservation decisions often involve determining if a given area should be protected or developed, irrespective of whether or not particular organisms of conservation concern might or might not occur there. In such cases, information on *endemicity*, the biogeographical uniqueness of the area's flora and fauna, is usually assessed before any phylogenetic information is taken into account. The concept of surrogacy has taken on special significance in the biodiversity-conservation literature because "complete" biotic inventories are difficult to assemble because of the time necessary to collect and identify all members of a biota, and are therefore not generally feasible. *Surrogacy* is the practice of choosing certain groups of organisms to be representative of the biota as a whole. This can be manifested as what we might call "transitive surrogacy," in which the diversity of group x is used to infer the diversity of groups y, z, and so on, and also what might be called "integral surrogacy," in which the diversity of higher taxa in a charismatic group, such as birds, is used as a blanket measure of species diversity for comparison of one area to another. To what degree surrogacy works may well be determined by the quality of choice or indicator organisms in each case under study. Knowledge derived from the study of biogeography would suggest that the concept should be used with care.

Consider, for example, transitive surrogate comparisons between flowering plants and insects. Australia may have slightly more than twice as many angiosperms as weevil species. It would certainly be much easier to sample the plant species for a given area than the weevils. Yet nothing is known about the degree of correspondence between plant and weevil endemicity. Even if such a correlation were assumed, we still might not choose plants as the sole surrogate group because weevils may not be uniformly distributed across flowering plants as hosts. Furthermore, phylogenetically unique groups (of either plants or weevils) might not occur in areas of highest endemism.

A very coarse-scale approach to examining diversity was taken by Williams et al. (1994). In this case study, the families of seed plants were used as integral surrogates for an overall measure of species richness. The ultimate value of their approach may best be tested by determining to what degree it measures diversity when compared with results of applying the same higher-taxon approach to other groups of organisms. Bertrand et al. (2006) criticized the use of higher taxa as surrogates for species diversity on the grounds that Linnaean ranks above the species level are not comparable. For example, the 8000 or so species in the class Aves with the 80,000 described species in the family Curculionidae. We might then conclude that, in general, the trouble with surrogacy is that its conclusions rest on assumptions that are either known to be false, because not all groups respond to vicariant events in the same way, or are untested—and the tests of which would eliminate the need for surrogacy in the first place.

Mapping and Distributional Predictions

In Chapter 8 we discussed the creation of extensive georeferenced specimen databases, many of which are available on the internet through exclusive portals or aggregation sites such as GBIF, DiscoverLife, the Australasian Virtual Herbarium, iDigBio, and others. These databases have greatly facilitated aspects of the taxonomic process and have virtually limitless applications in the fields of conservation and biodiversity assessment and for tracking and predicting distributional changes.

At the same time that databases have been proliferating on the internet, so has the ability to map the data found in them. The Australasian Virtual Herbarium, also mentioned in Chapter 8 and earlier in this chapter, uses outline maps to plot distributions of taxa on its public site. The ease of use of Google Earth as an interactive tool has allowed for its incorporation into several sites, providing rapid mapping on high-resolution images of the earth's surface. Sites such as DiscoverLife use a combination of satellite imagery and topographic maps to rapidly plot tens of thousands of points in real time, with a click of any given point revealing the specimen data from which it was plotted.

The ready availability of a wide variety of base maps now allows for plotting distributions against environmental variables such as rainfall, temperature regimes, vegetative cover, and other parameters. Such approaches have greatly increased the precision and accuracy with which actual distributions can be predicted relative to known distributions. Note that unlike phylogenetic models, one can test a distributional model by going out and finding a population where it is predicted but not known to occur.

The use of ecological niche models that employ physical environmental variables to assess biotic distributions has burgeoned, and in the view of some investigators holds great promise for the prediction of current and future distributions in the face of the aforementioned climate change. This is a subject beyond the scope of the present volume but one for which there is now a large and ever-growing literature (cf. Peterson et al. 2011).

Concluding Thoughts

The above discussion provides a glimpse into the ways systematic information can be used in our efforts to protect and preserve the world's endangered biota. The refinement of approaches is a process in an active state of discussion and development. We might note, in closing, that even a complete knowledge of diversity, phylogenetic relationships, and endemism would not guarantee that decisions about conservation would be made on the basis of science. The track record of conservation efforts through modern history has shown that although protecting the environment and its inhabitants may be informed by science, such choices stem fundamentally from a political process, driven by the wills of people, the policies of governments, and the forces of economics.

As is the case for geographical space on the planet, the global productivity budget, in terms of the amount of solar energy available for conversion into biomass, is a zero-sum game. The more resources humans harness for their own devices, the fewer are available for the remaining biota. It is regrettable but perhaps inevitable, as our populations, and therefore our food, water, and energy needs, continue to burgeon, that conservation is likely to shift its emphasis even further from the charitable (or decadent, depending on one's perspective) preservation of "nature" to the self-interested protection of resources for human consumption.

In the twenty-first century, we are confronted with a global environmental crisis—human-induced climate change, driven, ultimately, by overpopulation of the planet by our species—that threatens to negate all prior and ongoing conservation efforts. We are on the verge of losing not just a wealth of species but entire biomes, such as tropical reefs and rainforests. Sadly, most of the knowledge of "Recent" biodiversity accessible to future generations of humans may come from the collections and publications of systematists produced in the less than half a millennium it has taken our species to overrun and devastate the natural world. Rhinoceroses, whales, and tigers may be the *Triceratops*, plesiosaurs, and *Tyrannosauri* of a not-too-distant future. For this reason alone, the work of systematists to discover, document, and preserve evidence of biotic diversity for future study is a matter of immediate critical concern. The window is closing, the future is unlikely to resemble the past, and what is, will not be.

SUGGESTED READINGS

Forey, P.L., C.J. Humphries, and R.I. Vane-Wright, eds. 1994. Systematics and conservation evaluation. Systematics Association Special Volume 50. Oxford: Clarendon. [A collection of papers related to systematics and conservation]

Gaston, K.J., ed. 1996. Biodiversity: A biology of numbers and difference. Oxford: Blackwell Science. [A treatment of general biodiversity issues]

Hodkinson, T.R., and J.A.N. Parnell, eds. 2007. Reconstructing the Tree of Life: Taxonomy and systematics of species rich taxa. Boca Raton, FL: CRC Press. [A collection of papers reviewing a range of approaches, mostly using example taxa]

Mittermeier, R.A., P. Robles Gil, M. Hoffmann, J. Pilgrim, T. Brooks, C.G. Mittermeier, J. Lamoreux, and G.A.B. da Fonseca. 2005. Hotspots revisited: Earth's biologically richest and most endangered terrestrial ecoregions. Arlington, VA: Conservation International. [An easy-to-read single source on the importance of hot spots]

Myers, N. 1988. Threatened biotas: "hot spots" in tropical forests. Environmentalist 8:187–208. [The classic paper on hot spots]

Pellens, R., and P. Grandcolas, eds. 2016. Biodiversity conservation and phylogenetic systematics: Preserving our evolutionary heritage in an extinction crisis. Cham, Switzerland: SpringerOpen. [A relatively recent symposium volume]

Purvis, A., J.C. Gittleman, and T. Brooks, eds. 2005. Phylogeny and conservation. Conservation Biology 10. Cambridge: Cambridge University Press. [A less recent symposium volume]

Scherson, R.A., and D.P. Faith, eds. 2018. Phylogenetic diversity: Applications and challenges in diversity science. Cham, Switzerland: Springer. [An edited volume on phylogenetic diversity measures]

POSTSCRIPT
Parsimony and the Future of Systematics

Sober (1988) observed that evolution is both a pattern-creating and a pattern-destroying process. Systematics is the study of that pattern. Since we cannot know, in any given instance, whether the pattern we perceive is a pattern reflecting evolutionary history, or a pattern reflecting the evolutionary overwriting of history, we assume that the data convey information about the evident natural hierarchy of life (evident to Aristotle, evident to Linnaeus, evident to Darwin, evident to Hennig, evident to us). This is the assumption that any scientist makes when addressing any pattern in nature—that the apparent pattern is intelligible and meaningful and that evidence is pertinent to explanation. That could be false, but the alternative is chaos and ignorance.

Under the view of systematics advocated in this book, the exuberantly messy data of biological diversity are organized into a clear and coherent explanatory framework through the application of the principle of parsimony. As we have emphasized repeatedly, whether the chronicle of history has or has not unfolded in a parsimonious manner is immaterial to the epistemological lens through which we view it. In this view, parsimony is the underlying principle that allows the interpretation of evidence in a scientific manner. The principle of common cause, the principle of cause and effect, and the principle of uniformitarianism are all applications of the principle of parsimony to the explanation of events unfolding in time. Thus, parsimony is not merely an old-fashioned phylogenetic method that has been superceded by purportedly more powerful and sophisticated statistical tools: it is the epistemological key to evaluating empirical evidence and discovering orderly patterns in the world to the extent that our perceptions allow. If the future does not resemble the past, then the coin does not flip consistently, there are no identically distributed samples, and the predictive power of statistical inference is mute. Therefore, the success of every scientific inference and prediction relating to empirical phenomena in the world (as opposed to those confined to the Platonic realm of mathematics) hinges upon parsimony. Hopefully, we have been able to convince the reader of this point.

As we have discussed several times in this book, the claim against parsimony regarding statistical inconsistency is just a highfalutin way of saying that parsimonious interpretations of evidence could be wrong. David Hume established that skeptical point almost 300 years ago. Of course, parsimonious interpretations of evidence could be wrong, but so could unparsimonious interpretations of evidence, and the point is moot from an empirical perspective. The foundationalist appeal to greater realism through modeling is therefore nothing more than a philosophically naive rationale for departing from parsimonious interpretation of evidence. The Kantian dualism between noumena (things as they exist independent of human perception) and phenomena (things as we perceive them; recall our critique of the "evolutionary species concept" in Chapter 7) represents another venerable epistemological conundrum (Kant 1781) that is exacerbated, rather than resolved, by adding auxiliary assumptions to background knowledge. Because you cannot know what is real, you can hardly know what is "realistic"— you can only assume, based on parsimony in the guise of uniformitarianism, that your model characterizes some attribute of future observations that has been divined to be a reliable pattern from past observations.

These arguments show that opponents of the cladistic approach possess no justification for their statistical flights of fancy other than foundational parsimony assumptions—plainly not a rational basis for condemning parsimony!

We now turn to the positive side of the balance sheet. Since it relies on parsimony and not deterministic models, the cladistic approach does not need to assume anything about the process of evolution. The freedom of cladistics from evolutionary assumptions means that the pattern of phylogenetic relationships inferred using that method provides an independent source of evidence to corroborate evolutionary theory. The same cannot be said for models that make explicit evolutionary assumptions. The independence of systematics should be a compelling desideratum for any researcher who views evolution as a scientific theory based on evidence, rather than a foundational metaphysical truth.

In addition, cladograms have the advantage of directly and transparently representing the distribution of characters that support them. If one cares to do so, one can map every individual character-state transformation onto a tree, or reconstruct the data matrix from a tree that has the characters so mapped. We consider cladistics' clarity concerning assumptions and evidence to embody the fundamental values of the scientific endeavor: explanatory rigor, intelligibility, and repeatability. Alternative approaches suggest alternative values and motivations: idiosyncratic complexity and obscurantism.

John Huelsenbeck tweeted (January 2016), "The clade wars ended more than 20 years ago. Felsenstein won." Perhaps, in the relativistic Kuhnian arena of competing research programs, this is so (Sterner and Lidgard 2018). The hearts and minds of the generation of phylogenomicists following us do appear to have been smitten by a paradigm opposed to the one we advocate. But, as Ronald Brady (1985) said, "The province of science is not a democracy." Although ascendant for a time, cladists have never been in the majority, and we are long accustomed to the perception that cladists are "outside the "mainstream" and perceived by some as rude or nasty. Ironically, it might be argued that cladists have been victims of their own success, having awakened the sleeping giant of population genetics to the empirical accessibility of evolutionary history as revealed through cladistic methods.

We harbor no illusions that this book will turn the tide of model-based phylogenetics or quell the exuberance of those who reject the philosophy of science (as we apply it) and embrace statistical operationalism. Indeed, we anticipate derision from some quarters that we are merely blinkered reactionaries peddling obsolete dogma. As we described in Chapter 1, the history of systematics has been fraught for hundreds of years with acrimonious struggles among its participants over their differing systems and methods, and as this book attests, we embrace that tradition by defending our own. It is frustrating to watch one's preferred methodological worldview slide out of fashion and somewhat infuriating when the sole justification for its abandonment is, as we have shown, baseless rhetorical legerdemain. However, speaking truth to power is necessary for checking folly: there are fundamental roles in science for criticism, challenges to the status quo, and calls for circumspection in the face of irrational exuberance (Zimring 2019).

It is neither a talent nor a virtue to contrive scientific methods that are needlessly complex. While the modelers are busy adding squirrels and spring-loaded boxing gloves to the Rube Goldberg machine that phylogenetics has become, we are betting on parsimony in the long game. We remain skeptical, not dogmatic but unconvinced, that any epistemological framework for phylogenetic inference demonstrably superior to cladistics has been proposed to date. We suspect that, in time, owing to the lack of a coherent epistemological foundation, the elaborate contraption of model-based methods will collapse under its own weight, much like phenetics did in the 1970s. We hold out optimism that a parsimonious paradigm will prevail, either through a renaissance of cladistics or by the advancement of compelling and substantive arguments in support of alternative approaches less burdened with untestable metaphysical baggage.

"The method of science is the method of bold conjectures and ingenious and severe attempts to refute them" (Popper 1979:81).

"*Entia non sunt multiplicanda praeter necessitatem*" (William of Ockham).

NOTES ON PHYLOGENETIC SOFTWARE

Computer assisted phylogenetic analysis consists of several distinct operations. These include the following:

Data matrix preparation
Nucleotide sequence alignment
Inference/computation of phylogenetic trees
Tree analysis and visual representation

There are now so many different methods and programs available to perform the various activities listed above that a comprehensive review would consume another book's worth of documentation. Further, this is a continually shifting landscape, with new versions and tweaks and entirely new packages appearing on a continuous basis. Therefore, we simply offer a categorized list of some of the more recent and familiar programs. Many of these are syncretic toolkits that encompass multiple, somewhat incompatible, approaches.

Data matrix preparation
Mesquite
WINCLADA

Nucleotide/sequence alignment
Blast
ClustalW
Malign
Muscle

Inference/computation of phylogenetic trees—parsimony
PAUP*
POY
TNT

Inference/computation of phylogenetic trees—distance-based methods
MEGA

Inference/computation of phylogenetic trees—maximum likelihood
Garli
PhyML
Phylip
RaxML

Inference/computation of phylogenetic trees—Bayesian inference
BAMBe
BEAST
Mr. Bayes
Phangorn

Tree analysis and visual representation
BioGeoBears
FigTree
TreeGraph

The best way to discover information about, and gain access to, these programs is via the web. A number of them can be accessed or downloaded for free.

Many more programs/packages are listed on Wikipedia under the following:

https://en.wikipedia.org/wiki/List_of_phylogenetics_software
https://en.wikipedia.org/wiki/List_of_phylogenetic_tree_visualization_software
https://en.wikipedia.org/wiki/List_of_sequence_alignment_software

See also the following:

http://research.amnh.org/users/koloko/softlinks/phylogeny.html
http://evolution.gs.washington.edu/phylip/software.html

Glossary

"When *I* use a word," Humpty Dumpty said in rather a scornful tone,
"it means just what I choose it to mean—neither more nor less."
—Carroll (1960:188)

As with many aspects of systematics, the definitions of terms are sometimes controversial, and the definitions we provide here may not concur completely with usage of the same word by others. Our aims in the text and in this glossary are to use these words consistently and to make our intended meanings clear

ACCTRAN   accelerated transformation; optimization of character-state transformations as close as possible to the base of a tree, implying early origins and later reversals of derived states

additive binary coding   a method of recoding multistate characters that allows for representation of branching patterns through the use of multiple two-state variables (*see* nonredundant linear coding)

additive character   (Farris transformation) a multistate character in which differential costs for state-to-state transformations are specified and changes hypothesized during cladistic analysis must conform to the specified transformation series for the character if additional steps are not to be added (*see* nonadditive character)

ad hoc hypothesis   an assumption invoked to dispose of observations that do not conform to some preferred theory; in cladistics, used in reference to a priori invocations of homoplasy to explain similarity as nonhomologous

adjacency   the relative position of character states, one to another, for a multistate character, without the implication of directionality of transformation (*see* polarity)

advanced   *see* derived

algorithm (for phylogenetic analysis)   a decision-making process for computing cladograms, as in the "Wagner algorithm"

alignment   the process of arranging sequence data from different organisms so that the putatively homologous nucleotide positions correspond to one another across taxa; if sequences differ in length, this requires the insertion of gaps

allopatric   separated in space by distance or a geographical barrier

allopatric speciation   the origin of a new species lineage by geographical subdivision of the parental species and the resultant interruption of gene flow

allozyme   proteins used to examine genetic variation within and among species; differences among alleles are due to different net charges of the proteins, as visualized by their different rates or directions of migration in gel electrophoresis

anagenesis   change within a single lineage over evolutionary time; as opposed to cladogenesis, the spitting of lineages

ancestor   *see* hypothetical common ancestor

ancestral   *see* primitive

ancestral polymorphism a potential source of gene tree incongruence due to genetic diversity that survives cladogenesis and persists in related lineages

apomorphic advanced, as opposed to primitive, or special, as opposed to general; a derived feature unique to a group and therefore group defining

apomorphy an advanced or derived character state; a group-defining feature

area cladogram in studies of historical biogeography, a cladogram in which the areas where taxa occur are substituted for the taxa themselves

area of endemism the congruent distributional limits of two or more species (Platnick)

asymmetrical (classification) a hierarchic scheme of relationships in which branching always occurs in just one of the lineages arising from each successive node (level in the hierarchy); a completely asymmetrical cladogram is *pectinate*

autapomorphy a derived feature (character state) unique to a terminal taxon in a given data set (*see* apomorphy)

availability criteria specified in the codes of nomenclature that a name must meet in order to enter into formal use, such as having been published in an acceptable manner, having been accompanied by a description of the biological material on which it was based, and others

background knowledge testable facts and theories taken to be well established enough to be assumed as plausible initial conditions or premises for subsequent investigation

barcode a term applied in systematics to refer to the use of machine-readable unique specimen identifiers (labels) (*see also* DNA barcode)

Bayesian estimation a method of phylogenetic inference that expresses a degree of subjective belief in one or more topologies, given an a priori evolutionary model and some data

bilateral symmetry repetition of similar ("mirror image") parts on opposite sides of the sagittal body axis

binary character *see* two-state character

binomen the two-part name of a species (*see* binominal nomenclature)

binominal nomenclature the system of naming codified by Carolus Linnaeus, in which each species is recognized by a name composed of two words, the generic name and the specific name (or epithet), as, for example, *Homo sapiens*

biodiversity a term with varied meanings; often used with reference to taxon richness, as numbers of species or higher taxa in a given area

biogeography the study of patterns of geographic relationship among taxa; in the present work with strict reference to historical patterns

bootstrap(ping) a technique that uses pseudoreplicated subsamples of characters (sampled with replacement) in an attempt to estimate the degree to which a data set can stably reproduce a given tree topology

branch a segment of a cladogram or other phylogenetic tree between two branching points or between a branching point and a terminal taxon; sometimes referred to as an "edge" in graph-theoretical literature

branch-and-bound algorithm a phylogenetic algorithm that determines an exact result (guaranteed most parsimonious tree) by examining only a portion of the universe of all possible trees, having discarded those portions of that universe longer than any of the trees in the portion examined in detail

branch length the number of character-state transformations expected or inferred to have taken place between two nodes on a phylogenetic tree; this is an assumption-laden and epistemologically ambiguous concept

branch support a measure of evidentiary support for branches on a cladogram, determined by assessing the length difference or relative cost between shortest inferred trees that do and do not include the clade of interest

branch swapping a technique used in numerical phylogenetic computations that improves the chances of finding most parsimonious trees; "branches" from potentially useful trees are moved to different locations in an attempt to find shorter trees, the length of the tree being recomputed with each move

branching character a multistate character in which adjacency relationships in a given direction are multiple (*see* linear character)

Bremer support *see* branch support

center of origin in biogeography, the area in which a group presumably originated and from which it later spread

character a feature that can be compared among taxa—thus a theory, rather than an empirical observation (*see* character state)

character coding the formal process of converting raw observations of features of exemplar specimens into concepts of characters and character states with specified transformation patterns in a data matrix

character state one of the various conditions of a feature (character) that may be observed among members of a taxon

character-state identity the hypothesis that two or more taxa share the same character state, which is tested by congruence

character-state transformation the hypothetical link between adjacent states of a given character; the number (or total cost) of character-state transformations is the quantity that is minimized in cladistic analysis

character-state tree the graphic representation of the coding of an ordered multistate character; the topological relationships of the states of a character as coded

chorological progression within a group of taxa, the progressive advancement of character states with increase in distance from the geographic center of origin

circumscription the boundaries of a taxon, as determined by systematic study (changes of circumscription are one of the chief sources of nomenclatural synonymy)

clade concordance index a measure of intercladogram character conflict for all characters among a set of cladograms

cladist one who practices systematics based on the methods of cladistics (i.e., grouping by synapomorphy alone, through the application of the parsimony criterion)

cladistics a school of systematics that espouses grouping by synapomorphy through the application of the parsimony criterion and recognizing only monophyletic groups in classifications

cladogenesis the spitting of lineages over time, with the consequent increase in numbers of taxa (*see* anagenesis)

cladogram a depiction of hierarchic relationships among taxa in the form of a treelike diagram, which shows relative recency of relationship and on which character-state transformations may be mapped, but without the connotation of amount of difference or time since divergence (*see* dendrogram, phenogram, phylogram)

class in philosophy, a universal category defined by its properties; in biological classification, a taxonomic rank less inclusive than a phylum and more inclusive than an order

classification (biological) subordinated list of names of taxa, usually assigned to ranks of the Linnaean hierarchy; often representing a formalization of the results of phylogenetic analysis

clique the subset of perfectly congruent characters supporting the optimal result in a compatibility analysis

clustering-level distance an ultrametric, or Euclidean, distance of the type employed in phenetics; such a distance implies a uniform rate of divergence among taxa that cluster at the same level (*see* path-length distance)

codivergence *see* cospeciation

coevolution   the proposition that taxa have evolved in concert, as, for example, hosts and their parasites, with the expectation of congruent branching patterns in respective clades

coherence   an empiricist philosophical perspective in which theories are accepted to the degree that they are supported by evidence and a commitment is not made to their metaphysical truth (*see* correspondence)

combined analysis   *see* simultaneous analysis

compatibility   noncontradiction; two characters are compatible if they do not imply conflicting hypotheses of grouping

compatibility analysis   a technique for reconstructing relationships among taxa, whereby no character contributing to the result of the analysis may be incongruent with the result; the largest set of compatible characters is called a "clique"

component   a clade; a node and the branches descending from it in a cladogram (*see* hypothetical ancestor, term)

composite coding   the approach of coding character data in a multistate format where possible, as opposed to presence–absence (reductionist) coding

compromise tree   a tree derived from techniques that allow trees to be combined, but the results do not summarize exactly groupings from all of the input trees (*see* majority-rule "consensus")

concatenation   *see* simultaneous analysis

concordance   congruence

congruence   the property of two or more characters or trees supporting the same hierarchic pattern of relationships

conjunction   multiple occurrences of a structure hypothesized to be the same in the same organism (conjunction indicates the initial hypothesis of homology needs reassessment) (*see* serial homology)

consensus   the collection of groups (components) contained exactly in all (most parsimonious) trees resulting from a phylogenetic analysis (*see* compromise tree; consensus tree)

consensus tree   the tree depicting the pattern of relationships shared in common among all fundamental trees and collapsing into polytomies all incongruent nodes among them

consilience   agreement, usually among results from different analyses; often implying a whole greater than the sum of its parts, or increase in confidence; referred to by Hennig as "reciprocal illumination"

consistency   in cladistics, the degree to which variation in a character unambiguously supports a particular hierarchic scheme (*see* consistency index); in statistics, the property of gaining increased support for a "correct" inference with accumulation of evidence

consistency index (ci)   a measure of the cladistic informativeness of a character, maximal inconsistency having a value approaching 0.00, perfect consistency having a value of 1.00, those values being computed as the ratio of the minimum possible number of changes in a character on a tree divided by the observed number of changes in the character (see also *ensemble consistency index*, CI)

convergence   independent gain (or loss) of similar features in two or more lineages, often explained as a result of similar selection pressures (*see* homoplasy)

correspondence   a realist philosophical commitment to the idea of an external reality against which observations may be determined to be accurate or inaccurate (*see* coherence)

corroboration   in the hypothetico-deductive approach, the discovery of evidence that supports the prediction of a hypothesis (*see* falsification)

cospeciation   a pattern of parallel phylogenetic branching shared between two (or more) symbiotic taxa

cost   the value assigned to a state-to-state change during cladistic analysis; when these values are not assigned as equal, the most parsimonious tree is the one with the smallest overall cost (*see* length)

crown group   a clade; paleontologists often describe a crown group as "the clade inclusive of the most recent common ancestor of some extant taxon" (*see* stem group)

data matrix   evidence on characters for a set of taxa in tabular form, with the rows representing the taxa and the columns representing the characters

data partition   a subset of all available data that the investigator identifies as worthy of reification and investigation (such as molecules versus morphology)

decay index   *see* branch support

deduction   the approach to science in which predictions implied by theories are tested by observation (in the deductivist approach, hypotheses can be falsified or corroborated but not confirmed) (*see* induction)

definition   in philosophy, the necessary and sufficient attributes that determine the members of a class; in systematics, an enumeration of the apomorphies for a taxon (*see* diagnosis)

DELTRAN   delayed transformation; optimization of character-state transformations as far as possible toward the terminals in a tree, with the implication of independent gains of derived states

dendrogram   a general term for a branching diagram intended to represent hierarchic relationships among entities (e.g., taxa, individuals, alleles, areas)

derived   used in reference to character data for describing a relative condition, namely as opposed to primitive; the apomorphic condition of a feature

description   in systematics, a detailed written statement of the attributes possessed a given taxon, sometimes including multiple life stages

diagnosis   in systematics, a summary statement of attributes that allows recognition of a taxon and separation of that taxon from other taxa

dichotomy   two branches arising from a node on a cladogram (*see* polytomy, trichotomy)

direct optimization   the combination of sequence alignment and phylogenetic inference into a single operation, potentially resulting in shorter trees than may be discovered by the alternative two-step process (*see* alignment)

discordance   *see* hierarchical discordance

disjunction   a fragmented geographical distribution of a taxon, separated by large areas where the taxon is absent

dispersal   in biogeography, the method by which individual taxa increase the size of their ranges, yielding distributions that may or may not be shared in common with other taxa

distance   a measure of similarity (or divergence) among taxa (*see* path-length distance, clustering-level distance)

DNA barcode   a short region of DNA intended to serve as a universal indicator of species identity (for animals, a 640-nucleotide region of the mitochondrial cytochrome oxidase subunit I gene)

empiricism   the philosophical stance that limits scientific knowledge to theories constructed upon observations that may or may not reflect the true nature of the world independent of experience (*see* realism)

endemism   in biogeography, the idea of a taxon (taxa) being restricted to a place

ensemble consistency index (CI)   the consistency index for the suite of all characters used in computing a cladogram

epistemology   the branch of philosophy concerned with the nature of knowledge, asking questions such as, "How do we know what we know?" (*see* ontology)

essentialism   in systematics, a claim about the existence of fundamental qualities that unite diverse individuals into natural kinds (Sober); the school of thought that assigns innate attributes or essences to taxa without regard to the inherent variability of biological systems (after Mayr)

Euclidian distance   a straight line between two points in N-dimensional space (*see* clustering-level distance, Manhattan distance)

event-based methods   model-based approaches to inferring patterns of biogeographical or coevolutionary congruence

evolutionary taxonomy   the school of systematic (taxonomic) practice that recognizes taxa on the basis of combinations of apomorphies and plesiomorphies (frequently characterized as branching order and amount of divergence), often depending on the taxonomist's judgment

exact solution (to a phylogenetic problem)   the set of most parsimonious trees discovered by algorithms that examine all possible trees for a given data set or that produce equivalent results by means of branch-and-bound algorithms

exemplar   a species, other lower-level taxon, or individual organism, chosen to represent a higher-level taxon in a phylogenetic study

extrinsic data   data derived from sources not subject to genetic inheritance, for example, geographical range, or the host plant associations of phytophagous insects (*see* intrinsic data)

fact   a piece of knowledge presumed to be so universally accepted or overwhelmingly supported by evidence as to be uncontroversial (*see* intersubjective corroboration)

falsification   in the hypothetico-deductive approach, the discovery of evidence that refutes a hypothesis; in systematics, the most parsimonious tree for a given data set is the tree with the smallest number of falsifiers (instances of homoplasy) (*see* corroboration)

Farris transformation   *see* additive character

fit   consistency; the degree to which characters conform to (define) a cladogram they have been used to compute; often measured with the consistency index

Fitch transformation   *see* nonadditive character

fittest tree   the tree(s) computed according to the "implied weights" of the characters, so as to maximize total fit (*see* implied weighting)

fundamental tree   the single or set of multiple equally parsimonious (or otherwise optimal) trees discovered in a phylogenetic analysis

gap   in DNA sequence alignment, space inserted into the alignment to achieve correspondence of putatively homologous nucleotide positions across a group of taxa that possess putatively homologous gene regions with unequal numbers of nucleotides

gap coding   a method for converting continuously distributed variables into discrete character states by arbitrarily breaking them at perceived "gaps" in the distribution

genealogy   tokogeny and/or phylogeny; historical patterns of relationship of ancestry and descent either among individuals within a population or among taxa

general area cladogram   a cladogram showing the resolution of area relationships among a variety of taxon-area cladograms (*see* area cladogram)

gene tree   a branching scheme of relationships inferred from data comprising exclusively amino-acid or DNA sequence data from a single gene or gene region (*see* species tree)

genotype   the genetic complement of an organism, equivalent to the genome

geodispersal   biotic dispersal, for example, the "great faunal exchange" of terrestrial animals between North and South America after the emergence of the Isthmus of Panama

gradist   one who espouses the legitimacy of paraphyletic groups in classification (*see* evolutionary taxonomy)

ground plan   the set of attributes (character states) possessed by the hypothetical common ancestor of a group of taxa; may be postulated, as in a composite taxon used as an outgroup, or inferred, as for a hypothetical taxonomic unit (node) on a cladogram optimized from actual character data

heritable   possessing a genetic basis; having the potential to be passed from parents to offspring during reproduction

heterobathmy (of synapomorphy)   the phenomenon whereby different characters define groups at different levels in the taxonomic hierarchy

heuristic (phylogenetic solution)   the set of most parsimonious trees inferred through the use of algorithms applied to data sets so large that all possible trees cannot be examined (*see* exact solution)

hierarchical discordance   disagreement; as, for example, the degree to which an additive multistate character does not agree with a cladogram it helps define

holomorph   the totality of all character states, including DNA, exhibited by different life stages and/or sexes of a taxon (*see* semaphoront)

holophyletic   monophyletic; a term coined by evolutionary taxonomists to allow for the use of "monophyletic" in reference to paraphyletic groups

holophyly   the property of being holophyletic

holotype   the unique specimen designated to represent the concept for a named species; the name bearer

homology   the recognition across taxa of identity among structures (including genes) and behaviors, on the basis of similarity and position; the relationship among parts of organisms that provides evidence for common ancestry

homonymy   in homology determination, the occurrence of multiple similar structures in a single organism, such as leaves on a plant; in nomenclature, the same name applied to two or more groups (e.g., species, genus, family) within botany or zoology, but not between the two fields

homoplasy   incongruence of a character state on a given hypothesis of relationships; convergence; parallelism as implied by reversal or two or more independent gains of a character state on a cladogram; the inferred multiple origin, reduction, or re-evolution of structures or behaviors

horizontal gene transfer   a potential source of incongruence due to the incorporation of foreign genetic material into the genome, often effected by viruses

hot spot   in biodiversity studies, a restricted biogeographical region containing a large number of endemic species

HTU   *see* hypothetical taxonomic unit

hybrid zone   a geographical region where two species' ranges overlap and interspecific interbreeding takes place, potentially enabling genetic introgression

hypothetical common ancestor   an internal node in a cladogram representing the inferred set of attributes common to two or more terminal taxa

hypothetical taxonomic unit (HTU)   internal node on a cladogram (*see* hypothetical common ancestor, node)

hypothetico-deductivism   a philosophy of science that emphasizes falsification of hypotheses rather than their verification

identification   the process of assigning specimens to names

implied weighting   an optimality criterion whereby weights (values) of characters are determined by the degree to which the characters "fit" the tree, the optimal tree being the one with the greatest sum for the weights of all characters (Goloboff)

inapplicable data   data that cannot be known (or coded) for a given taxon because the relevant feature is not present or is so modified that it is unrecognizable

incertae sedis   literally, "of uncertain position"; applied to taxa that, due to lack of available evidence, cannot be placed with certainty in a classification, as for example, incomplete fossils

incongruence   lack of agreement, as usually applied to the fit of characters to a clado-gram (*see* congruence) or the disagreement between topologies of separate cladograms

incongruence length difference (ILD)   a measure of the degree of congruence between data partitions, measured as the length of the shortest tree from the combined data minus the sum of the tree lengths for the partitions analyzed separately

inconsistency   the undesirable statistical property in which increasing the amount of data provides increasing levels of support for an incorrect conclusion; to determine that a result is inconsistent requires knowledge of the truth

indel   "insertion/deletion"; hypothesized events leading to changes in length of homolo-gous DNA or RNA sequences

independence   the quality of characters that makes them appropriate for conjoint analy-sis, whereby variation in one character is not tied to variation in other characters; the ability of different characters in phylogenetic analysis to serve as separate sources of evidence

individual   in philosophy, a metaphysical reference to a thing, as opposed to a group of similar things. As defined by advocates of the concept, an individual is a spatiotem-porally bounded entity with no necessary and sufficient defining properties. Some systematists are convinced that taxa are individuals. Other systematists recognize that the tools we use to do systematics—such as data matrices—treat taxa as though they are classes (*see* class)

induction   the approach to science through which knowledge accrues over the course of continued observation, general theories are built up from assembled facts, and sup-porting evidence confirms hypotheses; the Baconian method (*see* deduction)

ingroup   the focal taxon of interest, among whose members relationships are to be inferred; the ingroup is assumed to be a clade with respect to the outgroup

instrumentalism   in systematics, the performance of data analysis without consider-ation of necessary and sufficient theoretical background knowledge

internal rooting   a method of ordered character-state transformation in which the pre-sumed outgroup state occupies an internal position in a transformation series

internode   a segment between two branching points on a cladogram

intersubjective corroboration   an empiricist proxy for truth; since all observations are inherently subjective, if observers agree upon definitions and conditions of an exis-tential statement, it may be accepted as uncontroversial and adopted as background knowledge for subsequent investigation (*see* fact)

intrinsic data   data subject to genetic inheritance; for example, morphological features or DNA sequences (*see* extrinsic data)

introgression   gene flow across taxonomic boundaries that results in incorporation of foreign genes into the genome of a taxon and therefore acts as a potential source of gene tree incongruence

invalid name   in nomenclature, an available name that is superceded by a valid name

isomorphy   identity of hierarchic relationships as portrayed by a cladogram and a for-mal classification

jackknife (jackknifing)   a technique that uses pseudoreplicated subsamples of char-acters or taxa (sampled without replacement) in an attempt to understand to what degree a data set can stably reproduce a given tree topology

lectotype   a specimen designated from among the members of a syntype series to serve the function of a holotype

length (of a tree)   the number of steps (character-state transformations) on a given cladogram; the quantity minimized in standard cladistic analysis (*see* cost)

lineage   a theoretical evolutionary unit persisting through time; the empirical equivalent is a terminal taxon or monophyletic group

linear character   a multistate character in which adjacency relationships are sequential; an ordinal character (Pimentel and Riggins) (*see* branching character)

Linnaean hierarchy   the nested system of named ranks for taxa codified by Linnaeus and elaborated by later workers

long-branch attraction   the theory that sequence data in rapidly evolving lineages may contain large numbers of characters evolving in parallel, and which may therefore empirically support groups even though the lineages are not each other's closest relatives; often invoked as an ad hoc reason to reject an unexpected phylogenetic result

macroevolution   phylogenetic and other processes, such as diversification and extinction, taking place above the species level

majority-rule "consensus"   a compromise tree that includes all nodes found in 50 percent or more of most parsimonious trees or search results. Often used to represent the output of bootstrap analyses

Manhattan distance   *see* path-length distance

mapping (of characters)   the practice of optimizing "extrinsic" character data on a cladogram produced from a matrix of "intrinsic" data; more generally, determining the distribution on a cladogram of any character not used in a prior analysis

matrix   *see* data matrix

maximum likelihood estimation   in phylogenetics, a method for inferring patterns of relationships among organisms through computation of probabilities of character distributions (evolution) on the basis of an a priori model of character evolution

meristic   serially homologous structures that exhibit variability in their numbers, such as fin rays in fish or antennal segments in insects; in effect, countable

metaphysics   in systematics, the set of beliefs and statements about the condition of things in the world that provides the underpinnings for the growth of knowledge, which advocates of parsimony argue in the practice of systematics should be limited to those necessary and sufficient to infer phylogenetic patterns

metric   one of a variety of mathematical properties of measurement; in systematics, metrics satisfying the triangle inequality are used as the basis for computing distances in cladistics and phenetics (*see* clustering-level distance, path-length distance)

microevolution   changes in allele frequency and other genetic events that take place at the population level (*see* tokogenetic relationships)

minimum spanning tree   a branching diagram in which observed entities can occur at internal nodes; often used in phylogeographical studies to represent unrooted networks of relationship among DNA sequences sampled among populations within a species

missing taxon   in biogeography, a taxon absent from one of the areas in one of the area cladograms being compared, the result being noncorrespondence of distributions across taxa

molecular data   in the twenty-first century, DNA sequence data; molecules such as cuticular hydrocarbons, and even proteins, are viewed as part of the phenotype

monophyletic group   a clade; a group defined by synapomorphies; a group containing a hypothetical common ancestor and all of its descendants; a group connected to the remainder of the Tree of Life by a single branch (*see* paraphyletic, polyphyletic)

morphocline analysis   ordering the states of a character on the basis of relative similarity alone, without regard for congruence with other characters

mosaic evolution   differential rates of evolution in different characters

most parsimonious tree (MPT)   for a given data set, the tree(s) of minimum length as computed under the parsimony criterion

multistate character   a feature for which there are three or more conditions (character states) in a set of three or more taxa

natural group   a monophyletic group

Neighbor Joining   a phenetic method that sequentially clusters most similar terminal taxa, frequently used for the analysis of DNA sequence data

neomorphic character   a synapomorphy ("evolutionary novelty") unique to a clade and absent in other taxa (Sereno)

neotype   a specimen selected to serve the function of the holotype when the holotype, lectotype, or syntypes have been lost or destroyed

nesting   the property of hierarchy, whereby smaller, less inclusive groups are subsets of (i.e., completely included in) larger, more inclusive groups

network   an undirected and therefore incomplete hierarchic arrangement of taxa; an unrooted result of numerical phylogenetic analysis

Newick format   *see* parenthetical notation

node (on a cladogram)   a branching point in a hierarchy that identifies a component; a hypothetical entity (not a common ancestor)

nomenclature   *see* binominal nomenclature

nomen nudum   (plural **nomina nuda**) a taxonomic name published without a description that de facto does not meet the criteria of availability; it has no standing in nomenclature

nonadditive character   (Fitch transformation) a multistate character in which the same cost (weight) is assigned to all state-to-state changes during cladistic analysis (*see* additive character)

nonredundant linear coding   a method of character coding that uses multiple variables to code complex multistate characters, but those variables need not all be two state, as in additive binary coding (*see* additive binary coding)

NP-completeness   a quality of computational problems for which there is no direct solution and for which the number of possible solutions increases disproportionately as the number of entities compared increases; in phylogenetics, sequence alignment and tree search are both NP-complete problems

numerical taxonomy   the name originally attached to phenetics; the theory and practice of grouping by overall similarity with the attendant assumption of uniform rates of change; sometimes used in reference to any analytic approach to classification that applies quantitative techniques

objective character weighting   weighting on the basis of the observed attributes of characters, such as the consistency index (*see* subjective character weighting)

objectivity   a naive desideratum in science that one may observe nature as it actually is, without a filter of sensory and mental perception (*see* intersubjective corroboration)

ontogeny   the process of organismal development; the stages of the life cycle of an organism

ontology   the area of philosophy concerned with the substance of knowledge, reality, and truth; existential statements, things observed, objects of hypothesis tests, resultant theories, and "facts" are ontological claims about the nature of the world (*see* epistemology)

operational (operationalism)   in systematics, when results are determined by the method used to generate them without a clear grounding of necessary and sufficient background knowledge; held as a virtue by pheneticists, often used pejoratively elsewhere

operational taxonomic unit (OTU)   a terminal taxon used in an analysis of relationships, especially as understood in the practice of phenetics

optimality criterion   the decision-making rule used to select among alternative topologies in phylogenetic analysis; for example, the optimality criterion for inferring most parsimonious trees is minimization of required character-state changes

optimization   in phylogenetic inference, the process of discovering the best tree under a given optimality criterion by examining the fit of the characters to alternative hierarchic topologies (Chapter 5); in biogeography and ecological studies, the method employed to assign character states to hypothetical taxonomic units (internal cladogram nodes) and to map extrinsic character data (Chapter 10; *see* ACCTRAN, DELTRAN)

ordered character   *see* additive character

ordinal character   *see* linear character

orthology   homology; the relationship between gene copies resulting from taxonomic divergence (*see* paralogy)

outgroup   the taxon (or taxa) used to determine the position of the root of a cladogram (or other phylogenetic hypothesis), and thus the polarities of the characters used to infer it

overall similarity   the concept of relationship based on character-state identity and difference, averaged across all characters; the quantity used for grouping by the phenetic approach, under which the shared absence of derived character states is deemed informative

paralogy   serially homologous duplication; originally used to describe duplicated gene regions that exist in the same genome and that are potentially subject to independent sequence divergence; by analogy, in biogeography and coevolution, the same area occurring more than once on an area cladogram

paraphyletic group   a group containing a hypothetical common ancestor and some, but not all, of its descendants; a group recognized by a combination of synapomorphies and symplesiomorphies; a group connected to the Tree of Life by a single branch but with one or more subordinate branches removed from it (*see* monophyletic, polyphyletic)

parenthetical notation   a linear representation of a Venn diagram, used by computer phylogenetics programs to describe the nesting of taxa in a cladogram, for example, (A(B,C)); also referred to as the Newick format

parsimony   simplicity of explanation; minimization of ad hoc hypotheses; the approach applied in cladistics whereby similarities are assumed to be homologous, in the absence of evidence to the contrary; in cladistic analysis, parsimony selects the cladogram with the minimum cost (number of steps) as the preferred hypothesis of relationships; parsimony is the epistemological principle underlying all scientific inference

partitioned branch support   a measure of the contribution of a data partition to the overall support for a given branch in a combined analysis

path-length distance   distance along a path between two points measured in "steps" without assumptions about rates of divergence among taxa (*see* clustering-level distance)

pectinate (classification)   a cladogram that branches asymmetrically, like the teeth in a comb

pheneticist   one who practices phenetics

phenetics   the method(s) of inferring relationships among organisms whereby groups are formed on the basis of overall similarity (i.e., considering both derived and ancestral similarities) and often assuming uniform rates of change

phenogram   a hierarchic diagrammatic representation of overall similarity derived from the application of phenetic techniques, which may or may not reflect phylogenetic relationships, depending on constancy of evolutionary rates

phenotype   the physical attributes of an organism resulting from the expression of the genotype

phylogenetic hypothesis   a statement about the hierarchical pattern of relationships among three or more taxa, usually based on empirical data that have been analyzed by some method to yield a branching diagram; often interpreted as representing the history of evolutionary divergence within a group

phylogenetic nomenclature   a system of names for taxonomic groups that has been proposed as a replacement for the Linnaean system and the codes of zoological and botanical nomenclature

phylogenetic relationship   the bifurcating hierarchical pattern of relationships among taxa

phylogenetic systematics   cladistics, often more particularly as outlined by Willi Hennig

phylogenetics   generally, the aspect of systematics focused on inference of branching patterns of relationship; however, the term often connotes a statistical, model-based approach antagonistic to cladistics

phylogenomics   inference of phylogenetic relationships based on large quantities of DNA sequence data

phylogeny   the process of evolutionary diversification; misused by many to refer to the inferred hierarchical pattern of relationships among a set of taxa (*see* phylogenetic hypothesis)

phylogeography   the study of population-level genetic variability using phylogenetic tools, often in the context of geographical variation

phylogram   a hierarchic diagrammatic representation of relationships that includes relative branching order and some indication of relative branch length

plesiomorphic   general; primitive, as opposed to derived; complementary to apomorphic; the quality of a character being group defining only at a more inclusive level

plesiomorphy   a general or primitive character state, not group defining at the level at which it is being observed

polarity   the inferred direction of character-state change from primitive to derived

polyphyletic   a group of taxa not including their hypothetical common ancestor; an assemblage of taxa connected to the Tree of Life by more than a single branch (*see* monophyletic, paraphyletic)

polytomy   three or more branches arising from a single node on a cladogram, reflecting lack of resolution (*see* dichotomy, trichotomy)

positivism   the largely rejected twentieth-century school of philosophy that argued that all propositions could be reduced to sense statements (i.e., statements implying no metaphysical ideas or theoretical claims); strongly associated with phenetics (*see* induction)

prediction   deduction from a theory; the anticipated nature of a future observation; in cladistics, the ability to hypothesize the distributions of previously unobserved (unstudied) characters among taxa

presence–absence coding   reductionist coding; an approach to character coding whereby all data are rendered in a two-state format

primitive   plesiomorphic, relatively general or earlier; used in reference to the appearance of character states on a cladogram (plesiomorphic states are those found in the outgroup and some members of the ingroup and therefore closer to the root)

priority   the rule of biological nomenclature stipulating that the name first applied to a taxon is the one that will be treated as valid

progression rule   Hennig's now largely discredited biogeographical theory that members of a taxon in the oldest part of its range are the most plesiomorphic and those in more recently colonized areas are progressively more derived

pseudoreplicate   unlike an independent sample drawn from a population, a subsample redrawn from the original sample

radial symmetry   repetition of similar structures around the central body axis of an organism, such as ray flowers in Compositae or tentacles of a sea anemone

ranking   the assignment of hierarchic position

realism   the philosophical stance that scientific knowledge bears upon and has access to things and processes in the world as they actually are, independent of the lenses of cognition and experience (*see* empiricism, objectivity)

reciprocal illumination   consilience; the capacity of different kinds of data to inform a result

reductionist coding   the practice of coding all character data in a presence–absence format (*see* composite coding)

redundant (paralogous) distributions   the sequential or multiple occurrence of the same area on an area cladogram; a phenomenon observed to occur frequently among the basal taxa (areas) on cladograms

retention index   (ri) the fraction of potential synapomorphy retained as synapomorphy for a character on a cladogram

robustness   a general term for well-resolved branches in a phylogenetic tree; in cladistics, stable or strongly supported branches; in model-based approaches, descriptive of models that produce the "correct" tree even when the data do not fit the model (*see* stability, support)

root   the point at which a cladogram is hypothetically connected to the remainder of the Tree of Life and is thereby given a temporal trajectory; the basal node used to determine the polarity of the characters used to infer a cladogram

rooting   determining the position of the basal node of a phylogenetic tree

scattering   the pattern of character-state distributions observed when a two-state character does not fit a cladogram parsimoniously

scenario   a story describing the evolution of organisms, structures, and so on; usually not based on critical analysis of, or falsifiable with, available evidence

secondary calibration   an inferred molecular clock-based node age from an independently calibrated tree that is subsequently used as a calibration point to infer ages for nodes on a different tree

sectorial search   a fast heuristic search method that analyzes subclades of a large data set separately and, upon discovery of shorter solutions for that group, reconnects them to the entire tree

semaphoront   literally, sign-bearer; the particular life-stage of an organism from which character states are observed, critical to consider when comparing males and females or ontogenetic stages (e.g., larval versus adult forms of holometabolous insects); all character states of all semaphoronts compose the *holomorph*

sensitivity   a measure of the capacity to recover the same hypothesis of relationships under alternate weighting schemes

sequencing (in classifications)   the treatment of successive sister-group relationships as being of equal rank; a method for minimizing the number of ranks required to convert a pectinate cladogram into a formal classification (*see* subordination)

sequencing (of nucleotides)   the determination of nucleotide composition and order in a portion of the genome

serial homology   the repetition of parts in the body plan of an organism, for example, the sequential recurrence of similar segments in the postcephalic region of a centipede

simultaneous analysis   the inclusion of all available data (the "total evidence") in a global parsimony (or other) analysis; when multiple gene segments are combined in this way, the data are often said to be concatenated

sister group   the relationship between a pair of taxa united by one or more apomorphic characters, inferred to be more closely related to one another than to any other taxon

special similarity   *see* synapomorphy

species   in cladistics, often used in reference to the minimal-level taxon subject to analysis; "The smallest aggregation of populations (sexual) or lineages (asexual) diagnosable by a unique combination of character states in comparable individuals (semaphoronts)" (Nixon and Wheeler)

species tree   a frequently and inherently ambiguous term, often used in reference to a cladogram or other hierarchic hypothesis of relationships based on a preponderance of the evidence; used by some authors to refer to the "true" relationships for a group (*see* gene tree)

stability   in phylogenetic inference, the degree to which the same topology is produced by analysis of pseudoreplicated subsamples of the original data matrix; in nomenclature, the preservation of the relationship between a name and the taxon to which it refers

state   *see* character state

stem group   an extinct taxon or paraphyletic assemblage of extinct taxa with respect to a crown group that includes extant taxa, as, for example, dinosaurs excluding birds

step(s)   on a cladogram, the number of state changes for a character or characters; the measure of length used when computing a path-length (Manhattan) distance

subjective character weighting   determining the value of characters for phylogenetic analysis on the basis of criteria not related to the informativeness of the characters themselves, as for example, apparent morphological complexity or supposed adaptive significance

subjectivity   the perception of phenomena through a potentially distorted lens of theoretical interpretation or belief (*see* intersubjective corroboration)

subordination (in classifications)   the treatment of successive sister-group relationships as being of progressively lower rank (i.e., the nesting of ranks corresponding to the nesting of the hierarchy itself); the method requires the maximum number of ranks to convert a pectinate cladogram into a formal classification (*see* sequencing [in classifications])

subtree analysis   an approach in historical biogeography for analyzing congruence among area cladograms, whereby paralogous areas are removed before comparisons among the area cladograms are made

successive approximations weighting   a method of evaluating the strength of characters as they support a given cladogram; a most parsimonious tree(s) is first computed, then characters are weighted, usually according to their consistency index on that tree(s), and the tree recomputed; the weighting and recomputation process is repeated until a stable result is achieved (*see* implied weighting)

surrogacy   in some studies of biodiversity, the use of one or a few taxa as a way of assessing the characteristics of a biota without the necessity of studying all of its constituent members; selective sampling

symmetrical (classification)   a branching pattern in which subdivision occurs equally in all lineages at successive nodes, resulting in a pattern resembling, for example, sports playoff brackets (*see* asymmetrical classification)

symplesiomorphy   a shared, primitive trait that may define a monophyletic group at a more inclusive level or may represent a complementary absence that unites no group

synapomorphy   shared, derived character state; a feature that supports the recognition of a clade

syncretist   a taxonomist who combines parts of different, often philosophically incompatible methods (e.g., phenetics and cladistics) or employs multiple methods to infer hypotheses of relationship

synonym   one of two or more different names applied to the same taxon

syntype   two or more specimens examined by the original author of a species, none of which was uniquely designated to serve as the name bearer for the taxon (holotype)

systematics   the practice of recognizing and naming taxa, determining hierarchic relationships among those taxa, and formally specifying those relationships; frequently used in a sense roughly equivalent to taxonomy

taxic homology   *see* synapomorphy

taxon (taxa)   a group of organisms at any level in the systematic hierarchy; in systematics, the least inclusive taxon is the species (although subspecies are recognized in zoological nomenclature as well as additional infraspecific categories in botany)

taxon-area cladogram *see* area cladogram

taxonomic congruence the comparison of hypotheses of relationships (trees) from separate analyses of data partitions to assess the proportion of nodes in common among them; the property of different data partitions producing results that are compatible (not in conflict) with one another; also used to compare trees of parasites and their hosts

taxonomy the practice of recognizing, describing, and classifying organisms; frequently used in a sense equivalent to *systematics*

term a terminal taxon; in biogeography, a terminal area in an area cladogram (*see* component)

terminal taxon a taxon for which data are actually observed in a cladistic analysis; a group of organisms that for the purposes of a given study is assumed to be homogeneous with respect to other such groups

test comparison of observation with theory, as comparing the actual distribution of characters with a hypothetical scheme of phylogenetic relationships (*see* deduction)

three-taxon statements a method for recoding characters as composite homology statements that reweights their evidentiary value

time tree a dendrogram in which the branch lengths from root to tip represent the time dimension and the nodes represent inferred ages of divergence

tokogenetic relationships the reticulate pattern of genealogical connections among individual organisms existing below the species level as a result of interbreeding or other reproductive factors

topographical identity *see* transformational homology

topology (of a cladogram) the branching order; also, the geometric form of a cladogram (e. g. pectinate or symmetrical)

total evidence *see* simultaneous analysis

transformation series a specified ordering of changes among the states of a character

transformational homology as employed in this book, the quality of being "the same character" without reference to character polarity or implying groups (*see* Sidebar 6, Chapter 3)

transition in a DNA sequence, a nucleotide substitution of a purine for another purine (A ↔ G) or a pyrimidine for another pyrimidine (C ↔ T) (*see* transversion)

transversion in a DNA sequence, the substitution of a purine for a pyrimidine or vice versa (*see* transition)

tree generally, any branching diagram that specifies hierarchic relationships among taxa; sometimes, a branching diagram implying a specific pattern of ancestor-descendant relationships or of speciation (after Nelson)

tree drifting a heuristic search method that retains suboptimal trees as starting points for additional heuristic searches, thereby increasing the opportunity to discover more parsimonious solutions

tree fusing a fast heuristic search method for large data sets that swaps resolved subclades among most parsimonious trees resulting from a basic search to seek more parsimonious solutions

tree space the universe of tree topologies that an algorithm examines during the course of searching for a most parsimonious solution to a phylogenetic problem

tree thinking the ability to interpret a tree diagram as a graphical representation of phylogenetic relationships; an elementary skill of systematic literacy: tree thinking is to systematics as knowing the letters of the alphabet is to reading

triangle inequality the property of triangles stipulating that the sum of the lengths of two sides cannot be less than the length of the third side

trichotomy three branches arising from a single node on a cladogram (*see* dichotomy, polytomy)

two-state character   a character for which there are two conditions, for example, present and absent

type specimen   *see* holotype

ultrametric   *see* clustering-level distance; in an ultrametric tree, the path length from the root to every tip is the same

uniformitarianism   the parsimonious metaphysical assumption that the future resembles the past; invoked by historical geologist Charles Lyell to argue that the processes that formed ancient geological features were the same as those acting on the present-day Earth

unordered character   *see* nonadditive character

valid name   the available name that correctly applies to a taxon under the rules of nomenclature

variable   a character; the representation of a character (and its states) in a data matrix

vicariance   in biogeography, the subdivision of ancestral ranges of taxa

Wagner algorithm   a method of computing phylogenetic relationships using the parsimony criterion; the underlying basis of all phylogenetic algorithms

weighting (of characters)   assigning greater importance to some characters relative to others (*see* objective character weighting, subjective character weighting)

widespread taxon   a taxon that occurs in more than one area of endemism, as areas of endemism are interpreted on the basis of the distributions of other taxa

Literature Cited

Aanen, D.K., P. Eggleton, C. Rouland-Lefèvre, T. Guldberg-Frøslev, S. Rosendahl, and J.J. Boomsma. 2002. The evolution of fungus-growing termites and their mutualistic fungal symbionts. Proceedings of the National Academy of Sciences of the USA 99:14887–14892.

Adams, E.N., III. 1972. Consensus techniques and the comparison of taxonomic trees. Systematic Zoology 21:390–397.

Agnarsson, I., and J.A. Miller. 2008. Is ACCTRAN better than DELTRAN? Cladistics 24:1032–1038.

Alfaro, M.E., and J.P. Huelsenbeck. 2006. Comparative performance of Bayesian and AIC-based measures of phylogenetic model uncertainty. Systematic Biology 55:89–96.

Altshul, S.F., W. Gish, W. Miller, E. W. Myers, and D.J. Lipman. 1990. Basic local alignment search tool. Journal of Molecular Biology 215:403–410.

Amundson, R. 2005. The changing role of the embryo in evolutionary thought: Roots of evo-devo. New York: Cambridge University Press.

Andersen, N.M. 1997. Phylogenetic tests of evolutionary scenarios: the evolution of flightlessness and wing polymorphism in insects. In: The origin of biodiversity in insects: Phylogenetic tests of evolutionary scenarios, ed. P. Grandcolas, 91–108. Paris: Mémoires du Muséum national d'Histoire naturelle, vol. 173.

Anderson, F.E., and D.L. Swofford. 2004. Should we be worried about long-branch attraction in real data sets? Investigations using metazoan 18S rDNA. Molecular Phylogenetics and Evolution 33:440–451.

Angiosperm Phylogeny Group. 2016. An update of the Angiosperm Phylogeny Group classification for the orders and families of flowering plants: APG IV. *Botanical Journal of the Linnean Society* 181:1–20.

Apodaca, M.J., L. Katinas, and E.L. Guerrero. 2019. Hidden areas of endemism: Small units in the South-eastern Neotropics. Systematics and Biodiversity 17:425–438.

Appel, T.A. 1987. The Cuvier–Geoffroy debate: French biology in the decades before Darwin. Oxford: Oxford University Press.

Archie, J.W. 1985. Methods for coding variable morphological features in numerical taxonomic analysis. Systematic Zoology 34:326–345.

Archie, J.W. 1989a. A randomization test for phylogenetic information is systematic data. Systematic Zoology 38:239–252.

Archie, J.W. 1989b. Homoplasy excess ratios: new indices for measuring levels of homoplasy in phylogenetic systematics and a critique of the consistency index. Systematic Zoology 38:253–269.

Arias, J.S., C.A. Szumik, and P.A. Goloboff. 2011. Spatial analysis of vicariance: a method for using direct geographical information in historical biogeography. Cladistics 27:617–628.

Ashlock, P.D. 1971. Monophyly and associated terms. Systematic Zoology 20:63–69.

Ashlock, P.D. 1974. The uses of cladistics. Annual Review of Ecology and Systematics 5:81–99.

Ashlock, P.D. 1979. An evolutionary systematist's view of classification. Systematic Zoology 28:441–450.

Australasian Virtual Herbarium. 2020. Council of Heads of Australasian Herbaria, http://avh.chah.org.au/ [accessed 25 March 2020].

Avise, J.C. 2006. Evolutionary pathways in nature: A phylogenetic approach. New York: Cambridge University Press.

Avise, J.C. 2014. Conceptual breakthroughs in evolutionary genetics: A brief history of Shifting Paradigms. Boca Raton, FL: Academic Press.

Avise, J.C., J. Arnold, R.M. Ball, E. Bermingham, T. Lamb, J.E. Neigel, C.A. Reeb, and N.C. Saunders. 1987. Intraspecific phylogeography: the mitochondrial DNA bridge between population genetics and systematics. Annual Review of Ecology and Systematics 18:489–522.

Ax, P. 1985. Stem species and the stem lineage concept. Cladistics 1:279–287.

Axelius, B. 1991. Areas of distribution and areas of endemism. Cladistics 7:197–199.

Ayala, F. J. 1999. Molecular clock mirages. BioEssays 21:71–75.

Baker, P.A., S.C. Fritz, C.W. Dick, A.J. Eckert, B.K. Horton, S. Manzoni, C.C. Ribas, C.N. Garzione, and D.S. Battisti. 2014. The emerging field of geogenomics: constraining geological problems with genetic data. Earth-Science Reviews 135:38–47.

Baker, R.H., and R. DeSalle. 1997. Multiple sources of character information and the phylogeny of Hawaiian *Drosophila*. Systematic Biology 46:654–673.

Ball, I. R. 1975. Nature and formulation of biogeographical hypotheses. Systematic Zoology 24:407–430.

Barbadilla, A., L.M. King, and R.C. Lewontin. 1996. What does electrophoretic variation tell us about protein variation? Molecular Biology and Evolution 13:427–432.

Barkley, T.M., P. DePriest, V. Funk, R. W. Kiger, W.J. Kress, and G. Moore. 2004. Linnaean nomenclature in the twenty-first century: a report from a workshop on integrating traditional nomenclature and phylogenetic classification. Taxon 53:153–158.

Barrett, M., M.J. Donoghue, and E. Sober. 1991. Against consensus. Systematic Zoology 40:486–493.

Barry, M. 1837. On the unity of structure in the animal kingdom. Edinburgh New Philosophical Journal 22:117–141.

Bates, H.W. 1862. Contributions to an insect fauna of the Amazon Valley. Lepidoptera: Heliconidae. Transactions of the Linnean Society 23:495–566.

Baum, B. 1992. Combining trees as a way of combining data sets for phylogenetic inference, and the desirability of combining gene trees. Taxon 41:3–10.

Baum, D.A. 2017. Does the future of phylogenetic systematics really rest on the legacy of one mid-20th-century German entomologist? Quarterly Review of Biology 92:450–453.

Baum, D.A., and S.D. Smith. 2013. Tree thinking: An introduction to phylogenetic biology. Greenwood Village, CO: Roberts.

Beatty, J. 1982. Classes and cladists. Systematic Zoology 31:25–34.

Beck, R.M., and C. Baillie. 2018. Improvements in the fossil record may largely resolve current conflicts between morphological and molecular estimates of mammal phylogeny. Proceedings of the Royal Society B 285, https://doi.org/10.1098/rspb.2018.1632.

Belon, P. 1555. L'histoire de la Nature des Oyseaux. Paris: Guillaume Cavellat.

Beltrán, M., C.D. Jiggins, A.V.Z. Brower, E. Bermingham, and J. Mallet, 2007. Do pollen feeding, pupal-mating and larval gregariousness have a single origin in *Heliconius* butterflies? Inferences from multilocus DNA sequence data. Biological Journal of the Linnean Society 92:221–239.

Benton, M. J., and P. C.J. Donoghue. 2007. Paleontological evidence to date the tree of life. Molecular Biology and Evolution 24:26–53.

Bergsten, J. 2005. A review of long-branch attraction. Cladistics 21:163–193.

Bergsten, J., D.T. Bilton, T. Fujisawa, M. Elliott, M.T. Monaghan, M. Balke, L. Hendrich, et al. 2012. The effect of geographical scale of sampling on DNA barcoding. Systematic Biology 61:851–870.

Bertrand, Y., F. Pleijel, and G.W. Rouse. 2006. Taxonomic surrogacy in biodiversity assessments, and the meaning of Linnaean ranks. Systematics and Biodiversity 4:149–159.

Bininda-Emonds, O., J. Gittleman, and A. Purvis. 1999. Building large trees by combining phylogenetic information: a complete phylogeny of the extant Carnivora. Biological Reviews 74:143–175.

Biodiversity Heritage Library, http://www.biodiversitylibrary.org/ [accessed 30 March 2020].

Biodiversity Hotspots. ©2019 Conservation International, https://www.conservation.org/priorities/biodiversity-hotspots [accessed 30 July 2020].

Blackwelder, R.E. 1967. Taxonomy: A text and reference book. New York: John Wiley and Sons.

Bock, W.J. 1974. Philosophical foundations of classical evolutionary classification. Systematic Zoology 22:375–392.

Bock, W.J. 2007. Explanations in evolutionary biology. Journal of Zoological Systematics and Evolutionary Research 45:89–103.

Bogdanowicz, S.M., W.E. Wallner, T.M. Bell, and R.G. Harrison. 1993. Asian gypsy moths (Lepidoptera: Lymantriidae) in North America: evidence from molecular data. Annals of the Entomological Society of America 86:710–715.

Bouckaert, R.R., J. Heled, D. Kuehnert, T. Vaughan, C.-H. Wu, D. Xie, M.A. Suchard, A. Rambaut, and A.J. Drummond. 2014. BEAST 2: a software platform for Bayesian evolutionary analysis. PLOS Computational Biology 10:e1003537.

Bouckaert, R.R., and J. Heled. (Unpublished.) DensiTree 2: seeing trees through the forest. BioRxiv, https://doi.org/10.1101/012401 [posted 8 December 2014].

Boudreaux, H.B. 1979. Arthropod phylogeny with special reference to insects. New York: John Wiley and Sons.

Boyd, R.N. 1999. Homeostasis, species, and higher taxa. In: Species: New interdisciplinary essays, ed. R.A. Wilson, 141–185. Cambridge, MA: MIT Press.

Boyden, A. 1947. Homology and analogy: a critical review of the meaning and implication of these concepts in biology. American Midland Naturalist 37:648–669.

Boyer, S.L., R.M. Clouse, L.R. Benavides, P. Sharma, P.I. Schwendinger, I. Karunarathna, and G. Giribet. 2007. Biogeography of the world: a case study from cyphophthalmid Opiliones, a globally distributed group of arachnids. Journal of Biogeography 34:2070–2085.

Brady, R.H. 1983. Parsimony, hierarchy, and biological implications. In: Advances in cladistics. Vol. 2, Proceedings of the Second Meeting of the Willi Hennig Society, ed. N.I. Platnick and V.A. Funk, 49–60. New York: Columbia University Press.

Brady, R.H. 1985. On the independence of systematics. Cladistics 1:113–126.

Brady, R.H. 1994. Explanation, description, and the meaning of transformation. In: Models in phylogeny reconstruction, ed. R. Scotland, D. Seibert, and D.M. Williams, 11–29. Oxford: Clarendon.

Bravo, G.A., A. Antonelli, C.D. Bacon, K. Bartoszek, M.P.K. Blom, S. Huynh, G. Jones, et al. 2019. Embracing heterogeneity: coalescing the Tree of Life and the future of phylogenomics. PeerJ 7:e3699.

Bremer, K. 1988. The limits of amino acid sequence data in angiosperm phylogenetic reconstruction. Evolution 42:795–803.

Bremer, K. 1990. Combinable component consensus. Cladistics 6:369–372.

Bremer, K. 1994. Branch support and tree stability. Cladistics 10:295–304.

Bremer, K., and H.-E. Wanntorp. 1978. Phylogenetic systematics in botany. Taxon 27:317–329.

Britten, R. J. 1986. Rates of DNA sequence evolution differ between taxonomic groups. Science 231:1393–1398.

Bromham, L. 2016. Testing hypotheses in macroevolution. Studies in History and Philosophy of Science 55:47–59.

Bromham, L. 2019. Six impossible things before breakfast: assumptions, models, and belief in molecular dating. Trends in Ecology and Evolution 34:474–486.

Brooks, D.R. 1981. Hennig's parasitological method: a proposed solution. Systematic Zoology 30:229–249.

Brower, A.V.Z. 1994a. Phylogeny of *Heliconius* butterflies inferred from mitochondrial DNA sequences (Lepidoptera: Nymphalidae). Molecular Phylogenetics and Evolution 3:159–174.

Brower, A.V.Z. 1994b. Rapid morphological radiation and convergence among races of the butterfly *Heliconius erato* inferred from patterns of mitochondrial DNA evolution. Proceedings of the National Academy of Sciences of the USA 91:6491–6495.

Brower, A.V.Z. 1996. Parallel race formation and the evolution of mimicry in *Heliconius* butterflies: a phylogenetic hypothesis from mitochondrial DNA sequences. Evolution 50:195–221.

Brower, A.V.Z. 1997. The evolution of ecologically important characters in *Heliconius* butterflies (Lepidoptera: Nymphalidae): a cladistic review. Zoological Journal of the Linnean Society 119:457–472.

Brower, A.V.Z. 1999. Delimitation of phylogenetic species with DNA sequences: a critique of Davis and Nixon's population aggregation analysis. Systematic Biology 48:199–213.

Brower, A.V.Z. 2000a. Evolution is not an assumption of cladistics. Cladistics 16:143–54.

Brower, A.V.Z. 2000b. Homology and the inference of systematic relationships: Some historical and philosophical perspectives. In: Homology and systematics: coding characters for phylogenetic analysis, ed. R.W. Scotland and R.T. Pennington, 10–21. London: Taylor and Francis.

Brower, A.V.Z. 2002. Cladistics, phylogeny, evidence and explanation: a reply to Lee. Zoologica Scripta 31:221–223.

Brower, A.V.Z. 2006a. Problems with DNA barcodes for species delimitation: "ten species" of *Astraptes fulgerator* reassessed (Lepidoptera: Hesperiidae). Systematics and Biodiversity 4:127–132.

Brower, A.V.Z. 2006b. The how and why of branch support and partitioned branch support, with a new index to assess partition incongruence. Cladistics 22:378–386.

Brower, A.V.Z. 2012. The meaning of "phenetic". Cladistics 28:113–114.

Brower, A.V.Z. 2015. Transformational and taxic homology revisited. Cladistics 31:197–201.

Brower, A.V.Z. 2016a. Are we all cladists? In: The future of phylogenetic systematics: The legacy of Willi Hennig, ed. D.M. Williams, M. Schmitt, and Q.D. Wheeler, 88–114. Cambridge: Cambridge University Press.

Brower, A.V.Z. 2016b. What is a cladogram and what is not? Cladistics 32:573–576.

Brower, A.V.Z. 2018a. Fifty shades of cladism. Biology and Philosophy 33:1–8.

Brower, A.V.Z. 2018b. Going rogue. Cladistics 34:467–468.

Brower, A.V.Z. 2018c. Statistical consistency and phylogenetic inference: a brief review. Cladistics 34:562–567.

Brower, A.V.Z. 2019. Background knowledge: the assumptions of pattern cladistics. Cladistics 35:717–731.

Brower, A.V.Z. 2020. Dead on arrival: a postmortem assessment of "phylogenetic nomenclature" 20+ years on. Cladistics, doi: 10.1111/cla.12432.

Brower, A.V.Z., and M. C. C. de Pinna. 2012. Homology and errors. Cladistics 28:529–538.

Brower, A.V.Z., and R. DeSalle. 1994. Practical and theoretical considerations for choice of a DNA sequence region in insect molecular systematics, with a short review of published studies using nuclear gene regions. Annals of the Entomological Society of America 87:702–716.

Brower, A.V.Z., R. DeSalle, and A. P. Vogler. 1996. Gene trees, species trees, and systematics: a cladistic perspective. Annual Review of Ecology and Systematics 27:423–450.

Brower, A.V.Z., and M. G. Egan. 1997. Cladistics of *Heliconius* butterflies and relatives (Nymphalidae: Heliconiiti): the phylogenetic position of *Eueides* based on sequences from mtDNA and a nuclear gene. Proceedings of the Royal Society London B 264:969–977.

Brower, A. V. Z., and I. J. Garzón-Orduña. 2018. Missing data, clade support and "reticulation": the molecular systematics of *Heliconius* and related genera (Lepidoptera: Nymphalidae) reexamined. Cladistics 34:151–166.

Brower, A.V.Z., and V. Schawaroch. 1996. Three steps of homology assessment. Cladistics 12:265–272.

Brower, L.P., J.V.Z. Brower, and P.W. Westcott. 1960. Experimental studies of mimicry. 5. The reactions of toads (*Bufo terrestris*) to bumblebees (*Bombus americanorum*) and their robberfly mimics (*Mallophora bomboides*), with a discussion of aggressive mimicry. American Naturalist 94:343–356.

Brown, W.M. 1983. Evolution of animal mitochondrial DNA. In: Evolution of genes and proteins, ed. M. Nei and R.K. Koehn, 62–88. Sunderland, MA: Sinauer.

Brundin, L.Z. 1966. Transantarctic relationships and their significance, as evidenced by chironomid midges. Kungliga Svenska Vetenskapsakademiens Handlingar, Fjarde Series, 11:1–472.

Brundin, L.Z. 1981. Croizat's panbiogeography versus phylogenetic biogeography. In: Vicariance biogeography: A critique, ed. G. Nelson and D.E. Rosen, 94–138. New York: Columbia University Press.

BugGuide.net. https://bugguide.net/node/view/15740 [accessed 6 March 2020].

Bull J.J., J.P. Huelsenbeck, C.W. Cunningham, D.L. Swofford, and P.J. Waddell. 1993. Partitioning and combining data in phylogenetic analysis. Systematic Biology 42:384–397.

Cain, A. J., and G. A. Harrison. 1960. Phyletic weighting. Proceedings of the Zoological Society of London 13:1–31.

Camin, J.H., and R.R. Sokal. 1965. A method for deducing branching sequences in phylogeny. Evolution 19:311–326.

Cantino, P.D., and K. de Queiroz. 2007. International code of phylogenetic nomenclature, version 4b, https://www.ohio.edu/phylocode/ [accessed 30 March 2020].

Cao, X., J. Liu, J. Chen, G. Zheng, M. Kuntner, and I. Agnarsson. 2016. Rapid dissemination of taxonomic discoveries based on DNA barcoding and morphology. Scientific Reports 6:1–13.

Carnap, R. 1950. Logical foundations of probability. Chicago: University of Chicago Press.

Carpenter, J.M. 1988a. Choosing among multiple equally most parsimonious cladograms. Cladistics 4:291–296.

Carpenter, J.M. 1988b. The phylogenetic system of the Stenogastrinae (Hymenoptera: Vespidae). Journal of the New York Entomological Society 96:140–175.

Carpenter, J.M. 1989. Testing scenarios: wasp social behavior. Cladistics 5:131–144.

Carpenter, J.M. 1990. On genetic distances and social wasps. Systematic Zoology 39:391–397.

Carpenter, J.M. 1991. Phylogenetic relationships and the origin of social behavior in the Vespidae. In: Social biology of wasps, ed. K.G. Ross and R.W. Matthews, 7–32. Ithaca, NY: Cornell University Press.

Carpenter, J.M. 1992. Random cladistics. Cladistics 8:147–153.

Carpenter, J.M. 1996. Uninformative bootstrapping. Cladistics 12:177–181.

Carpenter, J.M. 2003a. Critique of pure folly. Botanical Review 69:79–92.

Carpenter, J.M. 2003b. On "molecular phylogeny of Vespidae (Hymenoptera) and the evolution of sociality in wasps." American Museum Novitates 3389:1–17.

Carpenter, J.M., P.A. Goloboff, and J.S. Farris. 1998. PTP is meaningless, T-PTP is contradictory: a reply to Trueman. Cladistics 14:105–116.

Carroll, L. 1960. Alice's adventures in wonderland & through the looking-glass. New York: Penguin.

Carver, M., G.F. Gross, and T.E. Woodward 1991. Hemiptera. In: The insects of Australia, vol. 1, ed. I.D. Naumann, 429–509. Melbourne: Melbourne University Press.

Cassis, G., and G. F. Gross. 1995. Zoological catalog of Australia. 27.3A. Hemiptera: Heteroptera (Coleorrhyncha to Cimicomorpha). Australian Biological Resources Study. Melbourne: Commonwealth Scientific and Industrial Research Organisation.

Cavalli-Sforza, L.L., and A.W.F. Edwards. 1967. Phylogenetic analysis: models and estimation procedures. Evolution 21:550–570.

Cayley, A. 1856. Note sur une formule pour la reversion des séries. Journal für die reine und angewandte Mathematik 52:276–284.

Chen, M.-H., L. Kuo, and P.O. Lewis, eds. 2014. Bayesian phylogenetics: Methods, algorithms and applications. Boca Raton, FL: CRC Press.

China, W.E. 1933. A new family of Hemiptera: Heteroptera with notes on the phylogeny of the suborder. Annals and Magazine of Natural History, ser. 10, 12:180–196.

Chippindale, P. T., and J. J. Wiens. 1994. Weighting, partitioning, and combining characters in phylogenetic analysis. Systematic Biology 43:278–287.

Chor, B., M. D. Hendy, B. R. Holland, and D. Penny 2000. Multiple maxima of likelihood in phylogenetic trees: an analytic approach. Molecular Biology and Evolution 17:1529–1541.

Christenhusz, M.J.M., M.F. Fay, and M.W. Chase. 2017. Plants of the world: An illustrated encyclopedia of vascular plants. Richmond, UK, and Chicago: Royal Botanical Gardens, Kew, and University of Chicago Press.

Claridge, M.F., H.A. Dawah, and M.R. Wilson, eds. 1997. Species. Systematics Association Special Volume Series, no. 54. London: Chapman and Hall.

Cleland, C.E. 2002. Methodological and epistemic differences between historical science and experimental science. Philosophy of Science 69:474–496.

Coddington, J.A. 1988. Cladistic tests of adaptational hypotheses. Cladistics 4:3–22.

Cognato, A.I. 2006. Standard percent DNA sequence difference for insects does not predict species boundaries. Journal of Economic Entomology 99:1037–1045.

Coleman, K.A. and E.O. Wiley. 2001. On species individualism: a new defense of the species-as-individuals hypothesis. Philosophy of Science 68:498–517.

Cook, L., and M. Crisp. 2005. Directional asymmetry of long-distance dispersal and colonization could mislead reconstructions of biogeography. Journal of Biogeography 32:741–754.

Corrick, M.G., B.A. Fuhrer, and A.S. George. 1996. Wildflowers of southern western Australia. Noble Park, Australia: Five Mile Press.

Cotterill, F.P.D., P.J. Taylor, S. Gippoliti, J.M. Bishop, and C.P. Groves. 2014. Why one century of phenetics is enough: response to "Are there really twice as many bovid species as we thought?" Systematic Biology 63:819–832.

Cotton, J.A., and M. Wilkinson. 2009. Supertrees join the mainstream of phylogenetics. Trends in Ecology and Evolution 24:1–3.

Coyne, J.A., and H.A. Orr. 2004. Speciation. Sunderland, MA: Sinauer.

Cracraft, J. 1983. Species concepts and speciation analysis. Current Ornithology 1:159–187.

Cracraft, J. 1992. The species of the birds-of-paradise (Paradisaeidae): applying the phylogenetic species concept to a complex pattern of diversification. Cladistics 8:1–43.

Cranston, P.S., N.B. Hardy, G.E. Morse, L. Puslednik, and S.R. McCluen. 2010. When molecules and morphology concur: the 'Gondwanan' midges (Diptera: Chironomidae). Systematic Entomology 35:636–648.

Cranston, P.S., F.-T. Krell, K. Walker, and D. Hewes. 2015. Wiley's Early View constitutes valid publication for date-sensitive nomenclature. Systematic Entomology 40:2–4.

Craw, R.C. 1982. Phylogenetics, areas, geology, and the biogeography of Croizat: a radical view. Systematic Zoology 31:304–316.

Craw, R.C. 1983. Panbiogeography and vicariance cladistics: Are they truly different? Systematic Zoology 32:431–438.

Craw, R.C., J.R. Grehan, and M.J. Heads. 1999. Panbiogeography: tracking the history of life. New York: Oxford University Press.

Crisp, M.D. 1994. Evolution of bird-pollination in some Australian legumes (Fabaceae). In: Phylogenetics and ecology, ed. P. Eggleton and R. Vane-Wright, 281–309. Linnean Society Symposium Series, vol. 17. London: Academic Press.

Croizat, L. 1962. Space, time, form: The biological synthesis. Published by the author, Caracas.

Crother, B.I. 2002. Is Karl Popper's philosophy of science all things to all people? Cladistics 18:445.

Crowson, R.A. 1970. Classification and Biology. Chicago: Aldine.

Cuénot, L.C.M.J. 1940. Remarques sur un essai d'arbre généalogique du règne animal. Comptes Rendus de l'Académie des Sciences de Paris 210:23–27.

Darlington, P.J., Jr. 1957. Zoogeography: The geographical distribution of animals. New York: John Wiley and Sons.

Darlington, P.J., Jr. 1965. Biogeography of the southern end of the world: Distribution and history of far-southern life and land, with an assessment of continental drift. New York: McGraw-Hill.

Darwin, C. 1859. On the origin of species, 1st ed. London: John Murray.

Darwin, C. 1872. The origin of species, 6th ed. New York: A. L. Burt.

Darwin, F., ed. 1909. The foundations of the origin of species: Two essays written in 1842 and 1844 by Charles Darwin. Cambridge: Cambridge University Press.

Daston, L., and P. Galison. 2007. Objectivity. Cambridge, MA: MIT Press.

Davis, J.I. 1995. A phylogenetic structure for the monocotyledons, as inferred from chloroplast DNA restriction site variation, and a comparison of measures of clade support. Systematic Botany 20:503–527.

Davis, J.I., and K.C. Nixon. 1992. Populations, genetic variation, and the delimitation of phylogenetic species. Systematic Biology 41:421–435.

Day, W.H.E., D.S. Johnson, and D. Sankoff. 1986. The computational complexity of inferring rooted phylogenies by parsimony. Mathematical Bioscience 81:33–42.

Dayrat, B. 2005. Towards integrative taxonomy. Biological Journal of the Linnean Society 85:407–415.

de Candolle, A.-P. 1819. Theorie Elementaire de la Botanique, 2nd ed. Paris: Déterville.

de Carvalho, M.R., F.A. Bockmann, D.S. Amorim, C.R.F. Brandão, M. de Vivo, J.L. de Figueiredo, H.A. Britski, et al. 2007. Taxonomic impediment or impediment to taxonomy? A commentary on systematics and the cybertaxonomic-automation paradigm. Evolutionary Biology 34:140–143.

de Jong, R. 2017. Fossil butterflies, calibration points and the molecular clock (Lepidoptera: Papilionoidea). Zootaxa 4270:1–63.

de Jong, R., R.I. Vane-Wright, and P.R. Ackery. 1996. The higher classification of butterflies (Lepidoptera): problems and prospects. Entomologica Scandinavica 27:65–101.

De Laet, J., and E. Smets. 1998. On the three-taxon approach to parsimony analysis. Cladistics 14:363–381.

Deleporte, P. 1993. Characters, attributes, and tests of evolutionary scenarios. Cladistics 9:427–432.

De Moya, R.S., W.K. Savage, C. Tenney, X. Bao, N. Wahlberg, and R.I. Hill. 2017. Interrelationships and diversification of *Argynnis* Fabricius and *Speyeria* Scudder butterflies. Systematic Entomology 42:635–649.

de Pinna, M.C.C. 1991. Concepts and tests of homology in the cladistic paradigm. Cladistics 7:367–394.

de Queiroz, A. 2014. The monkey's voyage: How improbable journeys shaped the history of life. New York: Basic Books.

de Queiroz, A., M.J. Donoghue, and J. Kim. 1995. Separate versus combined analysis of phylogenetic evidence. Annual Review of Ecology and Systematics 26:657–681.

de Queiroz, K. 1988. Systematics and the Darwinian revolution. Philosophy of Science 55:238–259.

de Queiroz, K. 1999. The general lineage concept of species and defining properties of the species category. In: Species, ed. R.A. Wilson, 49–89. Cambridge, MA: MIT Press.

de Queiroz, K. 2004. The measurement of test severity, significance tests for resolution, and a unified philosophy of phylogenetic inference. Zoologica Scripta 33:463–473.

de Queiroz, K., and P.D. Cantino. 2020. International code of phylogenetic nomenclature (Phylocode). Boca Raton, FL: CRC Press.

de Queiroz, K., and J. Gauthier. 1990. Phylogeny as a central principle in taxonomy: phylogenetic definitions of taxon names. Systematic Zoology 39:307–322.

de Queiroz, K., and S. Poe. 2001. Philosophy and phylogenetic inference: a comparison of likelihood and parsimony methods in the context of Karl Popper's writings on corroboration. Systematic Biology 50:305–321.

de Queiroz, K., and S. Poe. 2003. Failed refutations: further comments on parsimony and likelihood methods and their relationship to Popper's degree of corroboration. Systematic Biology 52:352–367.

D'Erchia, A.M., C. Gissi, G. Pesole, C. Saccone, and U. Arnason. 1996. The guinea pig is not a rodent. Nature 381:597–600.

DeSalle, R., and A.V.Z. Brower. 1997. Process partitions, congruence, and the independence of characters: inferring relationships among closely-related

Hawaiian *Drosophila* from multiple gene regions. Systematic Biology 46:751–764.

Dingus, L., and T. Rowe. 1998. The mistaken extinction: Dinosaur evolution and the origin of birds. New York: W. H. Freeman.

DiscoverLife. https://www.discoverlife.org/ [accessed 24 March 2020].

Dobzhansky, T. 1973. Nothing in biology makes sense except in the light of evolution. American Biology Teacher 35:125–129.

Dominguez, E., and Q.D. Wheeler. 1997. Taxonomic stability is ignorance. Cladistics 13:367–372.

Donoghue, P.C.J. 2005. Matters of the record. Saving the stem group—a contradiction in terms? Paleobiology 31:553–558.

Doyle, J.A., and M.J. Donoghue. 1986. Seed plant phylogeny and the origin of the angiosperms: an experimental approach. Botanical Review 52:321–431.

Doyle, J.J. 1992. Gene trees and species trees: molecular systematics as one-character taxonomy. Systematic Botany 17:144–163.

Drummond, A.J., and A. Rambaut. 2007. BEAST: Bayesian evolutionary analysis by sampling trees. BMC Evolutionary Biology 7:214.

Dubois, A., R. Bour, and A. Ohler. 2015. Nomenclatural availability of preliminary electronic versions of taxonomic papers: in need of clear definition. Bulletin of Zoological Nomenclature 72:252–265.

Duchene, S., R. Lanfear, and S.Y.W. Ho. 2014. The impact of calibration and clock-model choice on molecular estimates of divergence times. Molecular Phylogenetics and Evolution 78:277–289.

Dyar, H.G., and F. Knab. 1906. The larvae of Culicidae classified as independent organisms. Journal of the New York Entomological Society 14:169–230, plates IV–XVI.

Ebach, M.C., and C. Holdrege. 2005. More taxonomy, not DNA barcoding. BioScience 55:822–823.

Ebach, M.C., D.M. Williams, and T.A. Vanderlaan. 2013. Implementation as theory, hierarchy as transformation, homology as synapomorphy. Zootaxa 3641:587–594.

Edwards, A.W.F., and L.L. Cavalli-Sforza. 1964. Reconstruction of evolutionary trees. In: Phenetic and phylogenetic classification, ed. V.H. Heywood and J. McNeill, 67–76. Systematics Association Special Volume 6. London: Systematics Association.

Edwards, S.V. 2009. Is a new and general theory of molecular systematics emerging? Evolution 63:1–19.

Edwards, S.V., Z. Xi, A. Janke, B.C. Faircloth, J.E. McCormack, T.C. Glenn, B. Zhong, et al. 2016. Implementing and testing the multispecies coalescent model: a valuable paradigm for phylogenomics. Molecular Phylogenetics and Evolution 94:447–462.

Eldredge, N. 1979. Alternative approaches to evolutionary theory. Bulletin of the Carnegie Museum of Natural History 13:7–19.

Eldredge, N., and J. Cracraft. 1980. Phylogenetic patterns and the evolutionary process. New York: Columbia University Press.

Eldredge, N., and S.J. Gould. 1972. Punctuated equilibria: an alternative to phyletic gradualism. In: Models in paleobiology, ed. T.J.M. Schopf, 82–115. San Francisco: Cooper, Freeman.

Elias, M., R.I. Hill, K.R. Willmott, K.K. Dasmahapatra, A.V.Z. Brower, J. Mallet, and C.D. Jiggins. 2007. Limited performance of DNA barcoding in a diverse community of tropical butterflies. Proceedings of the Royal Society of London B 274:2881–2889.

Encyclopedia of Life. https://eol.org/ [accessed 31 March 2020].

Endler, J.A. 1982. Problems of distinguishing historical from ecological factors in biogeography. American Zoologist 22:441–452.

Engelmann, G.F., and E.O. Wiley. 1977. The place of ancestor-descendent relationships in phylogeny reconstruction. Systematic Zoology 26:1–11.

Enghoff, H. 1995. Historical biogeography of the Holarctic: area relationships, ancestral areas, and dispersal of non-marine animals. Cladistics 11:223–263.

Ereshefsky, M., ed. 1992. The units of evolution: Essays on the nature of species. Cambridge, MA: MIT Press.

Ereshefsky, M. 2007. Foundational issues concerning taxa and taxon names. Systematic Biology 56:295–301.

Erwin, T.L. 1991. An evolutionary basis for conservation strategies. Science 253:750–752.

Eschmeyer, W.N. 1990. Catalog of the genera of recent fishes. San Francisco: California Academy of Sciences.

Estabrook, G.F., J.G. Strauch Jr., and K.L. Fiala. 1977. An application of compatibility analysis to the Blackith's data on orthopteroid insects. Systematic Zoology 26:269–276.

Faith, D.P. 1992. On corroboration: a reply to Carpenter. Cladistics 8:265–273.

Faith, D.P., and P.S. Cranston. 1991. Could a cladogram this short have arisen by chance alone?: On a permutation test for cladistic structure. Cladistics 7:1–28.

Faith, D.P., and J.W.H. Trueman. 2001. Towards an inclusive philosophy for phylogenetic inference. Systematic Biology 50:331–350.

Faivovich, J., C.F.B. Haddad, P.C.A. Garcia, D.R. Frost, J.A. Campbell, and W.C. Wheeler. 2005. Systematic review of the frog family Hylidae, with special reference to Hylinae: phylogenetic analysis and taxonomic revision. Bulletin of the American Museum of Natural History 294:1–240.

Farrell, B., and C. Mitter. 1990. Phylogenesis of insect/plant interactions: Have *Phyllobrotica* leaf beetles (Chrysomelidae) and the Lamiales diversified in parallel? Evolution 44:1389–1403.

Farris, J.S. 1969. A successive approximations approach to character weighting. Systematic Zoology 18:374–385.

Farris, J.S. 1970. Methods for computing Wagner trees. Systematic Zoology 19:83–92.

Farris, J.S. 1971. The hypothesis of nonspecificity and taxonomic congruence. Annual Review of Ecology and Systematics 2:277–302.

Farris, J.S. 1972. Estimating phylogenetic trees from distance matrices. American Naturalist 106:645–668.

Farris, J.S. 1974. Formal definitions of paraphyly and polyphyly. Systematic Zoology 23:548–554.

Farris, J.S. 1976a. Expected asymmetry of phylogenetic trees. Systematic Zoology 25:196–198.

Farris, J.S. 1976b. Phylogenetic classification of fossils with Recent species. Systematic Zoology 25:271–282.

Farris, J.S. 1977. Phylogenetic analysis under Dollo's Law. Systematic Zoology 26:77–88.

Farris, J.S. 1979a. On the naturalness of phylogenetic classification. Systematic Zoology 28:200–214.

Farris, J.S. 1979b. The information content of the phylogenetic system. Systematic Zoology 28:483–519.

Farris, J.S. 1981. Distance data in phylogenetic analysis. In: Advances in Cladistics. Proceedings of the First Meeting of the Willi Hennig Society, ed. V.A. Funk and D.R. Brooks, 3–23. New York: New York Botanical Garden.

Farris, J.S. 1982. Simplicity and informativeness in systematics and phylogeny. Systematic Zoology 31:413–444.

Farris, J.S. 1983. The logical basis for phylogenetic analysis. In: Advances in Cladistics. Vol. 2, Proceedings of the Second Meeting of the Willi Hennig Society, ed. N.I. Platnick V.A. Funk, 1–36. New York: Columbia University Press.

Farris, J.S. 1988. Hennig86. Unpublished documentation of computer program.

Farris, J.S. 1989. The retention index and rescaled consistency index. Cladistics 5:417–419.

Farris, J.S. 1995. Conjectures and refutations. Cladistics 11:105–118.

Farris, J.S. 1999. Likelihood and inconsistency. Cladistics 15:199–204.

Farris, J.S. 2012. 3ta sleeps with the fishes. Cladistics 28:422–436.

Farris, J.S. 2013. Popper: not Bayes or Rieppel. Cladistics 29:230–232.

Farris, J.S. 2014. Popper with probability. Cladistics 30:5–7.

Farris, J.S., V.A. Albert, M. Källersjö, D. Lipscomb, and A.G. Kluge. 1996. Parsimony jackknifing outperforms neighbor-joining. Cladistics 12:99–124.

Farris, J.S., M. Källersjö, V.A. Albert, M. Allard, A. Anderberg, B. Bowditch, C. Bult, et al. 1995a. Explanation. Cladistics 11:211–218.

Farris, J.S., M. Källersjö, A.G. Kluge, and C. Bult. 1994. Testing significance of congruence. Cladistics 10:315–320.

Farris, J.S., M. Källersjö, A.G. Kluge, and C. Bult. 1995b. Constructing a significance test for incongruence. Systematic Biology 44:570–572.

Farris, J.S., and A.G. Kluge. 1979. A botanical clique. Systematic Zoology 28:400–411.

Farris, J.S., A.G. Kluge, and J.M. Carpenter. 2001. Popper and likelihood versus "Popper*." Systematic Biology 50:438–444.

Farris, J.S., A.G. Kluge, and M.J. Eckhardt. 1970. A numerical approach to phylogenetic systematics. Systematic Zoology 19:172–191.

Felsenstein, J. 1973. Maximum likelihood and minimum-steps methods for estimating evolutionary trees from data on discrete characters. Systematic Zoology 22:240–249.

Felsenstein, J. 1978. Cases in which parsimony or compatibility methods will be positively misleading. Systematic Zoology 27:401–410.

Felsenstein, J. 1981. A likelihood approach to character weighting and what it tells us about parsimony and compatibility. Biological Journal of the Linnean Society 16:183–196.

Felsenstein, J. 1983. Parsimony in systematics: biological and statistical issues. Annual Review of Ecology and Systematics 14:313–333.

Felsenstein, J. 1985. Confidence limits on phylogenies: an approach using the bootstrap. Evolution 39:783–791.

Felsenstein, J. 2004. Inferring phylogenies. Sunderland, MA: Sinauer.

Fichman, M. 1977. Wallace: zoogeography and the problem of land bridges. Journal of the History of Biology 10:45–63.

Fitch, W.M. 1970. Distinguishing homologous from analogous proteins. Systematic Zoology 19:99–113.

Fitch, W.M. 1971. Toward defining the course of evolution: minimum change for a specific tree topology. Systematic Zoology 20:406–416.

Fitch, W.M. 1976. Molecular evolutionary clocks. In: Molecular evolution, ed. F. Ayala, 160–178. Sunderland, MA: Sinauer.

Fitzhugh, K. 2006a. The abduction of phylogenetic hypotheses. Zootaxa 1145:1–110.

Fitzhugh, K. 2006b. The philosophical basis of character coding for the inference of phylogenetic hypotheses. Zoologica Scripta 35:261–286.

Fitzhugh, K. 2016. Dispelling five myths about hypothesis testing in biological systematics. Organisms, Diversity and Evolution 16:443–65.

Foulds, L. R., M. D. Hendy, and D. Penny. 1979. A graph theoretic approach to the development of minimal phylogenetic trees. Journal of Molecular Evolution 13:127–149.

Frankham, R., J.D. Ballou, M.R. Dudash, M.D.B. Eldridge, C.B. Fenster, R.C. Lacy, J.R. Mendelson III, J. Porton, K. Ralls, and O.A. Ryder. 2012. Implications of different species concepts for conserving biodiversity. Biological Conservation 153:25–31.

Franz, N.M. 2005. Outline of an explanatory account of cladistic practice. Biology and Philosophy 20:489–515.

Franz, N.M., N.M. Pier, D.M. Reeder, M. Chen, S. Yu, P. Kianmajd, S. Bowers, and B. Ludascher. 2016. Two influential primate classifications logically aligned. Systematic Biology 65:561–582.

Franz, N.M., and B.W. Sterner. 2018. To increase trust, change the social design behind aggregated biodiversity data. Database 2017:1–12.

Freeman, S. 2008. Biological science, 3rd ed. Upper Saddle River, NJ: Pearson Prentice Hall.

Friç, Z, N. Wahlberg, P. Pech, and J. Zrzavy. 2007. Phylogeny and classification of the *Phengaris-Maculinea* clade (Lepidoptera: Lycaenidae): total evidence and phylogenetic species concepts. Systematic Entomology 32:558–567.

Fricke, R., W.N. Eschmeyer, and R. Van der Laan, eds. 2020. Eschmeyer's catalog of fishes: Genera, species, references, http://researcharchive.calacademy.org/research/ichthyology/catalog/fishcatmain.asp. [accessed 24 March 2020].

Frost, D.R. 2020 Amphibian species of the world 6.0: An online reference, https://amphibiansoftheworld.amnh.org/ [accessed 24 March 2020].

Frost, D.R., and A.G. Kluge. 1994. A consideration of epistemology in systematic biology, with special reference to species. Cladistics 10:259–294.

Futuyma, D.J., and M. Kirkpatrick. 2017. Evolution, 4th ed. Sunderland, MA: Sinauer.

Futuyma, D.J., and S.S. McCafferty. 1990. Phylogeny and evolution of host plant associations in the leaf beetle genus *Ophraella* (Coleoptera: Chrysomelidae). Evolution 44:1885–1913.

Gaffney, E.S. 1979a. An introduction to the logic of phylogeny reconstruction. In: Phylogenetic analysis and paleontology, ed. J. Cracraft and N. Eldredge, 79–111. New York: Columbia University Press.

Gaffney, E.S. 1979b. Tetrapod monophyly: a phylogenetic analysis. Bulletin of the Carnegie Museum of Natural History 13:92–105.

Garamszegi, L.Z. 2014. Modern phylogenetic comparative methods and their application in evolutionary biology: Concepts and practice. Heidelberg: Springer.

Garnett, S.T., and L. Christidis. 2017. Taxonomy anarchy hampers conservation. Nature 546:25–27.

Garzón-Orduña, I.J., K.L. Silva-Brandão, K.R. Willmott, A.V.L. Freitas, and A.V.Z. Brower. 2015. An alternative, plant-based time-tree implies conflicting dates for the diversification of ithomiine butterflies (Lepidoptera: Nymphalidae: Danainae). Systematic Biology 64:752–767.

Gaston, K.J., and M.A. O'Neill. 2004. Automated species identification: why not? Philosophical Transactions of the Royal Society of London. Series B: Biological Sciences 359:655–667.

Gatesy, J. 2007. A tenth crucial question regarding model use in phylogenetics. Trends in Ecology and Evolution 22:509–510.

Gatesy, J., and R.H. Baker. 2005. Hidden likelihood support in genomic data: can forty-five wrongs make a right? Systematic Biology 54:483–492.

Gatesy, J., C. Matthee, R. DeSalle, and C. Hayashi. 2002. Resolution of a supertree/supermatrix paradox. Systematic Biology 51:652–664.

Gatesy, J., M. Milinkovitch, V. Waddell, and M. Stanhope. 1999a. Stability of cladistic relationships between Cetacea and higher-level artiodactyl taxa. Systematic Biology 48:6–20.

Gatesy, J., P. O'Grady, and R. H. Baker. 1999b. Corroboration among data sets in simultaneous analysis: hidden support for phylogenetic relationships among higher-level artiodactyl taxa. Cladistics 15:271–313.

Gatesy, J., and M. S. Springer. 2017. Phylogenomic red flags: homology errors and zombie lineages in the evolutionary diversification of placental mammals. Proceedings of the National Academy of Sciences of the USA 114:E9431–9432.

Gauthier, J., A.G. Kluge, and T. Rowe. 1988. Amniote phylogeny and the importance of fossils. Cladistics 4:105–209.

GBIF (Global Biodiversity Information Facility). http://www.gbif.org/ [accessed 31 March 2020].

Geoffroy St. Hilaire, E. 1818. Philosophie Anatomique: des Organes Respiratoires sous le Rapport de la Determination et de l'Identité de Leurs Pieces Osseuses. Paris. (Available from Biodiversity Heritage Library.)

Ghiselin, M.T. 1966. On psychologism in the logic of taxonomic controversies. Systematic Zoology 15:207–215.

Ghiselin, M.T. 1969. The triumph of the Darwinian method. Berkeley: University of California Press.

Ghiselin, M.T. 1975. A radical solution to the species problem. Systematic Zoology 23:536–544.

Ghiselin, M.T. 1984. "Definition," "character," and other equivocal terms. Systematic Zoology 33:104–110.

Giangrande, A. 2003. Biodiversity, conservation, and the 'taxonomic impediment.' Aquatic Conservation: Marine and Freshwater Ecosystems 13:451–459.

Gippoliti, S. and C.P. Groves. 2013. "Taxonomic inflation" in the historical context of mammalogy and conservation. Hystrix, the Italian Journal of Mammalogy 23:8–11.

Gilbert, L.E. 2003. Adaptive novelty through introgression in *Heliconius* wing patterns: evidence for a shared genetic "tool box" from synthetic hybrid zones and a theory of diversification. In: Butterflies: Ecology and evolution taking flight, ed. C.L. Boggs, W.B. Watt, and P.R. Ehrlich, 281–318. Chicago: University of Chicago Press.

Gilmour, J.S.L. 1940. Taxonomy and philosophy. In: The new systematics, ed. J.S. Huxley, 461–74. Oxford: Oxford University Press.

Giribet, G., R. DeSalle, and W.C. Wheeler. 2002. 'Pluralism' and the aims of phylogenetic research. In: Molecular systematics and evolution: Theory and practice, ed. R. DeSalle, G. Giribet, and W.C. Wheeler, 141–46. Basel: Birkhäuser Verlag.

Giribet, G., and W.C. Wheeler. 2007. The case for sensitivity: a response to Grant and Kluge. Cladistics 23:294–296.

Gladstein, D. S. 1997. Efficient incremental character optimization. Cladistics 13:21–26.

Godfray, H.C.J. 2002. Challenges for taxonomy. Nature 417:17–19.

Godfray, H.C.J. 2007. Linnaeus in the information age. Nature 446:259–260.

Goldman, N. 1990. Maximum likelihood inference of phylogenetic trees, with special reference to a Poisson process model of DNA substitution and to parsimony analyses. Systematic Zoology 39:345–361.

Goldstein, P.Z., and R. DeSalle 2010. Integrating DNA barcode data and taxonomic practice: determination, discovery, and description. BioEssays 33:135–147.

Goloboff, P.A. 1991. Homoplasy and the choice among cladograms. Cladistics 7:215–232.

Goloboff, P.A. 1993. Estimating character weights during tree search. Cladistics 9:83–91.

Goloboff, P.A. 1994. Character optimization and calculation of tree lengths. Cladistics 9:433–436.

Goloboff, P.A. 1996. Methods for faster parsimony analysis. Cladistics 12:199–220.

Goloboff, P.A. 1998. Principios básicos de cladistica. Buenos Aires: Sociedad Argentina de Botanica.

Goloboff, P.A. 1999. Analyzing large data sets in reasonable times: solutions for composite optima. Cladistics 15:415–428.

Goloboff, P.A. 2003. Parsimony, likelihood, and simplicity. Cladistics 19:91–103.

Goloboff, P.A. 2005. Minority rule supertrees? MRP, compatibility, and minimum flip may display the least frequent groups. Cladistics 21:282–294.

Goloboff, P.A. 2016. NDM and VNDM: programs for the identification of areas of endemism, version 3.1. Program and documentation, available at http://www.lillo.org.ar/phylogeny.

Goloboff, P.A., J.M. Carpenter, J.S. Arias, and D.R.M. Esquivel. 2008. Weighting against homoplasy improves phylogenetic analysis of morphological data sets. Cladistics 24:758–773.

Goloboff, P.A., and S. A. Catalano. 2010. Phylogenetic morphometrics (II): algorithms for landmark optimization. Cladistics 27:42–51.

Goloboff, P.A., C. I. Mattoni, and A. S. Quinteros. 2006 Continuous characters analyzed as such. Cladistics 22:589–601.

Goodman, M., J. Czelusniak, G.W. Moore, A.E. Romero-Herrera, and G. Matsuda. 1979. Fitting the gene lineage into its species lineage: a parsimony strategy illustrated by cladograms constructed from globin sequences. Systematic Zoology 28:132–163.

Goodman, M., C.B. Olson, J.E. Beeber, and J. Czelusniak. 1985. New perspectives in the molecular biological analysis of mammalian phylogeny. Acta Zoologica Fennica 169:19–35.

Gould, S.J. 1965. Is uniformitarianism necessary? American Journal of Science 263:223–228.

Gould, S.J. 1974. The evolutionary significance of "bizarre" structures: antler size and skull size in the "Irish elk," *Megaloceros giganteus*. Evolution 28:191–220.

Gould, S.J. 1985. A clock of evolution: We finally have a method for sorting out homologies from "subtle as subtle can be" analogies. Natural History 94:12–25.

Grant, T. 2002. Testing methods: the evaluation of discovery operations in evolutionary biology. Cladistics 18:94–111.

Grant, T., and A.G. Kluge. 2003. Data exploration in phylogenetic inference: scientific, heuristic, or neither. Cladistics 19:379–418.

Grant, T., and A.G. Kluge. 2005. Stability, sensitivity, science and heurism. Cladistics 21:597–604.

Grant, T., and A.G. Kluge. 2008. Credit where credit is due: the Goodman–Bremer support metric. Molecular Phylogenetics and Evolution 49:405–406.

Gray, H. 1918. Anatomy of the human body. Philadelphia: Lea and Febiger.

Gries, C. E., E. Gilbert, and N. M. Franz. 2014. Symbiota—a virtual platform for creating voucher-based biodiversity information communities. Biodiversity Data Journal 2:e1114.

Groves, C.P., and P. Grubb. 2011. Ungulate taxonomy. Baltimore: Johns Hopkins University Press.

Guindon, S., and O. Gascuel. 2003. A simple, fast, and accurate algorithm to estimate large phylogenies by maximum likelihood. Systematic Biology 52:696–704.

Guttmann, W. F. 1977. Phylogenetic reconstruction: theory, methodology, and application to chordate evolution. In: Major patterns of vertebrate evolution, ed. M.K. Hecht, P.C. Goody, and B.M. Hecht, 645–669. New York: Plenum.

Haas, O., and G. G. Simpson. 1946. Analysis of some phylogenetic terms, with attempts at redefinition. Proceedings of the American Philosophical Society 90:319–349.

Hackett, S.J., R.T. Kimball, S. Reddy, R.C.K. Bowie, E.L. Braun, M.J. Braun, J.L. Chojnowski, et al. 2008. A phylogenomic study of birds reveals their evolutionary history. Science 320:1763–1768.

Haeckel, E. 1866. Generelle Morphologie der Organismen. Berlin: G. Reimer.

Haffer, J. 1997. Forward. Species concepts and species limits in ornithology. In: Handbook of the birds of the world. Vol. 4, Sandgrouse to cuckoos, ed. J. del Hoyo, A. Elliott, and J. Sargatal, 11–24. Barcelona: Lynx Ediciones.

Hafner, M.S., and S.A. Nadler. 1988. Phylogenetic trees support the coevolution of parasites and their hosts. Nature 332:258–259.

Hafner, M.S., and S.A. Nadler. 1990. Co-speciation in host-parasite assemblages: comparative analysis of rates of evolution and timing of cospeciation. Systematic Zoology 39:192–204.

Harold, A.S., and R.D. Mooi. 1994. Areas of endemism: definition and recognition criteria. Systematic Biology 43:261–266.

Harrison, R. G. 1998. Linking evolutionary pattern and process: the relevance of species concepts for the study of speciation. In: Endless forms: Species and speciation, ed. D.J. Howard and S.H. Berlocher, 19–31. Oxford: Oxford University Press.

Harvey, P. H., and M. D. Pagel. 1991. The comparative method in evolutionary biology. New York: Oxford University Press.

Hausmann, A., H.C.J. Godfray, P. Huemer, M. Mutanen, R. Rougerie, E.J. van Nieukerken, S. Ratnasingham, and P.D.N. Hebert. 2013. Genetic patterns in European geometrid moths revealed by the barcode index number (BIN) system. PLOS ONE 8:e84518.

Hay, O.P. 1902. Bibliography and Catalog of Fossil Vertebrata of North America. Bulletin of United States Geological Survey 179:1–239.

Heads, M. 2012. Molecular panbiogeography of the tropics. Berkeley: University of California Press.

Heath, T.A., J.P. Huelsenbeck, and T. Stadler. 2014. The fossilized birth-death process for coherent calibration of divergence-time estimates. Proceedings of the National Academy of Sciences of the USA 111:E2957–66.

Hebert, P.D.N., A. Cywinska, S. L. Ball, and J. R. deWaard. 2003a. Biological identifications through DNA barcodes. Proceedings of the Royal Society of London B 270:313–321.

Hebert, P.D.N., E.H. Penton, J.M. Burns, D.H. Janzen, and W. Hallwachs. 2004. Ten species in one: DNA barcoding reveals cryptic species in the neotropical skipper butterfly *Astraptes fulgerator*. Proceedings of the National Academy of Sciences of the USA 101:14812–14817.

Hebert, P.D.N., S. Ratnasingham, and J.R. deWaard. 2003b. Barcoding animal life: cytochrome c oxidase subunit 1 divergences among closely related species.

Proceedings of the Royal Society of London B (supplement), https://doi. org/10.1098/rsbl.2003.0025.

Hedges, S.B., and L.R. Maxson. 1996. Re: molecules and morphology in amniote phylogeny. Molecular Phylogenetics and Evolution 6:312–314.

Hedges, S.B., P.H. Parker, C.G. Sibley, and S. Kumar. 1996. Continental breakup and the ordinal diversification of birds and mammals. Nature 381:226–229.

Hedges, S.B., Q. Tao, M. Walker, and S. Kumar. 2018. Accurate timetrees require accurate calibrations. Proceedings of the National Academy of Sciences of the USA 115:E9510–9511.

Helfenbein, K.G., and R. DeSalle. 2005. Falsifications and corroborations: Karl Popper's influence on systematics. Molecular Phylogenetics and Evolution 35:271–280.

Heller, R., P. Frandsen, E.D. Lorenzen, and H.R. Siegismund. 2013. Are there really twice as many bovid species as we thought? Systematic Biology 62:490–493.

Heller, R., P. Frandsen, E.D. Lorenzen, and H.R. Siegismund. 2014. Is diagnosability an indicator of speciation? Response to "Why one century of phenetics is enough." Systematic Biology 63:833–837.

Hempel, C.G. 1966. Philosophy of natural science. Englewood Cliffs, NJ: Prentice Hall.

Hennig, W. 1950. Grundzüge einer Theorie der phylogenetischen Systematik. Berlin: Deutscher Zentralverlag.

Hennig, W. 1960. Die Dipteren-Fauna von Neuseeland als systematische und tiergeographisches Problem. Beiträge zur Entomologie 10:221–329. English translation by P. Wygodzinsky. 1966. The Diptera fauna of New Zealand as a problem in systematics and biogeography. Pacific Insects Monograph 9:1–81.

Hennig, W. 1965. Phylogenetic systematics. Annual Review of Entomology 10:97–116.

Hennig, W. 1966. Phylogenetic systematics. Urbana: University of Illinois Press.

Hennig, W. 1969. Die Stammesgeschichte der Insekten. Frankfurt am Mein: Waldemar Kramer.

Hennig, W. 1981. Insect phylogeny. New York: J. Wiley.

Henry, T.J., and A.G. Wheeler Jr. 1988. Family Miridae. In: Catalog of Heteroptera, or true bugs, of Canada and the continental United States, 251–520. Leiden: E. J. Brill.

Hey, J. 2001. Genes, categories, and species. Oxford: Oxford University Press.

Hibbard, T.N., M.S. Andrade-Díaz, and M.S. Díaz-Gómez. 2018. But they move! Vicariance and dispersal in southern South America: using two methods to reconstruct the biogeography of a clade of lizards endemic to South America. PLOS ONE 13(9):e0202339.

Higgins, D.G., and P.M. Sharp. 1988. CLUSTAL: a package for performing multiple sequence alignment on a microcomputer. Gene 73:237–244.

Hillis, D.M., C. Moritz, and B.K. Mable, eds. 1996. Molecular systematics, 2nd ed. Sunderland, MA: Sinauer.

Hinchliff, C.E., S.A. Smith, J.F. Allman, J.G. Burleigh, R. Chaudhary, L.M. Coghill, K.A. Crandall, et al. 2015. Synthesis of phylogeny and taxonomy into a comprehensive tree of life. Proceedings of the National Academy of Sciences of the USA. 112:12764–12769.

Hines, H.M., J.H. Hunt, T.K. O'Connor, J.J. Gillespie, and S.A. Cameron. 2007. Multigene phylogeny reveals eusociality evolved twice in vespid wasps. Proceedings of the National Academy of Sciences of the USA 104:3295–3299.

Ho, S.Y.W., K.J. Tong, C.S.P. Foster, A.M. Ritchie, N. Lo, and M.D. Crisp. 2015. Biogeographic calibrations for the molecular clock. Biology Letters 11:20150194.

Hoagland, K.E. 1996. The taxonomic impediment and the convention of biodiversity. Association of Systematics Collections Newsletter 24:61–62, 66–67.

Holder, M., and P.O. Lewis. 2003. Phylogeny estimation: traditional and Bayesian approaches. Nature Reviews Genetics 4:275–284.

Howard, D.J., and S.H. Berlocher, eds. 1998. Endless forms. Oxford: Oxford University Press.

Hudson, R.R. 1990. Gene genealogies and the coalescent process. Oxford Surveys in Evolutionary Biology 7:1–44.

Huelsenbeck, J.P. 1995. Performance of phylogenetic methods in simulation. Systematic Biology 44:17–48.

Huelsenbeck, J.P. 1997. Is the Felsenstein zone a fly trap? Systematic Biology 46:69–74.

Huelsenbeck, J.P., M.E. Alfaro, and M.A. Suchard. 2011. Biologically inspired phylogenetic models strongly outperform the no common mechanism model. Systematic Biology 60:225–233.

Huelsenbeck, J.P., B. Larget, R.E. Miller, and F. Ronquist. 2002. Potential application and pitfalls of Bayesian inference of phylogeny. Systematic Biology 51:673–688.

Hugot, J.-P. 2003. New evidence for hystricognath rodent monophyly from the phylogeny of their pinworms. In: Tangled trees: phylogeny, cospeciation, and coevolution, ed. R.D.M. Page, 44–173. Chicago: University of Chicago Press.

Hull, D.L. 1965. The effect of Essentialism on taxonomy: two thousand years of stasis. British Journal of the Philosophy of Science 15:314–326; 16:1–18. Reprinted in M. Ereshefsky, ed., 1992. The units of evolution. Essays on the nature of species. Cambridge, MA: MIT Press.

Hull, D.L. 1970. Contemporary systematic philosophies. Annual Review of Ecology and Systematics 1:19–54.

Hull, D.L. 1976. Are species really individuals? Systematic Zoology 25:174–191.

Hull, D.L. 1983. Karl Popper and Plato's metaphor. In: Advances in Cladistics. Vol. 2, Proceedings of the Second Meeting of the Willi Hennig Society, ed. N.I. Platnick and V.A. Funk, 177–89. New York: Columbia University Press.

Hume, D. 1748. An inquiry concerning human understanding. Reprint, Indianapolis: Bobbs-Merrill, 1955.

Humphries, C.J. 1981. Biogeographical methods and the southern beeches. In: The evolving biosphere, ed. P.L. Forey, 283–97. Cambridge: Cambridge University Press.

Humphries, C.J., and L.R. Parenti. 1986. Cladistic biogeography: Interpreting patterns of plant and animal distributions. Oxford: Oxford University Press.

Humphries, C.J., and L. R. Parenti. 1999. Cladistic biogeography: Interpreting patterns of plant and animal distributions, 2nd ed. Oxford: Oxford University Press.

Hurlbert, S.H. 1984. Pseudoreplication and the design of ecological field experiments. Ecological Monographs 54:187–211.

Hutchinson, G.E. 1965. The ecological theater and the evolutionary play. New Haven, CT: Yale University Press.

Huxley, J.S. 1957. The three types of evolutionary process. Nature 180:454–455.

Hymenoptera online. http://antbase.org/databases/hod.htm [accessed 31 March 2020].

Index animalium. Smithsonian Libraries, http://www.sil.si.edu/digitalcollections/indexanimalium/ [accessed 4 July 2019].

Index of organism names. ©2008 Thomson Reuters, http://www.organismnames.com [accessed 26 August 2008].

International code of nomenclature for algae, fungi and plants (Shenzhen Code). 2018. International Association for Plant Taxonomy, https://www.iapt-taxon.org/nomen/main.php [accessed 31 March 2020].

International code of nomenclature for cultivated plants. 2016. Scripta Horticulturae 18. Leuven, Belgium: International Society for Horticultural Science.

International code of nomenclature of prokaryotes (2008 revision). 2019. International Journal of Systematic and Evolutionary Microbiology 69:S1–111, http://www.ncbi.nlm.nih.gov/books/bv.fcgi?rid=icnb.TOC&depth=2 [accessed 5 January 2009].

International code of virus classification and nomenclature. 2018. International Committee on Taxonomy of Viruses, https://talk.ictvonline.org/information/w/ictv-information/383/ictv-code [accessed 31 March 2020].

International code of zoological nomenclature, 4th ed. 1999. International Trust for Zoological Nomenclature. London: Natural History Museum, https://www.iczn.org/the-code/the-international-code-of-zoological-nomenclature/the-code-online/ [accessed 31 March 2020].

International Commission on Zoological Nomenclature. 2008. Proposed amendment of the International Code of Zoological Nomenclature to expand and refine methods of publication. Zootaxa 1908:57–67.

International Commission on Zoological Nomenclature. 2012. Amendment of Articles 8, 9, 10, 21 and 78 of the International Code of Zoological Nomenclature to expand and refine methods of publication. Zookeys 2019:1–10.

International plant names index (IPNI). 2019. Royal Botanic Gardens, Kew, Harvard University Herbaria, and Australian National Herbarium, https://www.ipni.org/index.html [accessed 24 March 2020].

Janzen, D.H., W. Hallwachs, P. Blandin, J.M. Burns, J.M. Cadiou, I. Chacon, T. Dapkey, et al. 2009. Integration of DNA barcoding into an ongoing inventory of complex tropical biodiversity. Molecular Ecology Resources 9:1–26.

Jarvis, E.D., S. Mirarab, A.J. Aberer, B. Li, P. Houde, C. Li, S.Y. W.Ho, et al. 2014. Whole genomic analyses resolve early branches in the tree of life of modern birds. Science 346:1320–1331.

Jefferies, R.P.S. 1979. The origin of chordates: a methodological essay. In: The origin of major invertebrate groups, ed. M.R. House, 443–47. Systematics Association Special Volume 12. London: Academic Press.

Jetz, W., and R.A. Pyron. 2018. The interplay of past diversification and evolutionary isolation with present imperilment across the amphibian tree of life. Nature Ecology and Evolution 2:850–858.

Jobb, G., A. von Haeseler, and K. Strimmer. 2004. TREEFINDER: a powerful graphical analysis environment for molecular phylogenetics. BMC Evolutionary Biology 4:18, https://doi.org/10.1186/1471-2148-4-18.

Johnson, K., and D.H. Clayton. 2003. Coevolutionary history of ecology replicates: comparing phylogenies of wing and body lice to columbiform hosts. In: Tangled trees: Phylogeny, cospeciation, and coevolution, ed. R.D.M. Page, 262–86. Chicago: University of Chicago Press.

Johnson, W.E., D.P. Oronato, M.E. Roelke-Parker, E.D. Land, M. Cunningham, R.C. Belden, R. McBride, et al. 2010. Genetic restoration of the Florida panther. Science 329:1641–1645.

Judd, D.D. 1998. Exploring component stability using life-stage concordance in sabethine mosquitoes (Diptera: Culicidae). Cladistics 14:63–94.

Judd, W.S., C.S. Campbell, E.A. Kellogg, P.F. Stevens, and M.J. Donoghue. 2015. Plant systematics: A phylogenetic approach, 4th ed. Sunderland, MA: Sinauer.

Källersjö, M., V.A. Albert, and J.S. Farris. 1999. Homoplasy increases phylogenetic structure. Cladistics 15:91–94.

Kant, I. 1781. Critique of pure reason. 1965 edition., translated by N.K. Smith, New York: St. Martin's Press.

Kearney, M. 2002. Fragmentary taxa, missing data, and ambiguity: mistaken assumptions and conclusions. Systematic Biology 51:369–381.

Keller, R.A., R.N. Boyd, and Q.D. Wheeler. 2003. The illogical basis of phylogenetic nomenclature. Botanical Review 69:93–110.

Kew record of taxonomic literature. Royal Botanical Gardens in Kew, https://www.library.ucsb.edu/research/db/204 [accessed 4 July 2019].

Kidd, K.K., and L.A. Sgaramella-Zonta. 1971. Phylogenetic analysis: concepts and methods. American Journal of Human Genetics 23:235–252.

Kim, J. 1993. Improving the accuracy of phylogenetic estimation by combining different methods. Systematic Biology 42:331–340.

Kimball, R.T., C.H. Oliveros, N. Wang, N.D. White, F.K. Barker, D.J. Field, D.T. Ksepka, et al. 2019. A phylogenomic supertree of birds. Diversity 11:109.

Kimura, M. 1968. Evolutionary rate at the molecular level. Nature 217:624–626.

Kingman, J.F.C. 1982. On the genealogy of large populations. Journal of Applied Probability 19:27–43.

Kitching, I.J., P.L. Forey, C.J. Humphries, and D.M. Williams. 1998. Cladistics: The Theory and Practice of Parsimony Analysis. Oxford: Oxford University Press.

Kitching, R.L. 1993. Biodiversity and taxonomy: impediment or opportunity. In: Conservation biology in Australia and Oceania, ed C. Moritz and J. Kikkawa, 253–268. Chipping Norton, NSW: Surrey Beatty and Sons.

Kjer, K.M., C. Simon, M. Yavorskaya, and R.G. Beutel. 2016. Progress, pitfalls and parallel universes: a history of insect phylogenetics. Journal of the Royal Society Interface 13:20160363.

Kling, M.M., B.D. Mishler, A.H. Thornhill, B.G. Baldwin, and D.D. Ackerly. 2018. Facets of phylodiversity: evolutionary diversification, divergence and survival as conservation targets. Philosophical Transactions of the Royal Society of London B 374:20170397.

Klingenberg, C.P., and N.A. Gidaszewski. 2010. Testing and quantifying phylogenetic signals and homoplasy in morphometric data. Systematic Biology 59:245–261.

Kluge, A.G. 1983. Cladistics and the classification of the great apes. In: New interpretations of ape ancestry, ed. R.L. Ciochon and R.S. Corruccini, 151–77. New York: Plenum.

Kluge, A.G. 1985. Ontogeny and phylogenetic systematics. Cladistics 1:13–27.

Kluge, A.G. 1989. A concern for evidence and a phylogenetic hypothesis of relationships among *Epicrates* (Boidae, Serpentes). Systematic Zoology 38:7–25.

Kluge, A.G. 1997. Testability and the refutation and corroboration of cladistic hypotheses. Cladistics 13:81–96.

Kluge, A.G. 2001. Parsimony with and without scientific justification. Cladistics 17:199–210.

Kluge, A.G. 2003. The repugnant and the mature in phylogenetic inference: atemporal similarity and historical identity. Cladistics 19:356–368.

Kluge, A.G. 2005a. Testing lineage and comparative methods for inferring adaptation. Zoologica Scripta 34:653–663.

Kluge, A.G. 2005b. What is the rationale for 'Ockham's razor' (a.k.a. parsimony) in phylogenetic inference? In: Parsimony, phylogeny, and genomics, ed. V.A. Albert, 15–42. Oxford: Oxford University Press.

Kluge, A.G. 2007. Completing the neo-Darwinian synthesis with an event criterion. Cladistics 23:613–633.

Kluge, A.G. 2009. Explanation and falsification in phylogenetic inference: exercises in Popperian philosophy. Acta Biotheoretica 57:171–186.

Kluge, A.G., and J.S. Farris. 1969. Quantitative phyletics and the evolution of anurans. Systematic Zoology 18:1–32.

Kluge, A.G., and T. Grant. 2006. From conviction to anti-superfluity: old and new justifications of parsimony in phylogenetic influence. Cladistics 22:276–288.

Kluge, A.G., and A.J. Wolf. 1993. Cladistics: What's in a word? Cladistics 9:183–199.

Knowles, L.L. 2008. Why does a method that fails continue to be used? Evolution 62:2713–2717.

Kodandaramaiah, U. 2011. Tectonic calibrations in molecular dating. Current Zoology 57:116–124.

Kolaczkowski, B., and J.W. Thornton. 2004. Performance of maximum parsimony and likelihood phylogenetics when evolution is heterogeneous. Nature 431:980–984.

Kozak, K.M., N. Wahlberg, A.F. Neild, K.K. Dasmahapatra, J. Mallet, and C.D. Jiggins. 2015. Multilocus species trees show the recent adaptive radiation of the mimetic *Heliconius* butterflies. Systematic Biology 64:505–524.

Krell, F.-T. 2015. A mixed bag: when are early online publications available for nomenclatural purposes? Bulletin of Zoological Nomenclature 72:19–32.

Kristensen, N.P. 1981. Phylogeny of insect orders. Annual Review of Entomology 26:135–157.

Kubatko, L.S., and J.H. Degnan. 2007. Inconsistency of phylogenetic estimates from concatenated data under coalescence. Systematic Biology 56:17–24.

Kuhn, T.S. 1970. The structure of scientific revolutions, 2nd ed. Chicago: University of Chicago Press.

Kunz, W. 2012. Do species exist? Principles of taxonomic classification. Weinheim, Germany: Wiley-VCH.

Laamanen, T.R., R. Meier, M.A. Miller, A. Hille, and B.M. Wiegmann. 2005. Phylogenetic analysis of *Themira* (Sepsidae: Diptera): sensitivity analysis, alignment, and indel treatment in a multigene study. Cladistics 21:258–271.

Lakatos, I. 1974. Popper on demarcation and induction,. In: The philosophy of Karl Popper, ed. P.A. Schlipp, 241–73. La Salle, IL: Open Court.

Lamas, G., ed. 2004. Atlas of Neotropical Lepidoptera. Checklist: Part 4A Hesperioidea—Papiionoidea. Gainesville, FL: Scientific Publishers/Association of Tropical Lepidoptera.

Lanfear, R., P.B. Frandsen, A.M. Wright, T. Senfeld, and B. Calcott. 2017. PartitionFinder 2: new methods for selecting partitioned models of evolution for molecular and morphological phylogenetic analysis. Molecular Biology and Evolution 34:772–773.

Lanfear, R., S. Y. Ho, T.J. Davies, A.T. Moles, L. Aarssen, N.G. Swensen, N. Warman, A.E. Zanne, and A.P. Allen. 2013. Taller plants have lower rates of molecular evolution: the rate of mitosis hypothesis. Nature Communications 4:1879.

Lankester, E.R. 1870. On the use of the term homology in modern zoology, and the distinction between homogenetic and homoplastic agreements. Annals and Magazine of Natural History, ser. 4, 6:34–43.

Lanyon, S.M. 1985. Detecting internal inconsistencies in distance data. Systematic Zoology 34:397–403.

Lanyon, S.M. 1993. Phylogenetic frameworks: towards a firmer foundation for the comparative approach. Biological Journal of the Linnean Society 49:45–61.

Lartillot, N. 2015. Probabilistic models of eukaryotic evolution: time for integration. Philosophical Transactions of the Royal Society of London B 370:20140338.

La Salle, J., Q.D. Wheeler, P. Jackway, S. Winterton, D. Hobern, and D. Lovell. 2009. Accelerating taxonomic discovery through automated character extraction. Zootaxa 2217:43–55.

Lawrence, G.H.M. 1951. Taxonomy of vascular plants. New York: Macmillan.

Lee, M.S.Y., and A.F. Hugall. 2003. Partitioned likelihood support and the evaluation of data set conflict. Systematic Biology 52:15–22.

Lemmon, A.R., J.M. Brown, K. Stanger-Hall, and E.M. Lemmon. 2009. The effect of ambiguous data on phylogenetic estimates obtained by maximum likelihood and Bayesian inference. Systematic Biology 58:130–145.

Leonelli, S. 2016. Data-centric biology: A philosophical study. Chicago: University of Chicago Press.

Le Quesne, W.J. 1969. A method of selection of characters in numerical taxonomy. Systematic Zoology 18:201–205.

Lewis, P.O. 2001. A likelihood approach to estimating phylogeny from discrete morphological character data. Systematic Biology 50:913–925.

Lewontin, R.C. 1978. Adaptation. Scientific American 239:213–230.

Li, B., and G. Lecointre. 2009. Formalizing reliability in the taxonomic congruence approach. Zoologica Scripta 38:101–112.

Lienau, E.K., and R. DeSalle. 2009. Evidence, content and corroboration and the Tree of Life. Acta Biotheoretica 57:187–199.

Lin, Y.-H., P.A. McLenachan, A.R. Gore, M.J. Phillips, R. Ota, M.D. Hendy, and D. Penny. 2002. Four new mitochondrial genomes and the increased stability of evolutionary trees of mammals from improved taxon sampling. Molecular Biology and Evolution 19:2060–2070.

Linnaeus, C. 1737. Critica Botanica. Translated by A. Hort. London: Ray Society, 1938.

Linnaeus, C. 1753. Species Plantarum. Stockholm: Laurentii Salvii.

Linnaeus, C. 1758. Systema Naturae, 10th ed. Stockholm: Laurentii Salvii.

Lipscomb, D.L. 1992. Parsimony, homology, and the analysis of multistate characters. Cladistics 8:45–65.

Liu, K., S. Raghavan, S. Nelesen, C.R. Linder, and T. Warnow. 2009. Rapid and accurate large-scale coestimation of sequence alignments and phylogenetic trees. Science 324:1561–1564.

Liu, L., J. Zhang, F.E. Rheindt, F. Lei, Y. Qu, Y. Wang, Y. Zhang, et al. 2017. Genomic evidence reveals a radiation of placental mammals uninterrupted by the KPg boundary. Proceedings of the National Academy of Sciences of the USA 114:E7282–7290.

Loconte, H., and D.W. Stevenson. 1990. Cladistics of the Spermatophyta. Brittonia 42:197–211.

Loconte, H., and D.W. Stevenson. 1991. Cladistics of the Magnoliidae. Cladistics 7:267–296.

Löytynoja, A., and N. Goldman. 2008. Phylogeny-aware gap placement prevents errors in sequence alignment and evolutionary analysis. Science 320:1632–1635.

Lundberg, J., 1972. Wagner networks and ancestors. Systematic Zoology 21:398–413.

Lynch, M., and J.S. Conery. 2003. The origins of genome complexity. Science 302:1401–1404.

Mabberley, D.J. 2017. Mabberley's plant book: A portable dictionary of plants, their classification and uses, 4th ed. Cambridge: Cambridge University Press.

Mace, G.M. 2004. The role of taxonomy in species conservation. Philosophical Transactions of the Royal Society of London B 359:711–719.

Maddison, D.R. 1991. The discovery and importance of multiple islands of most parsimonious trees. Systematic Zoology 40:315–328.

Maddison, D.R., and W.P. Maddison. 1992. MacClade, version 3: Analysis of phylogeny and character evolution. Sunderland, MA: Sinauer.

Maddison, D.R., and K.-S. Schulz, eds. 2007. The Tree of Life web project, http://tolweb.org [accessed 24 March 2020].

Maddison, W.P. 1997. Gene trees and species trees. Systematic Biology 46:523–536.

Maddison, W.P., M.J. Donoghue, and D.R. Maddison. 1984. Outgroup analysis and parsimony. Systematic Zoology 33:83–103.

Margush, T., and F.R. McMorris. 1981. Consensus n-trees. Bulletin of Mathematical Biology 43:239–244.

Marko, P. 2002. Fossil calibration of molecular clocks and the divergence times of geminate species pairs separated by the Isthmus of Panama. Molecular Biology and Evolution 19:2005–2021.

Martill, D.M., H. Tischlinger, and N.R. Longrich 2015. A four-legged snake from the Early Cretaceous of Gondwana. Science 349:416–419.

Martins, E.P., ed. 1996. Phylogenies and the comparative method in animal behavior. Oxford: Oxford University Press.

Maslin, T.P. 1952. Morphological criteria of phyletic relationships. Systematic Zoology 1:49–70.

May, R.M. 1990. Taxonomy as destiny. Nature 347:129–130.

Mayden, R.L. 1988. Vicariance biogeography, parsimony, and evolution in North American freshwater fishes. Systematic Zoology 37:329–355.

Mayden, R.L. 1997. A hierarchy of species concepts: the denouement in the saga of the species problem. In: Species: The units of biodiversity, ed. M.F. Claridge, H.A. Dawah, and M.R. Wilson, 383–424. London: Chapman and Hall.

Mayr, E. 1942. Systematics and the origin of species. New York: Columbia University Press.

Mayr, E. 1965. Numerical phenetics and taxonomic theory. Systematic Zoology 14:73–97.

Mayr, E. 1969. Principles of systematic zoology. New York: McGraw-Hill.

Mayr, E. 1974. Cladistic analysis or cladistic classification? Zeitschrift für Zoologische Systematik und Evolutionsforschung 12:94–128.

Mayr, E. 1982. The growth of biological thought. Diversity, evolution, and inheritance. Cambridge, MA: Belknap Press of Harvard University Press.

Mayr, E. 1988. Toward a new philosophy of biology. Observations of an evolutionist. Cambridge, MA: Harvard University Press.

Mayr, E. 1997. This is biology: The science of the living world. Cambridge, MA: Belknap Press of Harvard University Press.

Mayr, E., E.G. Linsley, and R.L. Usinger. 1953. Methods and principles of systematic zoology. New York: McGraw-Hill.

McGlone, M. 2005. Goodbye Gondwana. Journal of Biogeography 32:739–740.

McIntyre, L. 2019. The scientific attitude: Defending science from denial, fraud, and pseudoscience. Cambridge, MA: MIT Press.

McKenna, M.C., and S.K. Bell. 1997. Classification of mammals above the species level. New York: Columbia University Press.

Meier, R. 1994. On the inappropriateness of presence/absence coding for non-additive multistate characters in computerized cladistic analysis. Zoologischer Anzeiger 232:201–212.

Meier, R., K. Shiyang, G. Vaidya, and P.K.L. Ng. 2006. DNA barcoding and taxonomy in Diptera: a tale of high intraspecific variability and low identification success. Systematic Biology 55:715–728.

Meireles, C.M., J. Czelusniak, M.P.C. Schneider, J.A.P.C. Muniz, M.C. Brigido, H.S. Ferreira, and M. Goodman. 1999. Molecular phylogeny of ateline New World monkeys (Platyrrhini, Atelinae) based on c globin gene sequences: evidence that Brachyteles is the sister group of Lagothrix. Molecular Phylogenetics and Evolution 12:10–30.

Meyer, C.P., and G. Paulay. 2005. DNA barcoding: error rates based on comprehensive sampling. PLOS Biology 3:e422.

Michener, C.D. 1953. Life history studies in insect systematics. Systematic Zoology 2:112–118.

Michener, C.D. 1963. Some future developments in taxonomy. Systematic Zoology 12:151–172.

Michener, C.D. 1977. Discordant evolution and the classification of allodapine bees. Systematic Zoology 26:32–56.

Mickevich, M.F. 1978. Taxonomic congruence. Systematic Zoology 27:143–158.

Mickevich, M.F. 1982. Transformation series analysis. Systematic Zoology 31:461–468.

Mickevich, M.F., and J.S. Farris. 1981. The implications of congruence in *Menidia*. Systematic Zoology 30:351–370.

Mickevich, M.F., and D. Lipscomb. 1991. Parsimony and the choice between different transformations of the same character set. Cladistics 7:111–139.

Mickevich, M.F., and C. Mitter. 1981. Treating polymorphic characters in systematics: a phylogenetic treatment of electrophoretic data. In: Advances in Cladistics. Proceedings of the First Meeting of the Willi Hennig Society, ed. V.A. Funk and D.R. Brooks, 45–58. New York: New York Botanical Garden.

Mickevich, M.F., and N.I. Platnick. 1989. On the information content of classifications. Cladistics 5:33–47.

Mill, J.S. 1843. A system of logic—ratiocinative and inductive. London: Longman.

Millar, T.R., P.B. Heenan, A.D. Wilton, R.D. Smissen, and I. Breitwieser. 2017. Spatial distribution of species, genus and phylogenetic endemism in the vascular flora of New Zealand, and implications for conservation. Australian Systematic Botany 30:134–147.

Miller, J.S. 1987. Host-plant relationships in the Papilionidae (Lepidoptera): parallel cladogenesis or colonization? Cladistics 3:105–120.

Miller, J.S., A.V.Z. Brower, and R. DeSalle. 1997. Phylogeny of the neotropical moth tribe Josiini (Notodontidae: Dioptinae): comparing and combining evidence from DNA sequences and morphology. Biological Journal of the Linnean Society 60:297–316.

Miller, J.S., and J.W. Wenzel. 1995. Ecological characters and phylogeny. Annual Review of Entomology 40:389–415.

Milne-Edwards, H. 1844. Considérations sur quelques principes relatifs à la classification naturelle des animaux. Annales des Sciences Naturelles, ser. 3, 1:65–99.

Mindell, D. P. 1991. Aligning DNA sequences: homology and phylogenetic weighting. In: Phylogenetic analysis of DNA sequences, ed. M.J. Miyamoto and J. Cracraft, 73–89. New York: Oxford University Press.

Mitchell, P.C. 1901. On the intestinal tract of birds; with remarks on the valuation and nomenclature of zoological characters. Transactions of the Linnean Society of London, ser. 2, Zoology 8:173–275.

Mitchell, R.J. 1967. "Both Sides Now." Los Angeles, CA: Reprise Records.

Mitter, C., B. Farrell, and B. Wiegmann. 1988. The phylogenetic study of adaptive zones: Has phytophagy promoted insect diversification? American Naturalist 132:107–128.

Mittermeier, R.A., P. Robles Gil, M. Hoffmann, J. Pilgrim, T. Brooks, C.G. Mittermeier, J, Lamoreux, and G.A.B. da Fonseca. 2005. Hotspots revisited: Earth's biologically richest and most endangered terrestrial ecoregions. Arlington, VA: Conservation International.

Miyamoto, M. M. 1985. Consensus cladograms and general classifications. Cladistics 1:186–189.

Miyamoto, M.M., and W.M. Fitch. 1995. Testing species phylogenies and phylogenetic methods with congruence. Systematic Biology 44:64–76.

Moeller, A.H., A. Caro-Quinteiro, D. Mjungu, A.V. Georgiev, E.V. Lonsdorf, M.N. Muller, A.E. Pusey, M. Peeters, B.H. Hahn, and H. Ochman. 2016. Cospeciation of gut microbiota with hominids. Science 353:380–382.

Morris, J., D. Hartl, A. Knoll, R. Lue, A. Berry, A. Biewener, B. Farrell, N.M. Holbrook, N. Pierce, and A. Veil. 2013. Biology: How life works. New York: W. H. Freeman.

Morris, J.L., M.N. Puttick, J.W. Clark, D. Edwards, P. Kenrick, S. Pressel, C.H. Wellman, Z. Yang, H. Schneider, and P.C.J. Donoghue. 2018. The timescale of early land plant evolution. Proceedings of the National Academy of Sciences of the USA 115:E2274–2283.

Morrone, J.J., and J.M. Carpenter. 1994. In search of a method for cladistic biogeography: an empirical comparison of component analysis, Brooks parsimony analysis, and three-area statements. Cladistics 10:99–153.

Mueller, U.G., S.A. Rehner, and T.R. Schultz. 1998. The evolution of agriculture in ants. Science 281:2034–2038.

Myers, N. 1988. Threatened biotas: "hot spots" in tropical forests. Environmentalist 8:187–208.

Myers, N., R.A. Mittermeier, C.G. Mittermeier, G.A.B. da Fonseca, and J. Kent. 2000. Cspots for conservation priorities. Nature 403:853–858.

Nagel, E. 1961. The structure of science: Problems in the Logic of Scientific Explanation. London: Routledge & Kegan Paul.

Nascimento, F.F., M. dos Reis, M., and Z. Yang. 2017. A biologist's guide to Bayesian phylogenetic analysis. Nature Ecology and Evolution 1:1446–1454.

National Science Foundation. 2019. Dimensions of biodiversity: Projects 2010–2017. https://www.nsf.gov/pubs/2019/nsf19019/nsf19019.pdf [accessed 30 March 2020].

NCBI (National Center for Biotechnology Information). 2018. National Institutes of Health, http://www.ncbi.nlm.nih.gov/genbank/ [accessed 23 March 2020].

Neave, S.A. 1939–1996. Nomenclator Zoologicus. 10 vols. London: Zoological Society of London, http://ubio.org/NomenclatorZoologicus/ / [version 0.86, accessed 31 March 2020].

Nelson, G. 1970. Outline of a theory of comparative biology. Systematic Zoology 19:373–384.

Nelson, G. 1971. Paraphyly and polyphyly: redefinitions. Systematic Zoology 20:471–472.

Nelson, G. 1972. Phylogenetic relationship and classification. Systematic Zoology 21:227–231.

Nelson, G. 1973. The higher-level phylogeny of vertebrates. Systematic Zoology 22:87–91.

Nelson, G. 1974. Classification as an expression of phylogenetic relationships. Systematic Zoology 22:344–359.

Nelson, G. 1978a. From Candolle to Croizat: comments on the history of biogeography. Journal of the History of Biology 11:269–305.

Nelson, G. 1978b. Ontogeny, phylogeny, paleontology, and the biogenetic law. Systematic Zoology 27:324–345.

Nelson, G. 1979. Cladistic analysis and synthesis: principles and definitions, with a historical note on Adanson's Familles des Plantes (1763–1764). Systematic Zoology 28:1–21.

Nelson, G. 1985. Outgroups and ontogeny. Cladistics 1:29–45.

Nelson, G., and P.Y. Ladiges. 1996. Paralogy in cladistic biogeography and analysis of paralogy-free subtrees. American Museum Novitates 3167:1–44.

Nelson, G., and N.I. Platnick. 1980. Multiple branching in cladograms: two interpretations. Systematic Zoology 29:86–91.

Nelson, G., and N.I. Platnick. 1981. Systematics and biogeography: Cladistics and vicariance. New York: Columbia University Press.

Nelson, G., and N.I. Platnick. 1991. Three-taxon statements: a more precise use of parsimony? Cladistics 7:351–366.

Nixon, K.C. 1999. The parsimony ratchet, a new method for rapid parsimony analysis. Cladistics 15:407–414.

Nixon, K.C. 2000. WinClada. Computer program, available from http://www.diversity oflife.org/winclada/ [accessed 4 July 2019].

Nixon, K.C., and J.M. Carpenter. 1993. On outgroups. Cladistics 9:413–426.

Nixon, K.C., and J.M. Carpenter. 1996. On simultaneous analysis. Cladistics 12:221–241.

Nixon, K.C., and J.M. Carpenter. 2000. On the other "phylogenetic systematics." Cladistics 16:298–318.

Nixon, K.C., and J.M. Carpenter. 2011. On homology. Cladistics 28:160–169.

Nixon, K.C., J.M. Carpenter, and D.W. Stevenson. 2003. The PhyloCode is fatally flawed, and the "Linnaean" system can be easily fixed. Botanical Review 69:111–120.

Nixon, K.C., and J.I. Davis. 1991. Polymorphic taxa, missing values and cladistic analysis. Cladistics 7:233–241.

Nixon, K.C., and Q.D. Wheeler. 1990. An amplification of the phylogenetic species concept. Cladistics 6:211–223.

Nixon, K.C., and Q.D. Wheeler. 1992. Measures of phylogenetic diversity. In: Extinction and phylogeny, ed. M.J. Novacek and Q.D. Wheeler, 216–234. New York: Columbia University Press.

Norell, M.A. 1992. Taxic origin and temporal diversity: the effect of phylogeny. In: Extinction and phylogeny, ed. M.J. Novacek and Q.D. Wheeler, 89–118. New York: Columbia University Press.

Nylin, S., S. Agosta, S. Bensch, W.A. Boeger, M.P. Braga, D.R. Brooks, M.L. Forister, et al. 2017. Embracing colonizations: a new paradigm for species association dynamics. Trends in Ecology and Evolution 33:4–14.

O'Brien, S.J., M.E. Roelke, N. Yuhke, K.W. Richards, W.E. Johnson, W.L. Franklin, A.E. Anderson, O.L. Bass Jr., R.C. Belden, and J.S. Martenson. 1990. Genetic introgression within the Florida panther Felis concolor coryi. National Geographic Research 6:485–494.

Ogden, T.H., and M.F. Whiting. 2003. The problem with "the Paleoptera problem": sense and sensitivity. Cladistics 19:432–442.

O'Grady, R.T., and G.B. Deets. 1987. Coding multistate characters, with special reference to the use of parasites as characters of their hosts. Systematic Zoology 36:268–279.

O'Grady, R.T., G.B. Deets, and G.W. Benz. 1989. Additional observations on nonredundant linear coding of multistate characters. Systematic Zoology 38:54–57.

Ohta, T. 1973. Slightly deleterious mutant substitutions in evolution. Nature 246:96–98.

O'Leary, M.A., J.I. Bloch, J.J. Flynn, T.J. Gaudin, A. Giallombardo, N.P. Giannini, S.L. Goldberg, et al. 2013. The placental mammal ancestor and the post–K-Pg radiation of placentals. Science 339:662–667.

O'Meara, B.C. 2012. Evolutionary inferences from phylogenies: a review of methods. Annual Reviews of Ecology, Evolution and Systematics 43:267–285.

Oosterbroek, P., and G. Courtney. 1995. Phylogeny of the nematocerous families of Diptera (Insecta). Zoological Journal of the Linnean Society 115:267–311.

O'Reilly, J.E., M. dos Reis, and P.C.J. Donoghue. 2015. Dating tips for divergence time estimation. Trends in Genetics 31:637–650.

Oreskes, N., K. Shrader-Frechette, and K. Belitz. 1994. Verification, validation, and confirmation of numerical models in the earth sciences. Science 263:641–646.

Owen, R. 1843. Lecture on the comparative anatomy and physiology of the invertebrate animals. London: Longman, Brown, Green, and Longman.

Page, R.D.M. 1988. Quantitative cladistic biogeography: constructing and comparing area cladograms. Systematic Zoology 37:254–270.

Page, R.D.M. 1989. Comments on component compatibility in historical biogeography. Cladistics 5:167–182.

Page, R.D.M. 1990a. Component analysis: a valiant failure? Cladistics 6:119–136.

Page, R.D.M. 1990b. Temporal congruence and cladistic analysis of biogeography and cospeciation. Systematic Zoology 39:205–226.

Page, R.D.M. 1994. Parallel phylogenies: reconstructing the history of host-parasite assemblages. Cladistics 10:155–173.

Pape, T., and N.L. Evenhuis, eds. 2019. Systema Dipterorum, version 2.4, http://sd.zoobank.org/ [accessed 3 July 2019].

Parr, C.S., N. Wilson, P. Leary, K.S. Schulz, K. Lans, L. Walley, J.A. Hammock, et al. 2014. The encyclopedia of life v2: Providing global access to knowledge about life on Earth. Biodiversity Data Journal 2:e1079. http://eol.org [accessed 24 March 2020].

Patterson, C. 1981. Significance of fossils in determining evolutionary relationships. Annual Review of Ecology and Systematics 12:195–223.

Patterson, C. 1982. Morphological characters and homology. In: Problems in phylogenetic reconstruction, ed. K.A. Joysey and A.E. Friday, 21–74. London: Academic Press.

Patterson, C. 1987. Introduction. In: Molecules and morphology in evolution. Conflict or compromise?, ed. C. Patterson, 1–22. Cambridge: Cambridge University Press.

Patterson, C. 1988. Homology in classical and molecular biology. Molecular Biology and Evolution 5:603–625.

Patterson, C., and D.E. Rosen. 1977. Review of ichthyodectiform and other Mesozoic teleost fishes and the theory and practice of classifying fossils. Bulletin of the American Museum of Natural History 158:1–172.

Patterson, D.J., J. Cooper, P.M. Kirk, R.L. Pyle, and D.P. Remsen. 2010. Names are the key to the big new biology. Trends in Ecology and Evolution 25:686–691.

Pease, J B., D. Haak, M.W. Hahn, and L.C. Moyle. 2016. Phylogenomics reveals three sources of adaptive variation during a rapid radiation. PLOS Biology 14:e1002379.

Pelham, J.P. 2008. A catalogue of the butterflies of the United States and Canada with a complete bibliography of the descriptive and systematic literature. Journal of Research on the Lepidoptera 40.

Penz, C.M. 2007. Evaluating the monophyly and phylogenetic relationships of Brassolini genera (Lepidoptera, Nymphalidae). Systematic Entomology 32:668–689.

Peterson, A.T., J. Soberón, R.G. Pearson, R.P. Anderson, E. Martínez-Meyer, M. Nakamura, and M.B. Araújo. 2011. Ecological niches and geographic distributions (MPB-49). Princeton, NJ: Princeton University Press.

Philippe, H., E.A. Snell, E. Bapteste, P. Lopez, P.W.H. Holland, and D. Casane. 2004. Phylogenomics of eukaryotes: impact of missing data on large alignments. Molecular Biology and Evolution 21:1740–1752.

Pickett, K.M., and J.M. Carpenter. 2010. Simultaneous analysis and the origin of eusociality in the Vespidae (Insecta: Hymenoptera). Arthropod Systematics and Phylogeny 68:3–33.

Pickett, K.M., and C.P. Randle. 2005. Strange Bayes indeed: uniform topological priors imply non-uniform clade priors. Molecular Phylogenetics and Evolution 34:203–211.

Piekarski, P.K., J.M. Carpenter, A.R. Lemmon, E. Moriarty Lemmon, and B.J. Sharanowski. 2018. Phylogenomic evidence overturns current conceptions of social evolution in wasps (Vespidae). Molecular Biology and Evolution 35:2097–2109.

Piel, W.H., M.J. Donoghue, M.J. Sanderson, and M. Walsh. 2007. TreeBASE: a database of phylogenetic knowledge, https://treebase.org/treebase-web/home.html [accessed 23 March 2018].

Pimentel, R.A., and R. Riggins. 1987. The nature of cladistic data. Cladistics 3:201–209.

Platnick, N.I. 1978. Classifications, historical narratives, and hypotheses. Systematic Zoology 27:365–369.

Platnick, N.I. 1979. Philosophy and the transformation of cladistics. Systematic Zoology 28:537–546.

Platnick, N.I. 1981a. Discussion of: Croizat's panbiogeography versus phylogenetic biogeography, by L. Z. Brundin. In: Vicariance biogeography: A critique, ed. G. Nelson and D.E. Rosen, 144–150. New York: Columbia University Press.

Platnick, N.I. 1981b. Widespread taxa and biogeographic congruence. In: Advances in Cladistics. Proceedings of the First Meeting of the Willi Hennig Society, ed. V.A. Funk and D.R. Brooks, 223–227. New York: New York Botanical Garden.

Platnick, N.I. 1982. Defining characters and evolutionary groups. Systematic Zoology 31:282–284.

Platnick, N.I. 1985. Philosophy and the transformation of cladistics revisited. Cladistics 1:87–94.

Platnick, N.I. 1991. On areas of endemism. Australian Systematic Botany 4:xi–xii.

Platnick, N.I. 1992. Patterns of biodiversity. In: Systematics, ecology, and the biodiversity crisis, ed. N. Eldredge, 15–24. New York: Columbia University Press.

Platnick, N.I. 1997. Advances in spider taxonomy, 1992–1995. New York: New York Entomological Society.

Platnick, N.I., and H.D. Cameron. 1977. Cladistic methods in textual, linguistic, and phylogenetic analysis. Systematic Zoology 26:380–385.

Platnick, N.I., J.A. Coddington, R.R. Forster, and C.E. Griswold. 1991. Spinneret morphology and the phylogeny of haplogyne spiders (Araneae, Araneomorphae). American Museum Novitates 3016:1–73.

Platnick, N.I., and E.S. Gaffney. 1978. Systematics and the Popperian paradigm. Systematic Zoology 27:381–388.

Platnick, N.I., C.J. Humphries, G. Nelson, and D.M. Williams. 1996. Is Farris optimization perfect? Three-taxon statements and multiple branching. Cladistics 12:243–352.

Platnick, N.I., and G. Nelson. 1978. A method of analysis for historical biogeography. Systematic Zoology 27:1–16.

Platnick, N.I., and G. Nelson. 1984. Composite areas in vicariance biogeography. Systematic Zoology 33:328–335.

Pleijel, F. 1995. On character coding for phylogeny reconstruction. Cladistics 11:309–315.

Polhemus, D. A. 1996. Island arcs, and their influence on Indo-Pacific biogeography. In: The origin and evolution of Pacific Island biotas, New Guinea to Eastern Polynesia: Patterns and processes, ed. A. Keast and S.E. Miller, 51–66. Amsterdam: SPB Academic.

Popper, K.R. 1957. The poverty of historicism. Boston: Beacon.

Popper, K.R. 1968. The logic of scientific discovery, 2nd English ed. New York: Harper and Row.

Popper, K.R. 1975. The rationality of scientific revolutions. In: Problems of scientific revolution: Progress and obstacles to progress in science, ed. R. Harre, 72–101. Herbert Spencer Lectures, 1973. Oxford: Clarendon.

Popper, K.R. 1979. Objective knowledge—an evolutionary approach. Oxford: Clarendon.

Popper, K.R. 1980. Evolution. New Scientist 87:611.

Popper, K.R. 1983. Realism and the aim of science. New York: Routledge.

Posada, D., and K. A. Crandall. 1998. Modeltest: testing the model of DNA substitution. Bioinformatics 14:817–818.

Posada, D., and K.A. Crandall. 2001. Intraspecific gene genealogies: trees grafting into networks. Trends in Ecology and Evolution 16:37–45.

Prendini, L. 2001. Species or supraspecific taxa as terminals in cladistic analysis? Groundplans versus exemplars revisited. Systematic Biology 50:290–300.

Prendini, L. 2005. Comments on "Identifying spiders through DNA barcodes." Canadian Journal of Zoology 83:498–504.

Prum, R.O., J.S. Berv, A. Dornburg, D.J. Field, J.P. Townsend, E.M. Lemmon, and A.R. Lemmon. 2015. A comprehensive phylogeny of birds (Aves) using targeted next-generation DNA sequencing. Nature 526:569–573.

Pulawski, W.J. 2008. Catalog of Sphecidae sensu lato. California Academy of Sciences, https://www.calacademy.org/scientists/projects/catalog-of-sphecidae [accessed 29 May 2019].

Pyrcz, T.W., A.V.L. Freitas, P. Boyer, F.M.S. Dias, D.R. Dolibana, E.P. Barbosa, L.M. Magaldi, O.H.H. Mielke, M.M. Casagrande,and J. Lorenc-Brudecka. 2018. Uncovered diversity of a predominantly Andean butterfly clade in the Brazilian Atlantic forest: a revision of the genus *Praepedaliodes* Forster (Lepidoptera: Nymphalidae, Satyrinae, Satyrini). Neotropical Entomology 47:211–255.

Qu, X.-J., J.-J. Jin, S.-M. Chaw, D.-Z. Li, and T.-S.Yi. 2017. Multiple measures could alleviate long-branch attraction in phylogenomic reconstruction of Cupressoideae (Cupressaceae). Scientific Reports 7:41005.

Quinn, A. 2016. Phylogenetic inference to the best explanation and the bad lot argument. Synthese 193:3025–3039.

Quinn, A. 2017. When is a cladist not a cladist? Biology and Philosophy 32:581–598.

Quinn, A. 2019. Diagnosing discordance: signal in data, conflict in paradigms. Philosophy, Theory and Practice in Biology 11:17.

Rabosky, D. L. 2014. Automatic detection of key innovations, rate shifts, and diversity-dependence on phylogenetic trees. PLOS ONE 9:e89543.

Ramirez, M.J., J.A. Coddington, W.P. Maddison, P.E. Midford, L. Prendini, J. Miller, C.E. Griswold, et al. 2007. Linking of digital images to phylogenetic data matrices using a morphological ontology. Systematic Biology 56:283–294.

Ramsay, G.W. 1986. The taxonomic impediment to conservation. Weta 9:60–62.

Raven, P.H., G.B. Johnson, K.A. Mason, J.B. Losos, and S.R. Singer. 2014. Biology, 10th ed. Boston: McGraw Hill.

Ree, R.H., B.R. Moore, and M.J. Donoghue. 2005. A likelihood framework for inferring the evolution of geographic range on phylogenetic trees. Evolution 59:2299–2311.

Ree, R.H., and S.A. Smith. 2008. Maximum likelihood inference of geographic range evolution by dispersal, local extinction, and cladogenesis. Systematic Biology 57:4–14.

Reichenbach, H., 1930. Die philosophische Bedeutung der modernen Physik. Erkenntnis 1:49–71.

Remane, A. 1952. Die Grundlagen des Naturlichen Systems der Vergleichenden Anatomie und der Phylogenetik. Leipzig: Geest und Portig K. G.

Rensch, B. 1954. Neuere Probleme der Abstammungslehre, 2nd ed. Stuttgart: Ferdinand Enke Verlag.

Reuter, O.M. 1910. Neue Beiträge zur Phylogenie und Systematik der Miriden nebst einleitenden Bemerkungen über die Phylogenie der Heteropteren-Familien. Acta Societatis Scientiarum Fennicae 37:1–169, 1 pl.

Richards, R.A. 2010. The species problem: A philosophical analysis. Cambridge: Cambridge University Press.

Rieppel, O. 1988. Fundamentals of comparative biology. Basel: Birkhäuser.

Rieppel, O. 2003. Popper and systematics. Systematic Biology 52:259–270.

Rieppel, O. 2005a. A skeptical look at justification. Cladistics 21:203–207.

Rieppel, O. 2005b. The philosophy of total evidence and its relevance for phylogenetic inference. Papeis Avulsos de Zoologia 45:77–89.

Rieppel, O. 2006a. On concept formation in systematics. Cladistics 22:474–492.

Rieppel, O. 2006b. The PhyloCode: a critical discussion of its theoretical foundation. Cladistics 22:186–197.

Rieppel, O. 2007a. Parsimony, likelihood, and instrumentalism in systematics. Biology and Philosophy 22:141–144.

Rieppel, O. 2007b. The metaphysics of Hennig's phylogenetic systematics: substance, events, and laws of nature. Systematics and Biodiversity 5:345–360.

Rieppel, O. 2007c. The performance of morphological characters in broad-scale phylogenetic analyses. Biological Journal of the Linnean Society 92:297–308.

Rieppel, O. 2010. Species as a systemic process. In: Fur eine Philosophie der Biologie, ed. I. Jahn and A. Wessel, 43–61. Munich: Kleine Verlag.

Rieppel, O. 2016. Phylogenetic systematics: Haeckel to Hennig. Boca Raton, FL: CRC Press.

Rieppel, O., and M. Kearney. 2002. Similarity. Biological Journal of the Linnean Society 75:59–82.

Rieppel, O., and M. Kearney. 2007. The poverty of taxonomic characters. Biology and Philosophy 22:95–113.

Rieppel, O., M. Rieppel, and L. Rieppel. 2006. Logic in systematics. Journal of Zoological Systematics and Evolutionary Research 44:186–192.

Rindal, E., and A.V.Z. Brower. 2011. Do model-based phylogenetic analyses perform better than parsimony? A test with empirical data. Cladistics 27:331–334.

Ripplinger, J., and J. Sullivan. 2008. Does choice in model selection affect maximum likelihood analysis? Systematic Biology 57:76–85.

Roch, S., and M. Steel. 2015. Likelihood-based tree construction on a concatenation of aligned sequence data sets can be statistically inconsistent. Theoretical Population Biology 100:56–62.

Rodman, J.E. and J.H. Cody. 2003. The taxonomic impediment overcome: NSF's partnerships for enhancing expertise in taxonomy (PEET) as a model. Systematic Biology 52:428–435.

Rohlf, F.J. 1963. Congruence of larval and adult classifications in *Aedes* (Diptera: Culicidae). Systematic Zoology 12:97–117.

Rokas, A., B.L. Williams, N. King, and S.B. Carroll. 2003. Genome-scale approaches to resolving incongruence in molecular phylogenies. Nature 425:798–803.

Romer, A.S. 1962. Bibliography of fossil vertebrates exclusive of North America. Memoirs of the Geological Society of America 87:1509–1927 (2 vols.).

Ronquist, F. 1996. DIVA. Version 1.0. Computer program for MacOS and Win32, http://www.ebc.uu.se/systzoo/research/diva/diva.html [accessed 5 July 2019].

Ronquist, F. 1997. Dispersal-vicariance analysis: a new biogeographic approach to the quantification of historical biogeography. Systematic Biology 46:195–203.

Ronquist, F. 2002. TREEFITTER, version 1.3 b. Software and user manual, http://www. ebc. uu. se/systzoo/research/treefitter/treefitter. html [accessed 27 March 2020].

Ronquist, F. 2003. Parsimony analysis of coevolving species associations. In: Tangled trees: Phylogeny, cospeciation, and coevolution, ed. R.D.M. Page, 22–64. Chicago: University of Chicago Press.

Ronquist, F., S. Klopfstein, L. Vilhelmsen, S. Schulmeister, D.L. Murray, and A.P. Rasnitsyn. 2012. A total-evidence approach to dating with fossils, applied to the early radiation of the Hymenoptera. Systematic Biology 61:973–999.

Ronquist, F., and I. Sanmartin. 2011. Phylogenetic methods in biogeography. Annual Review of Ecology and Systematics 42:441–464.

Rosen, D.E. 1975. A vicariance model of Caribbean biogeography. Systematic Zoology 24:431–464.

Rosen, D.E. 1978. Vicariant patterns and historical explanation in biogeography. Systematic Zoology 27:159–188.

Rosen, D.E. 1979. Fishes from the uplands and intermontane basins of Guatemala: revisionary studies and comparative geography. Bulletin of the American Museum of Natural History 162:267–375.

Rosen, D.E. 1982. Do current theories of evolution satisfy the basic requirements of explanation? Systematic Zoology 31:76–85.

Ross, H. 1974. Biological systematics. Reading, MA: Addison-Wesley.

Rowe, D.L., K.A. Dunn, R.M. Adkins, and R.L. Honeycutt. 2010. Molecular clocks keep dispersal hypotheses afloat: evidence for trans-Atlantic rafting by rodents. Journal of Biogeography 37:305–324.

Rowe, T. 1987. Definition and diagnosis in the phylogenetic system. Systematic Zoology 36:208–211.

Russell, G.E.G. 1985. Analysis of the size and composition of the southern African flora. Bothalia 15:613–629.

Saitou, N., and M. Nei. 1987. The neighbor-joining method: a new method for reconstructing phylogenetic trees. Molecular Biology and Evolution 4:406–425.

Sanderson, M.J. 1997. A nonparametric approach to estimating divergence times in the absence of rate constancy. Molecular Biology and Evolution 14:1218–1231.

Sanderson, M.J., A. Purvis, and C. Henze. 1998. Phylogenetic supertrees: Assembling the tree of life. Trends in Ecology and Evolution 13:105–109.

Sankoff, D., and R. Cedergren. 1983. Simultaneous comparison of three or more sequences related by a tree. In: Time warps, string edits, and macromolecules: The theory and practice of sequence comparison, ed. D. Sankoff and J. Kruskall, 253–264. Reading, MA: Addison-Wesley.

Sankoff, D., and P. Rousseau. 1975. Locating the vertices of a Steiner tree in an arbitrary space. Mathematical Programming 9:240–246.

Sanmartin, I. 2007. Event-based biogeography: integrating patterns, processes, and time. In: Biogeography in a changing world, ed. M.C. Ebach and R.S. Tangney, 135–159. Boca Raton, FL: CRC Press.

Sanmartin, I., and F. Ronquist. 2004. Southern Hemisphere biogeography inferred from event-based models: plant versus animal patterns. Systematic Biology 53:216–243.

Sarich V.M., C.W. Schmid, and J. Marks. 1989. DNA hybridization as a guide to phylogenies: a critical analysis. Cladistics 5:3–32.

Särkinen T., L. Bohs, R.G. Olmstead, and S. Knapp. 2013. A phylogenetic framework for evolutionary study of the nightshades (Solanaceae): a dated 1000-tip tree. BMC Evolutionary Biology 13:214.

Sauquet, H. 2013. A practical guide to molecular dating. Comptes Rendue Palevol 12:355–367.

Sauquet, H., S.Y. W. Ho, M.A. Gandolfo, G.J. Jordan, P. Wilf, D.J. Cantrill, M.J. Bayly, et al. 2012. Testing the impact of calibration on molecular divergence times using a fossil-rich group: the case of *Nothofagus* (Fagales). Systematic Biology 61:289–313.

Schmidt-Lebuhn, A. N. 2012. Fallacies and false premises—a critical assessment of the arguments for the recognition of paraphyletic taxa in botany. Cladistics 28:174–187.

Schmitt, M. 2013. From taxonomy to phylogenetics—the life and work of Willi Hennig. Leiden: Brill.

Schmitz, J., and R.F.A. Moritz. 1998. Molecular phylogeny of Vespidae (Hymenoptera) and the evolution of sociality in wasps. Molecular Phylogenetics and Evolution 9:183–191.

Schoch, R.M. 1986. Phylogeny reconstruction in paleontology. New York: Van Nostrand Reinhold.

Schröder, E., 1870. Vier combinatorische probleme. Zeitschrift fuer Mathematik und Physik 15:361–376.

Schuh, R.T. 1991. Phylogenetic, host, and biogeographic analyses of the Pilophorini (Heteroptera: Miridae: Phylinae). Cladistics 7:157–189.

Schuh, R.T. 2003. The Linnaean System and its 250-year persistence. Botanical Review 69:59–78.

Schuh, R.T. 2006. Revision, phylogenetic, biogeographic, and host analyses of the endemic western North American *Phymatopsallus* group, with the description of 9 new genera and 15 new species (Insecta: Hemiptera: Miridae: Phylinae). Bulletin of the American Museum of Natural History 301:1–115.

Schuh, R.T. 2002–2013. On-line systematic catalog of plant bugs (Insecta: Heteroptera: Miridae), version 2.2, http://research.amnh.org/pbi/catalog/ [accessed 30 March 2020].

Schuh, R.T., G. Cassis, K. Seltmann, and C.A. Johnson. 2004–2014. Arthropod easy capture specimen database, http://research.amnh.org/pbi/locality/ [accessed 30 March 2020].

Schuh, R.T., and J.S. Farris. 1981. Methods for investigating taxonomic congruence and their application to the Leptopodomorpha. Systematic Zoology 30:331–351.

Schuh, R.T., and K.L. Menard. 2011. Santalalean-feeding plant bugs: ten new species in the genus *Hypseloecus* Reuter from Australia and South Africa (Heteroptera: Miridae: Phylinae): their hosts and placement in the Pilophorini. Australian Journal of Entomology 50:365–392.

Schuh, R.T., and J. T. Polhemus. 1980. Analysis of taxonomic congruence among morphological, ecological, and biogeographic data sets for the Leptopodomorpha (Hemiptera). Systematic Zoology 29:1–26.

Schuh, R.T., and G. M. Stonedahl. 1986. Historical biogeography in the Indo-Pacific: a cladistic approach. Cladistics 2:337–355.

Schuh, R.T., and C. Weirauch. 2020. True bugs of the world (Hemiptera: Heteroptera). Natural history and classification, 2nd ed. Manchester, UK: Siri Scientific.

Schuh, R.T., C. Weirauch, and W.C. Wheeler. 2009. Phylogenetic relationships within the Cimicomorpha (Hemiptera: Heteroptera): a total evidence analysis. Systematic Entomology 34:15–48.

Schulmeister, S., W.C. Wheeler, and J.M. Carpenter. 2002. Simultaneous analysis of the basal lineages of Hymenoptera (Insecta) using sensitivity analysis. Cladistics 18:455–484.

Schwartz, M.D., C. Weirauch, and R.T. Schuh. 2018. New genera and species of Myrtaceae-feeding Phylinae from Australia, and the description of a new species of *Restiophylus* (Insecta: Heteropera: Miridae). Bulletin of the American Museum of Natural History 424:1–157.

Sclater, P.L. 1858. On the general geographical distribution of the members of the class Aves. Journal of the Proceedings of the Linnean Society of London, Zoology 2:130–145.

Scotland, R.W., R.G. Olmstead, and J.R. Bennett. 2003. Phylogeny reconstruction: the role of morphology. Systematic Biology 52:539–548.

Sereno, P.C. 2007. Logical basis for morphological characters in phylogenetics. Cladistics 23:565–587.

Sharkey, M.J. 1989. A hypothesis-independent method of character weighting for cladistic analysis. Cladistics 5:63–86.

Sherborne, C.D. 1902–1932. Index Animalium. 9 vols. London: British Museum (Natural History), https://www.sil.si.edu/DigitalCollections/indexanimalium/taxonomicnames/ [accessed 31 March 2020].

Sibley, C.G., and J.E. Ahlquist. 1990. Phylogeny and classification of birds: A study in molecular evolution. New Haven, CT: Yale University Press.

Sibley, C.G., and B.L. Monroe Jr. 1990. Distribution and taxonomy of birds of the world. New Haven, CT: Yale University Press.

Siddall, M.E. 1995. Another monophyly index: revisiting the jackknife. Cladistics 11:33–56.

Siddall, M.E. 1998. Success of parsimony in the four-taxon case: long branch repulsion by likelihood in the Farris zone. Cladistics 14:209–220.

Siddall, M.E., and A.G. Kluge. 1997. Probabilism and phylogenetic inference. Cladistics 13:313–336.

Simmons, M.P., 2012. Misleading results of likelihood-based phylogenetic analyses in the presence of missing data. Cladistics 28:208–222.

Simmons, M.P., and P.A. Goloboff. 2014. Dubious resolution and support from sparse supermatrices: the importance of thorough tree searches. Molecular Phylogenetics and Evolution 78:334–348.

Simmons, M.P., K.M. Pickett, and M. Miya. 2004. How meaningful are Bayesian support values? Molecular Biology and Evolution 21:188–199.

Simonsen, T.J. 2006. Fritillary phylogeny, classification, and larval host plants: reconstructed mainly on the basis of male and female genitalic morphology (Lepidoptera: Nymphalidae: Argynnini). Biological Journal of the Linnean Society 89:627–673.

Simonsen, T., A.V.Z. Brower, N. Wahlberg, and R. de Jong. 2006. Morphology, molecules and fritillaries: approaching a stable phylogeny for Argynnini (Lepidoptera: Nymphalidae). Insect Systematics and Evolution 37:405–418.

Simpson, G.G. 1951. The species concept. Evolution 5:285–298.

Simpson, G.G. 1961. Principles of animal taxonomy. New York: Columbia University Press.

Slater, J.A. 1964. A catalogue of the Lygaeidae of the world. 2 vols. Storrs: University of Connecticut.

Slater, J.A., and D.B. Wilcox. 1973. The chinch bugs or Blissinae of South Africa (Hemiptera: Lygaeidae). Memoirs of the Entomological Society of Southern Africa 12:135.

Slowinski, J.B. 1998. The number of multiple alignments. *Molecular Phylogenetics and Evolution* 10:264–266.

Smith, V.S., R.D. Page, and K.P. Johnson. 2004. Data incongruence and the problem of avian louse phylogeny. Zoologica Scripta 33:239–259.

Sober, E.R. 1983. Parsimony methods in systematics. In: Advances in Cladistics. Vol. 2, Proceedings of the Second Meeting of the Willi Hennig Society, ed. N.I. Platnick and V.A. Funk, 37–47. New York: Columbia University Press.

Sober, E.R. 1985. A likelihood justification of parsimony. Cladistics 1:209–233.

Sober, E.R. 1988. Reconstructing the past: Parsimony, evolution, and inference. Cambridge, MA: MIT Press.

Sober, E.R. 1993. Experimental tests of phylogenetic inference methods. Systematic Biology 42:85–89.

Sober, E.R. 2015. Ockham's razors: A user's manual. Cambridge: Cambridge University Press.

Sokal, R.R., and F.J. Rohlf. 1981. Taxonomic congruence in the Leptopodomorpha re-examined. Systematic Zoology 30:309–325.

Sokal, R.R., and P.H.A. Sneath. 1963. Principles of numerical taxonomy. San Francisco: W.H. Freeman.

Soltis, D.E., P.S. Soltis, P.K. Endress, M.W. Chase, S. Manchester, W. Judd, L. Majure, and E. Madroviev. 2017. Phylogeny and evolution of the angiosperms. Chicago: University of Chicago Press.

Soltis, D.E., P.S. Soltis, M.E. Mort, M.W. Chase, V. Savolainen, S.B. Hoot, and C.M. Morton. 1998. Inferring complex phylogenies using parsimony: an empirical approach using three large DNA data sets for angiosperms. Systematic Biology 47:32–42.

Soltis, D.E., C.J. Visger, D.B. Marchant, and P.S. Soltis. 2016. Polyploidy: pitfalls and paths to a paradigm. American Journal of Botany 103:1–21.

Sorensen, J.T., B.C. Campbell, R.J. Gill, and J.D. Steffan-Campbell. 1995. Non-monophyly of Auchenorrhyncha (Homoptera), based upon 18S rDNA phylogeny: eco-evolutionary and cladistic implications within pre-Heteropterodea Hemiptera (s. l.) and a proposal for new monophyletic suborders. Pan-Pacific Entomologist 71:31–60.

SpeciesFile. http://software.speciesfile.org/HomePage/Software/SoftwareHomePage.aspx [accessed 24 March 2020].

Sperling, F.A.H. 2004. DNA barcoding: Deus ex machina. Newsletter of the Biological Survey of Canada (Terrestrial Arthropods) 22:50–53.

Sperling, F.A.H., G.S. Anderson, and D.A. Hickey. 1994. A DNA-based approach to the identification of insect species used for postmortem interval estimation. Journal of Forensic Sciences 39:418–427.

Springer, M.S., and J. Gatesy 2016. The gene tree delusion. Molecular Phylogenetics and Evolution 94:1–33.

Stamatakis, A. 2006. RAxML-VI-HPC: Maximum likelihood-based phylogenetic analyses with thousands of taxa and mixed models. Bioinformatics 22:2688–2690.

Stamatakis, A. 2014. RAxML version 8: a tool for phylogenetic analysis and post-analysis of large phylogenies. Bioinformatics 30:1312–1313.

Stamos, D.N. 1996. Popper, falsifiability, and evolutionary biology. Biology and Philosophy 11:161–191.

Standley, P.C. 1922. Trees and shrubs of Mexico. Contribution of the United States Herbarium, vol. 23. Washington, DC: Smithsonian Institution.

Steel, M. 1994. The maximum likelihood point for a phylogenetic tree is not unique. Systematic Biology 43:560–564.

Steel, M. 2002. Some statistical aspects of the maximum parsimony method. In: Molecular systematics and evolution: theory and practice, ed. R. DeSalle, G. Giribet, and W.C. Wheeler, 125–139. Basel: Birkhäuser Verlag.

Sterner, B., and S. Lidgard. 2018. Moving past the systematics wars. Journal of the History of Biology 51:31–67.

Stonedahl, G.M. 1988. Revisions of *Dioclerus, Harpedona, Mertila, Myiocapsus, Prodromus,* and *Thaumastomiris* (Heteroptera: Miridae, Bryocorinae: Eccritotarsini). Bulletin of the American Museum Natural History 187:1–99.

Stonedahl, G.M. 1990. Revision and cladistic analysis of the Holarctic genus *Atractotomus* Fieber (Heteroptera: Miridae: Phylinae). Bulletin of the American Museum Natural History 198:1–88.

Streicher, J.W., J.A. Schulte II, and J.J. Wiens. 2016. How should genes and taxa be sampled for phylogenomic analyses with missing data? An empirical study in iguanian lizards. Systematic Biology 65:128–145.

Strickland, H.E. 1837. On the inexpediency of altering established terms in natural history. Magazine of Natural History, ser. 2, 1:127–131.

Strickland, H.E. 1841. On the true method of discovering the natural system in zoology and botany. Annals and Magazine of Natural History, ser. 1, 6:184–194.

Strickland, H.E. 1842. Rules for zoological nomenclature. Report of the 12th meeting of the British Association held at Manchester in 1842. Reports of the British Association for the Advancement of Science 1842:105–121.

Suzuki, Y., G.V. Glazko, and M. Nei. 2002. Overcredibility of molecular phylogenies obtained by Bayesian phylogenetics. Proceedings of the National Academy of Sciences of the USA 99:16138–16143.

Swenson, N.G. 2019. Phylogenetic ecology: A history, critique and remodelling. Chicago, University of Chicago Press.

Swofford, D.L. 1991. When are phylogeny estimates from molecular and morphological data incongruent. In: Phylogenetic analysis of DNA sequences, ed. M.M. Miyamoto and J. Cracraft, 295–333. Oxford: Oxford University Press.

Swofford, D.L. 2003. PAUP* phylogenetic analysis using parsimony (*and other methods). Sunderland, MA: Sinauer. (As of June 2019, the current beta version was 4.0a165.)

Swofford, D.L., and W.P. Maddison. 1987. Reconstructing ancestral character states under Wagner parsimony. Mathematical Bioscience 87:199–229.

Swofford, D.L., and W.P. Maddison. 1992. Parsimony, character-state reconstructions, and evolutionary inferences. In: Systematics, historical ecology, and North American freshwater fishes, ed. R.L. Mayden, 186–223. Stanford, CA: Stanford University Press.

Swofford, D.L., and G. J. Olsen. 1990. Phylogeny reconstruction. In: Molecular systematics, ed. D.M. Hillis, and C. Moritz, 411–501. Sunderland, MA: Sinauer.

Swofford, D.L., G. J. Olsen, P. J. Waddell, and D. M. Hillis. 1996. Phylogenetic inference. In: Molecular systematics, 2nd ed., ed. D.M. Hillis, C. Moritz, and B.K. Mable, 407–514. Sunderland, MA: Sinauer.

Symonds, C.L., and G. Cassis. 2018. Systematics and analysis of the radiation of Orthotylini plant bugs associated with callitroid conifers in Australia. Bulletin of the American Museum of Natural History 422:1–226.

Szumik, C.A., F. Cuezzo, P.A. Goloboff, and A. Chalup. 2002. An optimality criterion to determine areas of endemism. Systematic Biology 51:806–816.

Szumik, C.A., and P.A. Goloboff. 2004. Areas of endemism: an improved optimality criterion. Systematic Biology 53:968–977.

Szumik, C.A., and P.A. Goloboff. 2015. Higher taxa and the identification of areas of endemism. Cladistics 31:568–572.

Szumik, C., V.V. Pereyra, and M.D. Casagranda. 2019. Areas of endemism: to overlap or not to overlap, that is the question. Cladistics 35:198–229.

Tajima, F. 1983. Evolutionary relationships of DNA sequences in finite populations. Genetics 105:437–460.

Tarver, J.E., M. Dos Reis, S. Mirarab, R.J. Moran, S. Parker, J.E. O'Reilly, B.L. King, et al. 2016. The interrelationships of placental mammals and the limits of phylogenetic inference. Genome Biology and Evolution 8:330–344.

Tchernov E., O. Rieppel, H. Zaher, M.J. Polcyn, and L. L. Jacobs. 2000. A fossil snake with limbs. Science 287:2010–2012.

Templeton, A.R. 2004. Statistical phylogeography: methods of evaluating and minimizing inference errors. Molecular Ecology 13:789–809.

Templeton, A.R. 2008. Nested clade analysis: an extensively validated method for strong phylogeographic inference. Molecular Ecology 17:1877–1880.

Thiele, K. 1993. The Holy Grail of the perfect character: the cladistic treatment of morphometric data. Cladistics 9:275–304.

Thiers, B. 2018 [continuously updated]. Index Herbariorum: A global directory of public herbaria and associated staff. New York Botanical Garden's Virtual Herbarium, http://sweetgum.nybg.org/science/ih/.

Thorpe, R.S. 1984. Coding morphometric characters for constructing distance Wagner networks. Evolution 38:244–255.

Toon, A., L.G. Cook, and M.D. Crisp. 2014. Evolutionary consequences of shifts to bird-pollination in the Australian pea-flowered legumes (Mirbelieae and Bossiaeeae). BMC Evolutionary Biology 14:43.

Toussaint, E.F A., J.W Breinholt, C. Earl, A.D. Warren, A.V.Z. Brower, M. Yago, K.M. Dexter, et al. 2018. Anchored phylogenomics illuminates the skipper butterfly tree of life. BMC Evolutionary Biology 18:101.

Townsend, J.P. 2007. Profiling phylogenetic informativeness. Systematic Biology 56:222–231.

Tree of Life. http://www.tolweb.org/ [accessed 15 September 2008].

Trueman, J.W.H., B.E. Pfeil., S.A. Kelchner, and D.K. Yeates. 2004. Did stick insects *really* regain their wings? Systematic Entomology 29:138–139.

Tuffley, C., and M. Steel. 1997. Links between maximum likelihood and maximum parsimony under a simple model of site substitution. Bulletin of Mathematical Biology 59:581–607.

Turner, J.R.G. 1967. Goddess changes sex, or the gender game. Systematic Zoology 16:349–350.

Tuxen, S.L., ed. 1970. Taxonomist's glossary of genitalia of insects, 2nd ed. Darien, CT: S-H Service Agency.

Ussher, J. 1650. Annales Veteris Testamenti, a prima mundi origine deducti, una cum rerum Asiaticarum et Aegyptiacarum chronico, a temporis historici principio usque ad Maccabaicorum initia producto. London: Flesher, Crook and Baker.

Uyeda, J.C., R. Zenil-Ferguson, and M.W. Pennell. 2018. Rethinking phylogenetic comparative methods. Systematic Biology 67:1091–1109.

VandeWall, H. 2007. Why water is not H_2O, and other critiques of essentialist ontology from the philosophy of chemistry. Philosophy of Science 74:906–919.

Vane-Wright, R.I., C.J. Humphries, and P.H. Williams. 1991. What to protect?—systematics and the agony of choice. Biological Conservation 55:235–254.

Van Fraassen, B.C. 1980. The scientific image. Oxford: Clarendon.

Varón, A., L.S. Vinh, I. Bomash, and W.C. Wheeler. 2007. POY 4.0 Beta 1908. American Museum of Natural History, http://research.amnh.org/scicomp/projects/poy.php.

Velasco, J.D. 2013. Philosophy and phylogenetics. Philosophy Compass 8:990–998.

Vergara-Silva, F. 2009. Pattern cladistics and the 'realism-antirealism debate' in the philosophy of biology. Acta Biotheoretica 57:269–294.

Vermeij, G.J. 1993. The biological history of a seaway. Science 260:1603–1604.

VertNet. 2019. http://www.vertnet.org/index.html [accessed 24 March 2020].

Vogt, L. 2007. A falsificationist perspective on the usage of process frequencies in phylogenetics. Zoologica Scripta 36:395–407.

Vogt, L. 2008. The unfalsifiability of cladograms and its consequences. Cladistics 24:62–73.

Vogt, L. 2014. Popper and phylogenetics, a misguided rendezvous. Australian Systematic Botany 27:85–94.

von Baer, K. 1828. Über Entwickelungsgeschichte der Thiere: Beobachtung und Reflexion. Königsberg: Bornträger.

Voris, H.K. 2000. Maps of Pleistocene sea levels in Southeast Asia: Shorelines, river systems and time durations. Journal of Biogeography 27:1153–1167.

Vrana, P., and W.C. Wheeler. 1992. Individual organisms as terminal entities: laying the species problem to rest. Cladistics 8:67–72.

Wagner, W.H., Jr. 1961. Problems in the classification of ferns. Recent Advances in Botany 1:841–844.

Wahlberg, N., M.F. Braby, A.V.Z. Brower, R. de Jong, M.-M. Lee, S. Nylin, N.E. Pierce, et al. 2005. Synergistic effects of combining morphological and molecular data in resolving the phylogeny of butterflies and skippers. Proceedings of the Royal Society of London B 272:1577–1586.

Wahlberg, N., J. Leneveu, U. Kodandaramaiah, C. Pena, S. Nylin, A. V.L. Freitas, and A.V.Z. Brower. 2009. Nymphalid butterflies diversify following near demise at the Cretaceous/Tertiary boundary. Proceedings of the Royal Society of London B 276:4295–4302.

Wallace, A.R. 1860. On the zoological geography of the Malay Archipelago. Journal of the Proceedings of the Linnean Society of London, Zoology 4:172–184.

Wallace, A.R. 1876. The geographical distribution of animals. London: McMillan.

Wanntorp, H.-E. 1983. Historical constraints in adaptation theory: traits and non-traits. Oikos 41:157–160.

Warnow, T. 2018. Computational phylogenetics: An introduction to designing methods for phylogeny estimation. Cambridge: Cambridge University Press.

Watrous, L.E., and Q.D. Wheeler. 1981. The out-group comparison method of character analysis. Systematic Zoology 30:1–11.

Weeks, P.J. and K.J. Gaston. 1997. Image analysis, neural networks, and the taxonomic impediment to biodiversity studies. Biodiversity & Conservation 6:263–274.

Wegener, A. 1966. The origin of continents and oceans. New York: Dover. English translation of the fourth edition, Braunschweig: Vieweg, 1929.

Weingartner, E., N. Walhberg, and S. Nylin. 2006. Dynamics of host plant use and species diversity in Polygonia butterflies (Nymphalidae). Journal of Evolutionary Biology 19:483–491.

Weirauch, C., R.T. Schuh, G. Cassis, and W.C. Wheeler. 2019. Revisiting habitat and lifestyle transitions in Heteroptera (Insecta: Hemiptera): insights from a combined morphological and molecular phylogeny. Cladistics 35:67–105.

Wenzel, J.W. 1992. Behavioral homology and phylogeny. Annual Review of Ecology and Systematics 23:361–381.

Wenzel, J.W. 1993. Application of the biogenetic law to behavioral ontogeny: a test using nest architecture in paper wasps. Journal of Evolutionary Biology 6:229–247.

Wenzel, J.W. 1997. When is a phylogenetic test good enough? In: The origin of biodiversity in insects: Phylogenetic tests of evolutionary scenarios, ed. P. Grandcolas, 31–54. Paris: Mémoires du Muséum national d'Histoire naturelle, vol. 173.

Wenzel, J.W., and J.M. Carpenter. 1994. Comparing methods: adaptive traits and tests of adaptation. In: Phylogenetics and ecology, ed. P. Eggleton and R.I. Vane-Wright, 79–101. London: Academic Press.

Weston, P.H. 1988. Indirect and direct methods in systematics. In: Ontogeny and systematics, ed. C.J. Humphries, 25–56. London: British Museum (Natural History).

Weston, P.H., and M.D. Crisp. 1994. Cladistic biogeography of waratahs (Proteaceae: Embothrieae) and their allies across the Pacific. Australian Systematic Botany 7:225–249.

Wheeler, Q.D. 1990. Ontogeny and character phylogeny. Cladistics 6:225–268.

Wheeler, Q.D. 1995. The "old systematics": classification and phylogeny. In: Biology, phylogeny, and classification of Coleoptera: Papers celebrating the 80th birthday of Roy A. Crowson, ed. J. Pakaluk and S.A. Slipinski, 31–62. Warsaw: Muzeum i Instytut Zoologii PAN.

Wheeler, Q.D. 2005. Losing the plot: DNA "barcodes" and taxonomy. Cladistics 21:405–407.

Wheeler, Q.D., and R. Meier, eds. 2000. Species concepts and phylogenetic theory. New York: Columbia University Press.

Wheeler, W.C. 1990. Nucleic acid sequence phylogeny and random outgroups. Cladistics 6:363–367.

Wheeler, W.C. 1992. Extinction, sampling, and molecular phylogenetics. In: Extinction and phylogeny, ed. M J. Novacek and Q.D. Wheeler, 205–215. New York: Columbia University Press.

Wheeler, W.C. 1993. The triangle inequality and character analysis. Molecular Biology and Evolution 10:707–712.

Wheeler, W.C. 1994. Sources of ambiguity in nucleic acid sequence alignment, 323–354. In: Molecular ecology and evolution: Approaches and applications, ed. B. Schierwater, B. Streit, G.P. Wagner, and R. DeSalle, 323–354. Basel: Birkhäuser Verlag.

Wheeler, W.C. 1995. Sequence alignment, parameter sensitivity, and the phylogenetic analysis of molecular data. Systematic Biology 44:321–331.

Wheeler, W.C. 1996. Optimization alignment: the end of multiple sequence alignment in phylogenetics? Cladistics 12:1–9.

Wheeler, W.C., L. Aagesen, C.P. Arango, J. Faivovich, T. Grant, C. D'Haese, D. Janies, W.L. Smith, A. Varon, and G. Giribet. 2006a. Dynamic homology and phylogenetic Systematics: A unified approach using POY. New York: American Museum of Natural History.

Wheeler, W.C., P. Cartwright, and C.Y. Hayashi. 1993. Arthropod phylogeny: a combined approach. Cladistics 9:1–39.

Wheeler, W.C., J.A. Coddington, L.M. Crowley, D. Dimitrov, P.A. Goloboff, C.E. Griswold, G. Hormiga, et al. 2017. The spider tree of life: phylogeny of Araneae based on target-gene analyses from an extensive taxon sampling. Cladistics 33:574–616.

Wheeler, W.C., and D.G. Gladstein. 1992. MALIGN: A multiple sequence alignment program. Program and documentation, version. 2.0. New York: American Museum of Natural History.

Wheeler, W.C., D.G. Gladstein, and J. De Laet. 1996–2003. POY, version 3.0. American Museum of Natural History, http://ftp.amnh.org/pub/molecular/poy.

Wheeler, W.C., M.J. Ramirez, L. Aagesen, and S. Schulmeister. 2006b. Partition-free congruence analysis: implications for sensitivity analysis. Cladistics 22:256–263.

Whiting, M.F., S. Bradler, and T. Maxwell. 2003. Loss and recovery of wings in stick insects. Nature 421:264–267.

Whiting, M.F., J.C. Carpenter, Q.D. Wheeler, and W.C. Wheeler. 1997. The Strepsiptera problem: phylogeny of the holometabolous insect orders inferred from 18s and 28s ribosomal DNA sequences and morphology. Systematic Biology 46:1–68.

Whiting, M. F., and A. S. Whiting. 2004. Is wing recurrence *really* impossible? A reply to Trueman et al. Systematic Entomology 29:140–141.

Wielgorskaya, T. 1995. Dictionary of the generic names of seed plants. Consulting editor Armen Takhtajan. New York: Columbia University Press.

Wiemers, M., and K. Fiedler. 2007. Does the DNA barcoding gap exist?—a case study in blue butterflies (Lepidoptera: Lycaenidae). Frontiers in Zoology 4:8.

Wiens, J.J. 1998. Does adding characters with missing data increase or decrease phylogenetic accuracy? Systematic Biology 47:625–640.

Wiens, J.J. 2003. Missing data, incomplete taxa, and phylogenetic accuracy. Systematic Biology 52:528–538.

Wiens, J.J., and J. Tiu. 2012. Highly incomplete taxa can rescue phylogenetic analyses from the negative impacts of limited taxon sampling. PLOS One 7:e42925.

Wikispecies. https://species.wikimedia.org/wiki/Main_Page [accessed 6 March, 2020].

Wiley, E.O. 1975. Karl R. Popper, systematics, and classification: A reply to Walter Bock and other evolutionary systematists. Systematic Zoology 24:233–243.

Wiley, E.O. 1979. An annotated Linnaean hierarchy, with comments on natural taxa and competing systems. Systematic Zoology 28:308–337.

Wiley, E.O. 1981. Phylogenetics. New York: John Wiley and Sons.

Wiley, E.O. 1988. Parsimony analysis and vicariance biogeography. Systematic Zoology 37:271–290.

Wiley, E.O. 1989. Kinds, individuals and theories. In: What the philosophy of biology is, ed. M. Ruse, 31–52. Dordrecht: Kluwer Academic.

Wiley, E.O., and B.S. Lieberman. 2011. Phylogenetics: Theory and practice of phylogenetic systematics, 2nd ed. Hoboken, NJ: John Wiley and Sons.

Wiley, E.O., and R.L. Mayden. 2000. The evolutionary species concept. In: Species concepts and phylogenetic theory: A debate, ed. Q.D. Wheeler and R. Meier, 70–89. New York: Columbia University Press.

Wilkins, J.S. 2009. Species: A history of the idea. Berkeley: University of California Press.

Wilkinson, M. 1995a. Coping with abundant missing entries in phylogenetic inference using parsimony. Systematic Biology 44:501–514.

Wilkinson, M. 1995b. A comparison of methods of character construction. Cladistics 11:297–308.

Wilkinson, M., J.A. Cotton, C. Creevey, O. Eulenstein, S.R. Harris, F.-J. Lapointe, C. Levasseur, J.O. McInerney, D. Pisani, and J.L. Thorley 2005. The shape of supertrees to come: tree shape-related properties of fourteen supertree methods. Systematic Biology 54:419–431.

Will, K.W., B.D. Mishler, and Q.D. Wheeler. 2005. The perils of DNA barcoding and the need for integrative taxonomy. Systematic Biology 54:844–851.

Will, K.W., and D. Rubinoff. 2004. Myth of the molecule: DNA barcodes for species cannot replace morphology for identification and classification. Cladistics 20:47–55.

Williams, D.M., and M.C. Ebach. 2008. Foundations of systematics and biogeography. New York: Springer.

Williams, D.M., and M.C. Ebach. 2017. What is intuitive taxonomic practice? Systematic Biology 66:637–643.

Williams, D.M., M. Schmitt, and Q.D. Wheeler, eds. 2016. The future of phylogenetic systematics: The Legacy of Willi Hennig. Cambridge: Cambridge University Press.

Williams, D.M., and D.J. Siebert. 2000. Characters, homology and three-item analysis. In: Homology and dystematics: Coding characters for phylogenetic analysis, ed. R.W. Scotland and R.T. Pennington, 183–208. London: Taylor and Francis.

Williams, P.H., C.J. Humphries, and K.J. Gaston. 1994. Centres of seed-plant diversity: the family way. Proceedings of the Royal Society of London B 256:67–70.

Willis, J.C. 1973. A dictionary of the flowering plants and ferns, 8th ed. Revised by H. K. Airy Shaw. Cambridge: Cambridge University Press.

Wilson, D.E., and D.M. Reeder, eds. 2005. Mammal species of the world: A taxonomic and geographic reference, 3rd ed. Baltimore: Johns Hopkins University Press.

Wilson, D.E., and D.M. Reeder, eds. 2020. Wilson and Reeder's mammal species of the world, https://www.departments.bucknell.edu/biology/resources/msw3/browse. asp [accessed 24 March 2020].

Wilson, R.A., ed. 1999. Species. Cambridge, MA: MIT Press.

Wimberger, P.H., and A. de Queiroz. 1996. Comparing behavioral and morphological characters as indicators of phylogeny. In: Phylogenies and the comparative method in animal behavior, ed. E.P. Martins, 206–33. Oxford: Oxford University Press.

Winkworth, R.C., S. J. Wagstaff, D. Glenny, and P.J. Lockhart. 2002. Plant dispersal N.E.W.S from New Zealand. Trends in Ecology and Evolution 17:514–520.

Winsor, M.P. 2003. Non-essentialist methods in pre-Darwinian taxonomy. Biology and Philosophy 18:387–400.

Winsor, M.P. 2006. The creation of the essentialism story: an exercise in metahistory. History and Philosophy of the Life Sciences 28:149–174.

Winter, M., V. Devictor, and O. Schweiger. 2013. Phylogenetic diversity and nature conservation: where are we? Trends in Ecology and Evolution 28:199–204.

Witteveen, J. 2016. Suppressing synonymy with a homonym: the emergence of the nomenclatural type concept in nineteenth century natural history. Journal of the History of Biology 49:135–189.

Wood, D.M., and A. Borkent. 1989. Phylogeny and classification of the Nematocera. In: Manual of Nearctic Diptera, vol. 3., 1333–1370. Ottawa: Research Branch, Agriculture Canada.

World Spider Catalog. 2020. World Spider Catalog, version 21.0. Natural History Museum Bern, http://wsc.nmbe.ch [accessed 24 March 2020].

Wray, G., J.S. Levinton, and L.H. Shapiro. 1996. Molecular evidence for deep Precambrian divergences among metazoan phyla. Science 274:568–573.

Wu, C.-I., and W.-H. Li. 1985. Evidence for higher rates of nucleotide substitution in rodents than in man. Proceedings of the National Academy of Sciences of the USA 82:1741–1745.

Wu, Y., N.F. Trepanowski, J.J. Molongoski, P.F. Reagel, S.W. Lingafelter, H. Nadel, S.W. Myers, and A.M. Ray. 2017. Identification of wood-boring beetles (Cerambycidae and Buprestidae) intercepted in trade-associated solid wood packaging material using DNA barcoding and morphology. Scientific Reports 7:40316.

Wygodzinsky, P., and S. Lodhi. 1989. Atlas of antennal trichobothria in the Pachynomidae and Reduviidae (Heteroptera). Journal of the New York Entomological Society 97:371–393.

Xi, Z., L. Liu, and C. C. Davis. 2016. The impact of missing data on species tree estimation. Molecular Biology and Evolution 33:838–860.

Yang, Z. 1997. How often do wrong models produce better phylogenies? Molecular Biology and Evolution 14:105-08.

Yang, Z. 2006. Computational molecular evolution. Oxford: Oxford University Press.

Yang, Z. 2007. PAML 4: phylogenetic analysis by maximum likelihood. Molecular Biology and Evolution 24:1586–1591.

Yang, Z. 2014. Molecular evolution: A statistical approach. Oxford: Oxford University Press.

Yeates, D.K. 1995. Groundplans and exemplars: paths to the tree of life. Cladistics 11:343–357.

Yeo, D., J. Puniamoorthy, R.W.J. Ngiam, and R. Meier. 2018. Towards holomorphology in entomology: rapid and cost-effective adult-larva matching using NGS barcodes. Systematic Entomology 43:678–691.

Yoder, M.J., I. Mikó, I., K.C. Seltmann, M.A. Bertone, and A.R. Deans. 2010. A gross anatomy ontology for Hymenoptera. PLOS ONE 5:e15991.

Yu, Y., A. J. Harris, C. Blair, and X. He. 2015. RASP (reconstruct ancestral states in phylogenies): a tool for historical biogeography. Molecular Phylogenetics and Evolution 87:46–49.

Yu, Y., A.J. Harris, and X. He. 2010. S-DIVA (statistical dispersal-vicariance analysis): a tool for inferring biogeographic histories. Molecular Phylogenetics and Evolution 56:848–850.

Zachos, F.E., M. Apollonio, E.V. Bärmann, M. Festa-Bianchet, U. Göhlich, J.C. Habel, E. Haring, et al. 2013a. Species inflation and taxonomic artefacts—a critical comment on recent trends in mammalian classification. Mammalian Biology 78:1–6.

Zachos, F.E., T.H. Clutton-Brock, M. Festa-Bianchet, S. Lovari, D.W. Macdonald, and G.B. Schaller. 2013b. Species splitting puts conservation at risk. Nature 494:35.

Zachos, F.E., and J.C. Habel, eds. 2011. Biodiversity hotspots: Distribution and protection of conservation priority areas. Berlin: Springer.

Zachos, F.E., and S. Lovari. 2014. Taxonomic inflation and the poverty of the phylogenetic species concept—a reply to Gippoliti and Groves. Hystrix 24:142–144.

Zandee, M., and M.C. Roos. 1987. Component-compatibility in historical biogeography. Cladistics 3:305–332.

Zelditch M.L., W.L. Fink, and D.L. Swiderski. 1995. Morphometrics, homology, and phylogenetics: quantified characters as synapomorphies. Systematic Biology 44:179–189.

Zhang, W., K.K. Dasmahapatra, J. Mallet, G.R.P. Moreira, and M.R. Kronforst. 2016. Genome-wide introgression among distantly-related *Heliconius* butterfly species. Genome Biology 17:25.

Zhou, B.B., M. Tarawneh, P. Wang, D. Chu, C. Wang, A.Y. Zomaya, and R.P. Brent. 2006. Evidence of multiple maximum likelihood points for a phylogenetic tree. In: Proceedings of the Sixth IEEE Symposium on Bioinformatics and BioEngineering, 193–200. Washington, DC: IEEE Computer Society.

Zimmerman, E.C. 1991–1994. Australian weevils (Coleoptera: Curculionidae), vols. 1–6. Melbourne: CSIRO (Commonwealth Scientific and Industrial Research Organisation) Publications and Australian Entomological Society.

Zimmermann, W. 1943. Die Methoden der Phylogenetik. In: Die Evolution der Organismen, ed. G. Heberer, 20–56. Jena, Germany: Gustav Fischer Verlag.

Zimring, J. 2019. What science is and how it really works. Cambridge: Cambridge University Press.

Zoological Record. https://clarivate.com/webofsciencegroup/solutions/webofscience-zoological-record/ [accessed 31 March 2020].

Zrzavý, J. 1997. Phylogenetics and ecology: all characters should be included in the cladistic analysis. Oikos 80:186–192.

Zuckerkandl, E., and L.B. Pauling. 1962. Molecular disease, evolution, and genic heterogeneity. Horizons in biochemistry, ed. K. Kasha and B. Pullman, 189–225. New York: Academic Press.

Author Index

Subject Index

Italicized page numbers refer to glossary entries.

Lightning Source UK Ltd.
Milton Keynes UK
UKHW020141101122
411927UK00006B/262